Deer Management for Forest Landowners and Managers

Deer Management for Forest Landowners and Managers

Edited by
David S. deCalesta
Michael C. Eckley

CRC Press
Taylor & Francis Group
Boca Raton London New York

CRC Press is an imprint of the
Taylor & Francis Group, an **informa** business

CRC Press
Taylor & Francis Group
6000 Broken Sound Parkway NW, Suite 300
Boca Raton, FL 33487-2742

© 2019 by Taylor & Francis Group, LLC
CRC Press is an imprint of Taylor & Francis Group, an Informa business

No claim to original U.S. Government works

Printed on acid-free paper

International Standard Book Number-13: 978-1-4665-8016-9 (Paperback)
978-0-367-19693-6 (Hardback)

Library of Congress Cataloging-in-Publication Data

Names: DeCalesta, David S., editor. | Eckley, Michael C., editor.
Title: Deer management for forest landowners and managers / editors: David S. DeCalesta, Michael C. Eckley.
Description: Boca Raton, FL : CRC Press, Taylor & Francis Group, 2019.
Identifiers: LCCN 2018056842| ISBN 9781466580169 (pbk. : alk. paper) |
ISBN 9780367196936 (hardback : alk. paper) | ISBN 9780429574634 (epub) |
ISBN 9780429572524 (mobi/kindle)
Subjects: LCSH: Deer--Control. | Forest management.
Classification: LCC SB994.D4 D44 2019 | DDC 639.97/965--dc23
LC record available at https://lccn.loc.gov/2018056842

Visit the Taylor & Francis Web site at
http://www.taylorandfrancis.com

and the CRC Press Web site at
http://www.crcpress.com

This book is dedicated to the memory of Pennsylvania Game Commission forester Robert Bauer (1946–2002) and Pennsylvania State University Extension forester Roe (Sandy) Cochran (1925–1991) for their contributions to deer and forest management. These men promoted deer management that integrated and weighed equally wildlife and forestry sciences and management practices—a new concept 25 years ago and critically important today.

Contents

SECTION II   Planning and Assessment

SECTION III   Managing Ecological and Human Factors

SECTION IV   Special Cases

SECTION V   Case Histories

Foreword

Deer species, particularly white-tailed deer (*Odocoileus virginianus*), have been a catalyst for an enormous system of traditions and beliefs. Throughout North America are millions of people who observe and hunt deer and have for generations. This interest has stimulated the development of a large body of science including ideas about harvestable surplus, carrying capacity, and the importance of hunters in managing wildlife. Indeed, public interest in hunting regulations, conservation ethics, and wildlife science drove restoration of deer across North America in what represents one of the premier examples of wildlife conservation.

At the same time, managing deer to meet the goals of many different stakeholder groups has been central to the story of deer. Perhaps more than the biology, this part of the story has produced enormous challenges and considerable controversy. One of these controversies was how best to integrate deer management and forest management for the benefit of both the deer hunter and the forest landowner. In large measure, the challenges and controversies were a consequence of the complexity of eastern ecosystems: the silvics of so many different tree species, the role of human economies in the exploitation of these systems, and ongoing changes in the natural ecological pressures affecting these systems. It took nearly a century to understand the biology of deer and the ecology of North American forests. Much of this knowledge was gained in the past several decades, and it encourages us that we may soon see the full dimension of this ecosystem.

That our search for complete understanding of deer and forest management has been a goal for a long time is evident from the writings of Aldo Leopold nearly a century ago. Trained as a forester, Leopold is remembered today as the father of wildlife management. His early management ideas about deer were shaped by the principle that having more deer is always better. As a scientist trying to understand the ecology of deer in forests, his ideas evolved and he grew to see the error in that principle. Leopold recognized that deer were part of a much more complicated natural system. His training in forestry and experiences with mule deer (*O. hemionus*) in the Southwest and later white-tailed deer in the upper Midwest began to show him that deer could have such impacts on forests in just a few years that they would shape the forests for centuries. Leopold was among the first to understand that removal of female deer by predators or through hunter harvest was key to managing not only deer populations but a forest ecosystem. While he struggled to get traction for his ideas during his lifetime, his essay on deer and their relationships with forests and the larger ecosystem, *Thinking Like a Mountain*, has become a classic in the wildlife conservation literature. Here he laid out his philosophy of a land ethic in terms that speak to the need to avoid focusing on a single species and engage the complexity of managing all elements of the ecological system.

The concept of deer interacting with vegetation was made concrete for me when I first saw the deer exclosures in Itasca State Park in northern Minnesota. Feeding preferences of deer were shifting the competition between white and red pine (*Pinus strobus* and *P. resinosa*). The comparative effects of the shift in composition of tree species inside and outside the exclosures were obvious in just a couple of decades. Across the Itasca region, these impacts would last more than a century. The impact deer could have on shifting the composition of a forest over centuries became even more clear when I saw the experimental trials underway on the Huntington Wildlife Forest in the central Adirondack Mountains of northern New York, a field station of the State University of New York College of Environmental Science and Forestry. Working together, wildlife biologists and foresters developed a series of experimental treatments in which deer densities and forest composition were manipulated to measure the effects of deer on forest regeneration. These trials showed that feeding preferences of deer were causing the shift from a diverse hardwood forest to a monoculture of American beech (*Fagus grandifolia*) by preferentially eating each year's crop of new sugar maple, yellow birch, and black cherry (*Acer saccharum, Betula alleghaniensis, Prunus serotina*) seedlings, allowing the beech to proliferate. Findings showed that shelterwood harvest of the forest and reduction of the deer population through aggressive hunter harvest from >20 to <10 deer per square kilometer resulted in a diverse regeneration including sugar maple, yellow birch, and black cherry.

The profound nature of the impacts of deer was evident in the northern hardwood forests of the Adirondack region. On the mesic soils of mid elevations throughout the entire region, the overstory appeared to be dominated by relatively even-aged sugar maple, yellow birch, and black cherry about 150 years old. These stands originated in the mid-nineteenth century when logging for red spruce on these sites produced widespread disturbance in the soils and heavy hunting of deer to feed loggers kept deer densities low. In the late twentieth century, disturbance from logging was not yielding a diverse regeneration because hunting was tightly regulated and deer populations throughout the region were much higher. What became obvious was that the diverse overstory was a product of regeneration occurring during a brief window of opportunity: a coincidence of extensive site disturbance, regionally low deer densities, and a series of good seed years.

My perspective on deer impacts on vegetation expanded as I was exposed to the ecology of the highly productive agricultural regions of New York State. Saratoga National Historical Park exists in a landscape dominated by agriculture, particularly alfalfa and corn. The park is closed to hunting, and as a result, in the 1980s when we first began research in the park, deer populations achieved densities of >40 deer per square kilometer. Large portions of the park were dominated by gray dogwood (*Cornus racemosa*). The dogwood grew as expanding clones with stem densities that were so great the deer could browse only the stems on the periphery. Here again, deer shifted the competitive advantage toward dogwood within the plant communities and affected the structure of the dogwood, allowing large portions of the park to be dominated by dogwood clones. The National Park Service attempted to portray the arrangement of forest and agriculture that was relevant to the historic battles that occurred on that site. Prolific dogwood clones created a problem because they were not present in 1777 when the battles took place, and our research was designed to recommend management solutions.

Similar situations existed at many of the small historical parks throughout the eastern United States. The Park Service was facing significant public challenges to its efforts to actively manage the deer to reduce their populations. Early on, the rationale for active management was based on the idea that these deer populations had exceeded carrying capacity, as evidenced by the impact of deer on the vegetation. However, using carrying capacity as a rationale for action proved problematic. There were multiple working definitions of the concept and there was no consensus among wildlife biologists on which was appropriate. There was no clear way to measure carrying capacity. Consequently, objective assessments of the deer population relative to carrying capacity were unlikely to withstand legal challenge. As an alternative, the Park Service elected to write specific management objectives for re-establishing the vegetation to promote visitor appreciation of the context for the historical period the park commemorated. These objectives were cast in terms foresters would appreciate: species composition, structure of the forest understory and overstory, and configuration of forest and fields. Where browsing by deer prevented the Park Service from achieving its vegetation management goals, reduction of deer populations was implemented.

In the early 1990s, my colleague, Dr. Brian Underwood, a wildlife biologist with the United States Geological Survey, and I summarized the experience of the National Park Service in an article in the journal *Ecological Applications*, published by the Ecological Society of America. We likened the challenges of deer and vegetation management to the ancient story of the blind man touching an elephant. For much of the twentieth century, we could see only a small portion of this complicated ecosystem. Furthermore, biologists and foresters, hunters and farmers each saw different parts of the system. And, their perceptions about the system were shaded by values each brought with them from life experience. These perceptions and values were implicit in the assumptions that filled in the blank spaces in knowledge of this system. As so often happens in human society, these assumptions grew to become strong beliefs, and each group was hard pressed to let go of them. The ensuing arguments about the management of deer, and forests and hunting and crops, were based on an amalgam of science and belief. Many of those people who were, and perhaps still are, part of those arguments can identify deeply with the words of John Godfrey Saxe's poem, *Blind Men and the Elephant,* and especially with one of the stanzas:

So, oft in theologic wars
The disputants, I ween,
Rail on in utter ignorance
Of what each other mean;
And prate about an Elephant
Not one of them has seen!

A generation from now, wildlife biologists and foresters, and those who study human behavior, will say the arguments over the management of deer and forests were shaped mostly by assumptions and beliefs, and that we railed in utter ignorance. But this is the fate of every generation. Each new generation will judge the scientific knowledge of its predecessor as insufficient. While the theologic debates about deer are not likely to end anytime soon, this book is proof that we are beginning to see a good deal of the elephant and with much greater clarity.

William F. Porter
Boone and Crockett Professor of Wildlife Conservation
Michigan State University

Preface

Like many, if not all, deer researchers, educators, managers, and consultants, my formal preparation for working with white-tailed deer (*Odocoileus virginianus*) was based on perceptions passed down from generations of deer hunters, my own hunting experiences, and especially on formal information presented in undergraduate and graduate collage classes on wildlife and big game management, including deer. My teaching and research on deer at the university level as a wildlife professor were based on these same sources of information, without the benefit/experience of working directly with actual deer managers on private and public forest lands. My responsibilities as Extension Wildlife Specialist for several universities introduced me to the nature and extent of problems deer managers faced, but I was buffered from those managers by an intermediate layer of Extension County Agents who looked to me to provide them with training materials for their use in responding to deer managers' requests for assistance. It was only when I began work with a USDA Forest Service research laboratory in northwestern Pennsylvania that I interacted directly with forest managers in developing, interpreting, and delivering deer research appropriate to their needs, and that was possible only because that particular research entity (Warren Forestry Sciences Laboratory) had an annual goal of translating research findings into practical management applications aimed at solving forest managers' problems. This information was presented in research publications and at annual workshops for forest and wildlife managers that were directed at managers tasked with resolving deer–forestry issues.

The underlying impetus for this book and the development of research aimed at resolving deer impacts on forest resources were created by the haphazard manner in which deer and forests were managed in the past. Cycles of high and low deer and forage abundance created by vast and unsustainable timber harvest in the 1800s and early 1900s created the basic conflict among hunters, foresters, and ecologists that persists and confounds deer management for managers. Present-day hunters and their fathers and grandfathers remember the days when deer density was so high, and the understory so depleted, that dozens of deer would be seen during the hunting season, especially opening day. They remember that the highs of deer density were preceded by restrictions in antlerless or doe hunting, and that population crashes followed liberal antlerless seasons as the game agencies attempted to reduce deer density and avoid such crashes and associated depletion of understory vegetation.

Not being foresters or ecologists, most hunters did not notice the depleted state of understory vegetation, lack of regenerating seedlings necessary for future forests, or lack of diversity of understory vegetation. Nor did they notice reductions in abundance and diversity of wildlife species affected by habitat degradation by deer. Hunters also did not understand that when deer density was high, deer quality (measured by body condition and size and quality of antlers) was much less than when deer were in balance with the forest vegetation. Many hunters still want deer densities at least as high as 15 deer/km^2 and are unaware that such densities are not sustainable and would inevitably lead to winter starvation die-offs of deer. Their response in the past to deer die-offs was to demand that game agencies conduct and support emergency feeding operations for deer and prohibit harvest of female deer.

On the other hand, foresters and ecologists are well aware of the effects of a century of overabundant deer herds on regeneration of seedlings and diversity and abundance of other understory vegetation. They have struggled to protect and enhance understory vegetation, but without reduction of deer density to sustainable levels, their efforts have been mainly palliative and limited to excluding deer from the vegetation with costly fencing. Vegetation outside fenced areas had no protection from overabundant deer herds and remained sparse and lacking in understory species richness and abundance. Scientists have documented severe reductions in diversity and abundance of herbaceous vegetation that never grew out of the reach of deer, and the effects on associated plant and animal communities, with some species being eliminated entirely.

McWilliams et al. (2018) highlighted the need and scope for providing useful information for managing deer impact to forestlands in midwestern and northeastern states in the United States. "Currently, 59% of the 182.4 million acres of forest land inventoried in the Midwest and Northeast was estimated to have moderate or high browse impacts … Three realities of forest regeneration management for forests under herbivory stress in the Midwest and Northeast (are): (1) the scope and persistence of large-ungulate herbivory has long-term wide-ranging implications for regeneration management; (2) less palatable tree species will continue to have a competitive advantage during the regeneration phase and are likely to be different species from the current canopy dominants; and, (3) successful regeneration management of these forests requires more emphasis on ungulate-compatible prescriptions, novel approaches, and adaptive science."

Managers of private and public forestlands are sandwiched between deer hunters and natural resource administrators in a disconnect alluded to by Sands et al. (2013) in their book *Wildlife Science—Connecting Research with Management*. The major premise was that wildlife management suffers from a disconnect between wildlife scientists developing management recommendations based on science and managers failing to acknowledge or use them. As a contributor in that book, I contended that the real disconnect was between wildlife administrators and the partnership formed by scientists and managers (deCalesta 2013). More recently, I realized that the disconnect is between two groups: (1) persons who actually manage deer and deer habitat on forestlands under their direction and (2) the scientists, natural resource administrators, and educators who provide these managers with management recommendations through research, teaching and outreach, and pressure by deer hunters. That disconnect is addressed in this book.

Where public deer hunting is a key element in reducing deer density and impact to acceptable levels, a critical requirement for the success of this partnership is that the two groups interact collaboratively and cooperatively with a third group—the deer hunters—to achieve deer and forest management goals. This book is designed to provide a comprehensive and unifying construct to unite the groups as equal partners in developing and delivering effective deer–forest management. To reduce needless repetition and definition, landowners of private forestlands, stewards of public forestlands, and wildlife/forest consultants are henceforth referred to as "managers."

In published comprehensive works on white-tailed deer biology and management (Taylor 1956, McCullough 1979, Halls 1984, Hewitt 2011), the emphasis was on deer ecology and deer management at regional, statewide, or resource agency management areas, each encompassing dozens to hundreds of individual managers of individual forestlands. These books focused on the what and why of deer management—basically, the biological underpinnings for management. As such, the books are useful primarily for educators and natural agency personnel without responsibility for managing deer on individual forestlands. Three other "W" questions as defined by Aristotle and Cicero, among others, related to solving (management) problems, such as on individual forestlands, are: (1) who (performs management aspects), (2) where (landscapes where management is actually carried out, and (3) when (seasonal timing and frequency of management application). There are a final two questions related to management: (1) how (do managers actually perform management activities) and (2) how much do they do (scope and scale of management activities and associated costs as related to management on individual forestlands).

In pointed contrast to earlier books on white-tailed deer management, the emphasis of this book is on providing background information *and* effective deer management practices for deer managers on discrete forestlands negatively impacted by white-tailed deer—in essence addressing all seven of the components mentioned above. This emphasis is reinforced by recounting successful deer management operations that addressed the components as detailed in case histories in Section V of this book.

I endorse the viewpoint promoted by Decker et al. (2004) regarding managing deer impacts: "To enable the continued management of white-tailed deer as a resource rather than as a pest, by articulating key dimensions of success when engaging in … (deer) management." My objectives in producing this book were fourfold: (1) provide managers of forestlands with requisite knowledge, guidance, and techniques for managing deer and deer impact; (2) help managers educate, motivate, and retain hunters and other stakeholders whose support they need to manage deer and forest resources; (3) help natural resource administrators, scientists, educators, and deer hunters understand the needs of deer managers and address them through development, delivery, and endorsement of research and supportive deer harvest regulations; and (4) provide a variety of representative case histories demonstrating how managers incorporated key concepts presented in this book to address and successfully manage deer density and impact.

Environmental factors that influence and determine deer density and impact vary as affected by weather and forest type and include precipitation, wind, temperature, humidity, and duration and intensity of sunlight. Geology and latitude, in concert with weather, produce a number of forest types, each providing differing qualities and quantities of deer habitat requirements. Forest types include the boreal forest of the northeastern United States and Canada, northern and Appalachian hardwood forests of the northeastern United States, central broad-leafed and oak forests of the midwestern and southeastern United States, and southeastern oak and pine forests (Young and Giese 2003).

The 1984 compendium on ecology and management of white-tailed deer (Halls 1984) described deer management within 18 different forested habitats; the most recent version (Hewitt 2011) cut that number to three, eastern North America, midwestern North America, and western North America, and identified an additional category based on high human population density—exurban, suburban, and urban. Within each of these categories, deer densities and impacts exhibit tremendous variability, necessitating a management approach that recognizes and addresses environmental, regulatory, and human dimension factors specific to individual forestlands regardless of which forested habitat category they fall within—geography is too broad a brush upon which to pin deer management strategy and practice. Indeed, the only way to manage deer and deer impacts effectively is to develop and deliver management programs tailored to individual forestlands.

As DeYoung (2011) noted, many of the management strategies developed for white-tailed deer have been based on research conducted in eastern and midwestern states and extant forest types. To minimize potentially confounding discrepancies engendered by using data from disparate sections of the white-tailed deer range in North America, the research and management bases and case histories presented in this book were drawn primarily from forestlands within the northeastern hardwood forest and oak-hickory types characteristic of eastern North America (parts or all of Connecticut, Massachusetts, Michigan, Minnesota, New Hampshire, New York, Ohio, Pennsylvania, Vermont, West Virginia, and Wisconsin).

Within whatever plans forest type managers develop for managing deer and other forest resources, the requirement is the same: make sure deer and forest science, and sources of information chosen for reference, are as good a match with their forest type and weather as possible, especially concerning deer density and impacted forest resources.

Principles and techniques espoused in Sections I through III, including monitoring and managing deer density and deer impacts, are applicable to all forest types. However, the differences of deer density and impacts among forest types mandate that forest managers develop representative strategies in consultation with forestry and deer management specialists familiar with the forest type, landscape context, and weather where their lands are located.

Management to reduce the effects of overabundant deer herds is one of the biggest challenges to producing sustained outputs of herbaceous, shrub, tree, vine, and wildlife communities within forested landscapes. Browsing by overabundant deer herds changes the dynamics of understory, midstory, and overstory vegetative structure and composition, resulting in simplified and less diverse plant and wildlife communities, rendering sustained production of diverse forest resources difficult. Affected forestlands run the gamut of property types and management objectives. Some are private woodlots of less than 4 ha; some are corporate woodlots in the hundreds to thousands of ha; and others are small-, medium-, or large-sized public forests such as community watersheds, parks, and state forests.

FIGURE P.1   Quality deer (left panel) vs. poor-quality deer (right panel). (Photos courtesy David S. deCalesta.)

Landowner values are often similar, but goals and objectives vary and may include: protection of sensitive plant and animal species and enhancement of biodiversity; protection of private/public property from deer impact; sustained production of forest products, including timber; production and harvest of quality deer; and capitalization on the deer resource via leasing forestlands to deer hunting groups. Deer hunting is allowed on many, but not all, of these different property types.

In several places throughout this book, the term "quality deer" is encountered without being described. To address this deficit, a quality deer (Figure P.1) is defined as one that exhibits the best physical attributes (body weight, antler development) obtainable from the nutrition afforded by local forage resources, including soil mineral content, when deer density is managed to allow full expression of species richness and abundance of understory vegetation.

Harvested quality deer exhibit high amounts of body fat (depth exceeding 1 cm) on the backstrap (the large muscle running along back from shoulders to rump) and surrounding the kidneys and heart, indicating sufficient quality and quantity of forage to carry them through severe winter weather in good condition using their fat reserves. Nutritive quality (and quantity) of deer forage determines the upper limit of "quality deer" characteristics. Depleted forest/farmlands on nonglaciated soils have the lowest potential optimal values for quality deer. Values should be higher where deer forages are not depleted and are located on glaciated soils. The highest values should occur on forestlands managed for trophy deer, where natural forages and soil minerals are supplemented with high-quality food plots and mineral blocks.

A major premise running through this book is that, under many circumstances, reducing deer density on forested properties is the most effective management step for reducing deer impact. In many cases, this can be accomplished by public hunting during established deer hunting seasons. In some cases, state natural resource agencies may increase season length, bag limit, and availability of permits to harvest additional antlerless deer on specified areas/properties if regular season deer hunting does not result in the desired reductions in deer density and impact. When this step does not result in the desired reductions in deer density and impact, managers may seek permission to remove additional deer by use of sharpshooters, sometimes outside of regular deer hunting season. This management option is referred to as "culling," as contrasted with public hunting, and is used primarily to reduce deer herd density, deer/vehicle collisions, property damage, and incidence of disease spread by deer (Kaatz 2015).

When I broached the idea of writing this book to a field forester of 40 years' experience and explained that it would be written for deer–forest managers and administrative types representing agencies that regulate how deer may be managed, his response was, "Give me a quick summary of the reasons for managing deer for basic understanding. But what I really need to know to manage deer and forests jointly is the how, who, when, where, and how much/how long." In accordance with his wishes, and in deference to field managers, each chapter in Section I has a "Manager Summary." These summaries provide managers and administrators with condensed descriptions of ecological and human dimension factors necessary for understanding how and why managers need to apply management steps to mitigate impacts of overabundant deer herds on forest resources. I do not provide similar summaries for chapters in Sections II through V because information presented in those chapters provides the nuts and bolts (and examples) for developing and delivering effective deer management programs—managers, as well as natural resource agency administrators, should read those chapters in their entireties.

Last, readers will note that Chapters 1 through 8 and 13 through 15 have more citations than Chapters 9 through 12, 16 through 19, and 22 through 25—all of which deal with human dimension factors in deer management. The paucity of citations in human dimension–related chapters reflects a missed opportunity and continuing need for (social) scientists to investigate the human dimensions aspects of deer management and the difficulty of designing research that addresses how humans interface with, and affect, deer management. The scarcity of information regarding management of deer related to human factors, and on realistic deer management areas, is a major driver of this book. Forest landowners have a real need for this information, which is scattered and scant in the literature. The book provides a starting point for forest landowners dealing with deer management requiring the use of ecological and human dimensions–related research and may provide incentive for researchers to delve into these neglected aspects of deer management research.

David S. deCalesta
Editor

REFERENCES

deCalesta, D. S. 2013. Collaboration among scientists, managers, landowners, and hunters: The Kinzua Quality Deer Cooperative. In: *Wildlife science—Connecting research with management*, eds. J. P. Sands, S. J. Demaso, M. J. Schnupp et al. 191–209. Boca Raton FL: CRC Press.

Decker, D. J., D. A. Raik, and W. F. Siemer. 2004. *Community-based deer management*. Ithaca NY: Northeast Wildlife Damage Management Research and Outreach Cooperative.

DeYoung, C. A. 2011. Population dynamics. In: *Biology and management of white-tailed deer*. D. G. Hewitt, ed. Boca Raton FL: CRC Press.

Halls, L. K. ed. 1984. *The white-tailed deer of North America*. New York: Stackpole Books.

Hewitt, D. G. ed. 2011. *Biology and management of white-tailed deer*. Boca Raton FL: CRC Press.

Kaatz, J. 2015. https://shootingtime.com/blog/deer-culling-illinois.

McCullough, D. R. 1979. *The George Reserve deer herd*. Ann Arbor MI: The University of Michigan Press.

McWilliams, W. H., J. A. Westfall, P. H. Brose et al. 2018. *Subcontinental-scale of large-ungulate herbivory and synoptic review of restoration management implications for midwestern and northeastern forests*. General Technical Report NRS-182. Newton Square PA: U. S. Department of Agriculture, Forest Science, Northern Research Station.

Sands, J. P., S. J. Demaso, M. J. Schnupp, and L. A. Brennan. 2013. *Wildlife science: Connecting research with management*. Boca Raton FL: CRC Press.

Taylor, W. P. 1956. *The deer of North America: Their history and management*. Harrisburg PA: Stackpole Books.

Young, R. A. and R. I. Giese. 2003. *Introduction to forest ecosystem science and management*. Hoboken NJ: Wiley & Sons.

Acknowledgments

In our decades-long work with the deer–forestry issue, our perspectives, experiences, insights, and philosophies have been shaped and broadened by mentors, peers, organizations, and stakeholders. We would be remiss if we did not recognize their influences and contributions to this book. Our initial exposures to comprehensive and analytical approaches to deer and forestry issues were influenced by the likes of James Bailey, Dale Hein, Dale McCullough, Aaron Moen, Scott Overton, and William Severinghaus. Mentors and peers who expanded our horizons and promoted critical thinking by example were Kip Adams, Gary Alt, Robert Bauer, Sandy Cochran, Michael Conover, John Dzemyan, Mark Ellingwood, David Marquis, William McShea, Ralph Nyland, Timothy Pierson, William Porter, James Redding, David Samuel, and Brian Shissler. Ralph Nyland provided helpful review of chapters involving deer and silviculture. Organizations that supported and enhanced our involvement with deer management issues were the Sand County Foundation, the Northeast Research Station, USDA Forest Service, the Allegheny National Forest, the Nature Conservancy, the Pennsylvania State University Extension Service, Forestry Investment Associates, Collins Pine, Generations Forestry/Keith Horn Forestry, and the Kinzua Quality Deer Cooperative. Last, our conversations and interactions with deer hunters have provided us with a critical perspective, and we emphatically acknowledge the indispensable role hunters play in managing deer density and impact by harvesting deer from forestlands.

Editors

David S. deCalesta received a BA in psychology from Dartmouth College and MS and PhD degrees in wildlife ecology from Colorado State University. He began a life-long study of deer with a PhD thesis on mule deer nutrition and physiology in 1970. Much of his work as Extension Wildlife Specialist and university teacher and researcher in zoology (North Carolina State University) and Wildlife Ecology and Forest Science (Oregon State University) focused on deer (mule, black-tailed, and white-tailed) interactions with forest vegetation and wildlife communities. His work as a research wildlife biologist with the USDA Forest Service featured interactions between white-tailed deer and forest plant and animal communities. He has spent the last third of his career as a wildlife consultant and forest certification specialist (Forest Stewardship Council and Sustainable Forestry Initiative), primarily working with deer–forest interactions on forest ownerships ranging from dozens to thousands of hectares on private and public forestlands, including state parks and forests and National Forests. With Timothy Pierson, he coordinated the activities of the Kinzua Quality Deer Cooperative, which demonstrated how public hunting could be managed to reduce deer density and impact to levels enhancing diversity and sustainability of forest products and wildlife and vegetative communities.

Michael C. Eckley received a BS in forest resources management with a minor in communications from West Virginia University and an MS in forestry at the University of Maine-Orono. His career has centered on eastern hardwood forest management and assisting private land ownerships throughout the eastern United States. Much of his time is devoted to outreach and education along with specialization in assessing forest conditions, planning, and promoting responsible forestry practices. Mike is a Society of American Foresters (SAF) Certified Forester and is currently employed by The Nature Conservancy, serving as their Forestry Manager for the Working Woodlands Program (www.nature.org/workingwoodlands).

Contributors

Daniel Aitchison
Westchester County Parks
Recreation and Conservation Division
Cross River, New York

Michael L. Ashdown
Department of Natural Resources
Cornell University
Ithaca, New York

Roy Brubaker
Small Woodlot Owner
East Waterford, Pennsylvania

Jim Chapman
Hyma Devore Lumber Mill
Youngsville, Pennsylvania

Delores Costa
Forestry Investment Associates
Coudersport, Pennsylvania

Paul D. Curtis
Department of Natural Resources
Cornell University
Ithaca, New York

David S. deCalesta
Halcyon-Phoenix Consulting
Venice, Florida

Michael C. Eckley
The Nature Conservancy
Williamsport, Pennsylvania

Jim Finley
Center for Private Forests
Pennsylvania State University
State College, Pennsylvania

Jeff Hamilton
Small Woodlot Owners
Sullivan County, Pennsylvania

Sue Hamilton
Small Woodlot Owners
Sullivan County, Pennsylvania

Marian Keegan
Hemlock Farm Community
Association
Lord's Valley, Pennsylvania

Jeffrey Kochel
Forest Investment Associates
Coudersport, Pennsylvania

Michael McEntire
Forestry Investment Associates
Smethport, Pennsylvania

Allyson Muth
Center for Private Forests
Pennsylvania State University
State College, Pennsylvania

Kevin Virkler
Adirondack League Club
Old Forge, New York

Mark Weckel
American Museum of Natural
History
Center for Biodiversity and
Conservation
New York City, New York

1 Introduction

David S. deCalesta

CONTENTS

> Sit down before fact like a little child, and be prepared to give up every preconceived notion, follow humbly wherever and to whatever abyss Nature leads, or you shall learn nothing.
>
> **T.H. Huxley (1903)**

MANAGER SUMMARY

Managing deer at densities required to sustain forest resources of interest to landowners is complex and difficult, owing to the multiple interactions between deer ecology and human dimensions—all must be addressed and integrated into comprehensive management plans. Multiple factors must be addressed, such as values among differing stakeholder groups, landscape and time scales, and financial and human resources available. Important components include monitoring, managing vegetation, managing hunters, and managing access and permits for harvesting antlerless deer. All must adjust to changing conditions (adaptive management). No one size fits all, as there are multiple categories of forest landowners besides the typical, and they include small woodlot owners, quality deer management adherents, forestlands leased for deer hunting, forestlands in human residential areas/public lands without hunting, and cooperatives. The book concludes with nine case histories illustrating successful deer management over a variety of the typical and special cases, including documentation of a case history that failed because it did not integrate ecology and human dimensions.

1.1 INTRODUCTION

The white-tailed deer (*Odocoileus virginianus*) is the most-researched and published wildlife species in North America, with thousands of scientific articles and three books (Taylor 1956, Halls 1984,

Hewitt 2011) devoted to their history, ecology, and management. McCullough's (1979) study of deer population dynamics is a cornerstone for understanding how deer populations grow (and decline). McShea et al. (1997) published a comprehensive treatise on management of overabundant deer populations. Frye (2006) detailed the conflict, controversy, and difficulty of managing white-tailed deer in Pennsylvania as affected by multiple stakeholders. This book draws upon these resources and the research and management experiences of the book's authors.

This is a "how-to" book for managing impacts of overabundant deer on forest resources. The book is arranged in five sections. Section I presents the ecological and human factors identified and quantified from research that influence deer density and deer impact on forest resources that are integrated by the unifying concept of carrying capacity. An important reality of comprehensive deer management is that actions based on ecological factors must be enhanced by application of human-based actions for success in deer management. Section II describes a framework for assessing deer density and impact as related to factors in Section I and details a process for establishing goals and objectives for deer management. Section III details a process for integrating ecological- and human-based factors into management activities, which includes incorporation of adaptive management for adjusting these activities based on assessment of progress toward goals. Section IV describes five "special" deer–forest management situations (small woodlot management, quality deer management, lease hunting management, residential/public forestland management without administrative support for public hunting to resolve deer impacts, and deer–forest management cooperatives) outside the realm of traditional management for forest products. Section V presents nine case histories representative of typical and special forest management situations.

The Northeast Section of the Wildlife Society (2015), in a white paper on managing impacts of overabundant deer, identified 11 components of management plans to address the issue. Although written in the context of deer management in developed (residential) areas, the components are a good starting point for describing steps for deer management and are referenced to specific chapters in this book.

1.2 PRIMER ON INFORMATION/DOCUMENTATION OF NEEDS FOR DEER MANAGEMENT

1. Identify deer impacts (addressed in Chapter 15)
2. Define objectives to measure progress to alleviating/eliminating negative deer impacts (addressed in Chapter 16)
3. Collect data on deer impacts (addressed in Chapter 17)
4. Review management options (addressed in Chapters 1, 16, 25)
5. Invoke decision-making process (addressed in Chapters 1, 25, and 26)
6. Develop and implement a communication plan (addressed in Chapters 22 and 23)
7. Ensure state wildlife agencies … authorize regulated harvest where special hunts may be needed (addressed in Chapters 12 and 24)
8. Identify permitting requirements (addressed in Chapter 24)
9. Implement management actions (addressed in Chapters 25 and 26)
10. Monitor changes in deer impact levels (addressed in Chapter 17)
11. Review and modify management options (addressed in Chapter 26)

Missing from needs identified in the white paper for deer management are: (1) emphasis on management planning/conduct at the actual management level (individual private/public forestland), as discussed in Chapter 9; (2) identifying and meeting needs of hunters (Chapters 10 and 23); (3) assessing and implementing financial and human resource needs for deer management (Chapter 19); and (4) addressing spatial and temporal factors in deer management (addressed in Chapter 3).

1.3 CAVEAT

Before we get started, it is important that we adjust our perspective on deer management to address three major disconnects between resource agencies charged with managing deer and managers who actually manage deer. The disconnects relate to: (1) management philosophy, (2) negative impact resulting from managing for maximum deer density, and (3) landscape-level deer management.

1.3.1 MANAGEMENT PHILOSOPHY

One of the notions of the developing field of wildlife management early in the twentieth century, as promulgated by Aldo Leopold (1933), was that of managing for a sustainable surplus (harvestable crop for game species). The emerging philosophy was: the higher the abundance of the game species, the better.

The concept was visualized as a bucket of water, which when full represented the maximum number of individuals existing forage and other life requirements would support in a local population. Mortality factors representing "leaks" in the bucket were predation, parasites, and diseases; forage and other cover requirements; overharvest by hunting; and accidents. The combination of mortality factors was seen as countering the contributions of reproduction and especially recruitment to population numbers with the result that the habitat was not carrying as many individuals of the target population as possible and not optimizing the huntable surplus. Reproduction and recruitment were thought of as compensating for the leaks, especially if the magnitude of the leaks could be reduced by management, for example, aggressive reduction of predators, improvement of forage including supplemental feeding, and reduction of hunter harvest of doe deer.

Elimination of historic predators, improvement of forage conditions, and limiting harvest with hunting regulations were highly successful deer management steps after the near-extirpation of deer at the end of the nineteenth century. But the resultant population explosion of whitetail deer beginning in the 1920s resulted in unforeseen economic and ecological impacts on forest resources. As McCullough (1979) stated regarding deer management and the surplus concept, "the concept of surplus has, indeed, become a real problem" and "an obstacle to further advancement of the art."

1.3.2 UNFORESEEN ECONOMIC AND ECOLOGICAL IMPACT OF MAXIMIZED DEER DENSITY

The problem with the initial philosophy regarding game management was that the emphasis was on maximizing population density (maximum sustained yield; MSY) and did not address the unanticipated negative impact on forest vegetation caused by high deer numbers. As we will see in the following chapters, deer do not intrinsically control their numbers. When deer population control by external mortality factors is relaxed by diminution of these factors, deer densities can skyrocket. The result, experienced throughout the white-tail's range, has been depletion of understory vegetation and negative impact on forest resources, including seedling regeneration required to replace removal of overstory trees from natural disturbances, mortality factors, and/or timber harvest. Reduced vertical structure and species composition of forest vegetation has led to reduction in abundance and species richness (biodiversity) of dependent plant and animal communities. Increase in abundance of other wildlife species managed to recovery and higher abundance has not been associated with unacceptably high negative impacts on forest resources. In this regard, deer are unique and their management must be based on bringing their abundance to levels compatible with the health of other forest resources rather than maximizing abundance as a goal that is common for other wildlife species.

1.3.3 LANDSCAPE-LEVEL MANAGEMENT

Protection of deer from overharvest was the overriding initial concern of natural resource agencies charged with managing deer. To this end, deer harvest was, and continues to be, regulated within landscapes (termed deer management units) of hundreds to thousands of hectares assumed to be relatively homogeneous regarding availability and abundance of habitat components for producing sustained yields of deer. If deer abundance within such units was perceived as being up, then season length and bag limits were adjusted upward by natural resource agencies to take advantage of the increased availability for harvest. If abundance was perceived as being down, harvest regulations for management units were adjusted downward. Bases for decisions included harvest records; reproductive information from road-killed does; and irregular, nonrepresentative, and nonrobust surveys for characterizing deer population trends within the management units. Information on deer impact on forest resources at the level of deer management units was limited to complaints from affected landowners within the units.

However, the plain fact is, deer are not comprehensively managed within large, administrative deer management units. Management involves more than adjusting season and bag limits to adjust harvest. Comprehensive management incorporates, in addition, monitoring of deer abundance and impacts, manipulation of structure and composition of habitat components including food and cover, management and maintenance of hunting access to hunting areas, and consideration of human and economic costs in managing hunting and negative deer impacts on forest resources. The actual deer management unit is the individual parcel of land owned and operated by the managers of private forestlands, state and national forests, parks, and other public lands, and is identified in this book as a deer forest management area (DFMA). Because deer and forest management in DFMAs are directed by landowners or stewards, these two groups are heretofore identified as managers, as contrasted to agency professionals overseeing deer management on deer management units (DMUs), who are administrators rather than managers. In fact, the term DMU is an unintended misnomer, because there is no unified deer or forest management universally applied to all forestlands within a DMU. However, because the only management activities conducted by state natural resource agencies on deer administrative units (DAUs; excepting those included in state government forestlands) consist of regulation of deer harvest, DMUs should properly be identified as deer administrative units and are so identified in the remaining portions of this book. A comprehensive discussion of the differences between DAUs and DFMAs is presented in Chapter 9, "Human Factors: Deer–Forest Management Areas vs. Deer Administrative Units." Management activities, including regulations for adjusting deer harvest, should be tailored to DFMAs rather than DAUs, which are artificial deer management units composed of multiple individual forestlands often managed for differing forest resource goals on private and public forestlands.

In summary, deer management philosophy regarding target deer density should incorporate the needs of forestland managers as well as those of persons hunting deer recreationally. Target deer density should reflect concerns for reducing/avoiding negative deer impact on forest resources, and management actions should be developed and applied for controlling deer density as facilitated by hunting regulations. Regulations and management activities affecting deer density and impact should be developed for and applied at the true deer management unit level, the DFMA, as managed by individual forest managers.

1.3.4 ECOLOGICAL AND HUMAN FACTORS AFFECTING DEER DENSITY AND MANAGEMENT

Ecological factors form the biological determinants of deer density and impact—things that may have been influenced by past human activities (e.g., removal of historic predators of white-tailed deer, including mountain lions, wolves, and American natives; creating vast amounts of deer forage by repetitive clearcutting of eastern forests) but are no longer of such magnitude to effect much change in deer density and impact. Ecological factors are quantitatively related to deer density and impact as discerned and described by research—there is little wiggle room for adjusting the nature of these relationships.

Human factors describe human-related activities that influence deer density and impact that, in concert with ecological factors, make up the essential components that must be addressed

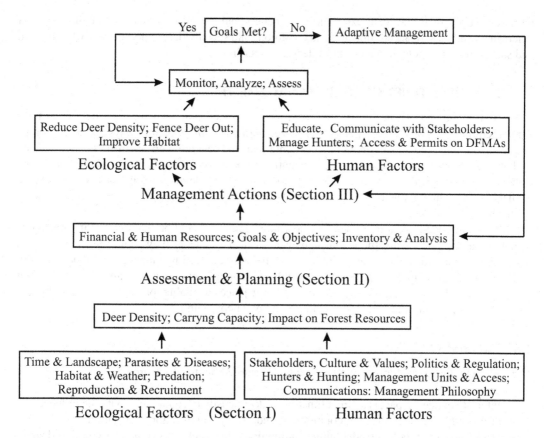

FIGURE 1.1 Factors and components involved in developing, delivering, and adjusting deer management actions to meet goals for forest resources as affected by deer.

and incorporated in development and delivery of proactive deer management programs. Deer management within forestlands is an exceedingly complex undertaking, as evidenced by the many and interconnected components of Sections I–III as arrayed in the following flow chart (Figure 1.1).

Ecological and human factors interact to determine deer density and impact. Such impacts may be categorized as they affect varying goals for forest resources, including enhancement/protection of biodiversity, regeneration of desirable tree species, production of quality deer, and maximization of deer abundance. These categories are associated with respective deer densities referred to as carrying capacities. Each carrying capacity is associated with a level of deer impact on specific forest resources and on management of structure and species composition of forest vegetation (silviculture). They are termed "cultural carrying capacities" (Ellingwood and Spignesi 1986) because they represent levels of deer density associated with impacts on the values and culture of different stakeholder groups (e.g., hunters, farmers, foresters, ecologists).

1.4 TENET, PREMISES, AND PRACTICALITIES

The book is grounded upon a guiding tenet and on premises, practicalities, and adaptiveness specific to deer management. Influencing variables, including perceptions of stakeholders, uniqueness of local habitats, and history of local/regional deer management efforts and issues specific to local landscapes, shape deer management. It is essential that managers interacting with local hunters and other stakeholders be well versed in the local history of deer management and stakeholder culture. This is where "art" and "craft" come into play, as components of integrated deer management must be based on practical and local experience. Managers and consultants are the face and voice of landowners and must relate credibly to local conditions and stakeholders to describe, justify, support, and successfully promote management of deer and deer impact.

1.5   PRINCIPAL TENET OF DEER MANAGEMENT

Individual public and private forestlands form the basic deer management unit—deer meet their habitat requirements on these properties, forest vegetation is manipulated in ways that affect forage and cover, hunting access roads are built and maintained, and deer numbers are (usually) manipulated by hunting. When deer are too abundant, negative impacts on forest resources cause economic, esthetic, and ecological losses to owners of these lands. These lands form the bulk of forestlands available to deer hunters who mostly hunt on the lands of others, as most do not own forestlands upon which to hunt deer.

The acknowledged overarching tenet of deer management should be that owners and managers of individual forestlands are the actual managers of deer. As such, they should receive as much responsible, informed, and regulatory latitude and assistance regarding management of deer density and impacts as municipal, state, and federal wildlife management agencies can provide. Without these lands and landowners, there would be far fewer places available for deer hunting, and much less quality habitat for other species in wildlife and plant communities.

1.6   PREMISES

The underlying premises for deer management as promoted by this book are:

- Intensity of deer impact is a direct function of deer density in conjunction with the influence of forest type and climate and condition of understory and overstory vegetation.
- Deer impact can be controlled either by manipulating deer density or by excluding deer from susceptible forest resources.
- For healthy and diverse forests. managers must strive for deer densities historically in harmony with balanced ecosystems and sustainable production of forest resources, tangible and intangible.
- For forest management to be sustainable, forest resources of interest may only be harvested at rates that are sustainable, requiring that: (1) only a prescribed percentage of trees may be harvested on a schedule that guarantees a succession of forest stages, each specific to a category of understory and overstory vegetation, and (2) only prescribed percentages of other forest vegetation and individual wildlife species may be removed/harvested on a schedule that prevents overharvest.
- VerCauteren and Hyngstrom (2011) noted that white-tailed deer density, while historically controlled by external mortality factors, including predation, disease, starvation, accidents, and old age, presently is almost completely controlled by hunting and collisions with vehicles. If predatory pressure (now primarily through hunting) is relaxed, deer density quickly escalates and may result in a population crash caused by massive starvation loss. Annual deer removals by hunting must be maintained or deer density will rapidly escalate to prehunting levels, overabundance, and negative impacts on forest resources, and may result in massive starvation deaths of deer.
- The best approach for achieving and maintaining desired deer density is public hunting, excepting situations where the landowners are unable or unwilling to allow public hunting, in which case deer exclusion (fencing), targeted deer removals by special permits, or provision of alternative forages may be viable options.
- The complexities of deer management require solid grounding and application of the science behind wildlife and forest management; social/cultural factors influencing stakeholders;

political and economic considerations; recruitment, education, motivation, and retention of hunters; and annual (or at least recurring) appraisals of the status of deer density and impacts on forest resources as required management actions.

- Management activities must be applied with persistence and in perpetuity—factors affecting deer density and impact are dynamic and must be monitored and addressed on a recurring (preferably annual) basis.
- All components of developed management systems should undergo recurring assessment/ monitoring to identify needed changes in management activities as determined by whether management goals have been achieved.
- Managers must be able and willing to adapt individual components of their management program for continual improvement and achievement of resource goals (e.g., adaptive management).

1.6.1 PRACTICALITIES

Certain realities governing deer management must also be recognized and incorporated into deer management:

- Who owns the deer and responsibility for deer impacts? A major legal reality that must be recognized and accepted by landowners, hunters, and natural resource administrators is that in the United States, wildlife species, including deer, are not the property of the persons or entities owning or managing public or private forestlands. Rather, in all states, deer are identified as a common resource owned collectively by all citizens. Deer management, including harvest, is regulated by state and/or federal resource agencies. As such, landowners whose habitat is used by deer are not responsible for damages those deer may inflict on abutting lands owned by others. Also, such landowners cannot indiscriminately reduce deer density on their properties by hunting, except as allowed by established hunting regulations. Respective state and local governing agencies are responsible for mitigating such damage by reducing deer density through public hunting or by special permit to landowners. Thus, managers may only reduce deer density and impact within their forestlands through management practices (such as hunting) as regulated/ permitted by their respective state natural resource agencies. The exceptions are licensed landowners who raise wildlife for recreation within private, fenced hunting preserves and/ or landowners who raise wildlife as domestic livestock for sale of meat and other products for human consumption.
- Who manages (and pays for) deer density and impact? Forestland managers often must reduce deer density and impact before they can meet goals for forest resources, including other wildlife species and timber. In many cases, deer impact can be managed by reducing deer density with public hunting. Deer habitat, road maintenance and road access to hunting areas, and deer density and impacts are managed at the individual property level. Costs of these activities are borne by property owners with little to no financial compensation provided by natural resource agencies or the hunters who hunt deer on their properties.
- Failure of effective communication between managers and deer hunters is a serious impediment to deer management. Granted, there is value in reduction of deer density and impact resulting from hunters under the implicit bargain that hunters are allowed to hunt on these properties in exchange for reducing deer density. However, many hunters become upset with reductions in deer density resulting from hunting and lobby natural resource agencies to enact stricter hunting regulations to reverse declining deer densities. This situation results from poor communication among, and education of, forestland owners, natural resource agencies, and hunters regarding landowner needs and management activities, including associated costs.

- In a perfect world, science would provide quantitative and universally accepted directives for controlling deer density and impact through hunting. In the real world, forestland managers have to address a number of cultural and political factors that often override science. The Kinzua Quality Deer Cooperative case history (Chapter 34) is one such example: managers wanted reduced deer density and impact so they could produce timber products economically, while hunters, who could reduce deer density by harvesting deer, wanted more deer. The case history on testing the quality deer management approach (Chapter 40) is another example: the owners wanted higher deer density, as occurred in the 1950s, while the deer biologist and forester advised them that they could not have higher deer density and a healthy forest and deer herd. In a test of wills, nature wins over human desires.

- Complicating deer management is the disconnect between natural resource agencies and landowners (deCalesta 2017). Local, state, and federal agencies determine conditions (season and bag limits) under which deer may be harvested and by whom. Such regulations are usually designed for large administrative units (often referred to as deer management units) consisting of hundreds to thousands of hectares. These management units encompass individual forestlands managed by a wide variety of owners with differing goals for deer and deer impact and render individualized deer management impossible. Forestland managers are often constrained by deer harvest regulations developed for the large management units the properties they manage lay within, wherein harvest regulations are too restrictive to produce needed reductions in deer density and impact.

1.7   ROLE OF VALUES AND CULTURE IN DEER MANAGEMENT

Before managers of deer-impacted forestlands can apply science to reach natural resource goals, they must identify and deal with values and cultures of key stakeholder groups that often disregard science as the basis for management. Values of stakeholders (especially hunters), politics, regulations, communications, and costs of deer management must be identified and integrated with science-based deer management to manage deer and other forest resources.

1.7.1   Requirement for Integrated Deer and Forest Management

The above-noted requirements for managing deer density and impact on individual forestlands (DMAs) can be overwhelming for managers/owners who may have limited training and experience in forestry, let alone deer or wildlife management. Seeking help from wildlife and forestry professionals separately, rather than in concert, often produces conflicting advice and unsatisfactory results. Considerably more assistance, communication, and cooperation among forestland owners, stakeholders, wildlife and forestry professionals, and resource agency administrations are required for deer management than for timber management—it is not enough to manage deer alone, because the multiple publics that influence deer management affect what can and cannot be done to manage deer and their impacts.

In addition to meeting the challenges posed by the factors identified above, successful deer management requires a thorough grounding in, and application of, deer and forest science. Equally important are: application of quantitative methodology for monitoring deer density, deer impact, and hunter values and satisfaction; collaboration and cooperation among landowners, deer hunters, biologists, and natural resource agency field and administrative personnel; resolution of the disconnect between quantitative deer science and qualitative hunter values; and development, implementation, and maintenance of comprehensive management plans.

To address these multiple needs and challenges, this book integrates the processes of goal setting, inventory, monitoring, interpretation, and development and implementation of plans for managing deer in eastern forestlands. The emphasis is on working with the multiple publics/stakeholder groups that define and determine the success of managing deer and impacts. Establishing and maintaining effective communications that foster productive and enduring interactions between forest/deer managers and multiple stakeholders is featured as a major process for achieving management goals. The overarching need is for education of stakeholders and property owners. Deer and forest science must be understood and promoted as the quantitative basis for management of both resources. Equally so, the values, culture, and goals of all stakeholder groups regarding deer and deer management must be communicated among groups, given credence and respect, and integrated with relevant science. These tasks must all be addressed if managers are to be successful in managing deer and deer impacts.

1.8 ADAPTIVE MANAGEMENT

Managers of deer and forests learned long ago that research and resulting management recommendations constantly evolve, creating new information that sometimes conflicts with established management practices and protocols. They also learned that monitoring could uncover needed changes in management applications as they pursued goals for management of their forest resources. For survival, their operating mantra was to adapt new approaches to problems that were either too specific to their properties to be addressed by current recommendations, or that simply did not produce desired improvements in progress to their goals for forest resources. This trial-and-error approach has been identified and described formally as *adaptive (resource) management,* a necessity that managers recognize, support, and recommend incorporating into evolving deer and forest management planning and decision-making. More than just a recommendation for adapting ongoing management practices, we advocate adjusting and improving, as needed, *all* components of deer management, including: monitoring; vegetation manipulation; educating, recruiting, rewarding, retaining, and communicating with hunters; identifying and addressing stakeholder needs and realities; controlling deer density and impact; and addressing political and economic factors affecting deer management.

REFERENCES

deCalesta, D. S. 2017. Bridging the disconnect between agencies and managers to manage deer impact. *Human Wildlife Interactions Journal* 11:112–115.

Ellingwood, M. and J. V. Spignesi. 1986. Management of an urban deer herd and the concept of cultural carrying capacity. *Transactions of the Northeast Deer Technical Committee* 22:42–55.

Frye, R. 2006. *Deer wars: Science, tradition, and the battle over managing whitetails in Pennsylvania.* State College PA: Pennsylvania State University.

Halls, L. K. (ed.) 1984. *The white-tailed deer of North America.* New York: Stackpole Books.

Hewitt, D. (ed.) 2011. *Biology and management of white-tailed deer.* Boca Raton FL: CRC Press.

Huxley, T. 1903. *The Life and Letters of Thomas Henry Huxley,* Volume 1. Teddington United Kingdom. Echo Library.

Leopold, A. 1933. *Game management.* New York: Charles Scribner's Sons.

McCullough, D. R. 1979. *The George Reserve deer herd.* Ann Arbor MI: The University of Michigan Press.

McShea, W. J., H. B. Underwood, and J. H. Rappole. 1997. *The science of overabundance: Deer ecology and population management.* Washington: Smithsonian Institute Press.

Northeast Section of the Wildlife Society. 2015. *Position statement on managing chronically overabundant deer.*

Taylor, W. P. 1956. *The deer of North America: Their history and management.* Harrisburg PA: Stackpole Books.

VerCauteren, K. and S. E. Hyngstrom. 2011. *Managing White-Tailed Deer: Midwest North America. Papers in natural resources 380.* Lincoln NE: University of Nebraska.

Section I

Ecological and Human Factors

In reality, we haven't escaped the gravity of life at all. We are still beholden to ecological laws, the same as any other life-form

Janine M. Benyus (1997)

The problem of game management is not how we should handle the deer ... the real problem is one of human management

Aldo Leopold (1943)

I.1 INTRODUCTION

Forest managers must become familiar with the many interacting ecological and human factors affecting deer and how these factors influence deer, their density, and their impact on forest resources before developing and executing deer management plans. This section describes the ecological influences on deer as derived from science and the additional influences exerted by human interactions on deer, their ecology, and their management. Both must be incorporated as equals in developing and delivering effective deer management plans to achieve goals for forest resources.

I.2 ECOLOGICAL FACTORS

Ecological factors associated with deer life-cycle requirements have been the primary go-to sources of information managers have relied upon to develop plans for managing deer density and impact. These factors are divided into autecological and synecological. Deer autecology, or population ecology, is the study of deer and their passive relationships/interactions with the physical environment, including weather, food and water, habitat, and the temporal and spatial arrangement of all as related to deer use of habitat and deer reproduction and recruitment. Deer synecology, or community ecology, refers to active relationships/interactions between deer and other organisms (plant and animal) in their environment, including impacts by deer on these other organisms. Deer autecology and synecology are reviewed in Chapters 2 through 8.

Autecological and synecological factors are considered limiting when they "determine or negatively affect a species in an ecosystem" (https://www.reference.com/science/limiting-factor-ecology). Managers must understand and incorporate components of autecology and synecology as limiting factors if they are to conduct deer management comprehensively and successfully.

I.3   HUMAN (DIMENSION) FACTORS

Beginning in the 1970s, social scientists began describing an additional and essential component of wildlife management—that of human dimensions. Decker et al. (2012) prefaced their book on human dimensions of wildlife management by stating that, "Wildlife professionals can more readily manage species and social-ecological systems by fully considering the role that humans play in every stage of the process." Decker et al. (2004) earlier stated that, "When the human and biological dimensions of wildlife management are well-informed by research and effectively integrated, wildlife management can move forward," and they identified four ways that wildlife management utilized the field of social science—by understanding: (1) how people value wildlife, (2) the benefits people seek from wildlife management, (3) how people accept elements of wildlife management, and (4) how different stakeholders affect or are affected by wildlife and wildlife management decisions.

Human dimension factors are reviewed in Chapters 9 through 15. Regarding the four categories identified by Decker et al. (2004), how people value deer is addressed by Chapters 10 and 12, benefits people seek from deer management are covered by Chapters 9 and 20, acceptance of stakeholders of deer management is covered in Chapters 15 and 23, and how stakeholders are affected/affect deer management decisions is covered in Chapters 13 and 15.

Integrating ecological and human dimension factors in deer management is presented in Chapter 25.

REFERENCES

Benyus, J. 1997. *Biomimicry: Innovation Inspired by Nature*. New York: Harper Perennial.
Decker, D. J., T. L. Brown, J. J. Vaske 2004. Human dimensions of wildlife management. In *Society and Natural Resources: A Summary of Knowledge*. 10th International Symposium on Society and Natural Resources. ed. M. J. Manfredo, J. K. Vaske, D. Field 187–198. Jefferson, MO: Modern Litho.
Decker, D. J., S. J. Riley, and W. F. Siemer, eds. 2012. *Human Dimensions of Wildlife Management*, 2nd edition. Baltimore, MD: Johns Hopkins University Press.
Leopold, A. 1943. In Flader, S. L. 1974. *Thinking Like a Mountain: Aldo Leopold and the Evolution of an Ecological Attitude toward Deer, Wolves, and Forests*. Columbia, MO: University of Missouri Press.

2 Autecology
Weather, Forest Type, and Habitat

David S. deCalesta and Paul D. Curtis

CONTENTS

Know thy enemy.

SunTzu (Fifth century BC Chinese military treatise)

MANAGER SUMMARY

Deer density and associated impact on forest resources are directly influenced by weather; forest type; and presence, absence, and/or condition of deer habitat requirements as affected by species composition, vertical forest structure, and proportion of different habitat components provided by forest succession. These factors vary by geographic region and forest management practices, requiring the combined inputs from deer and forest biologists to help assess and manage habitat conditions affecting deer density and impact.

2.1   INTRODUCTION

Deer life-cycle requirements are food and water, regulation of body temperature (thermal regulation), and reproduction and recruitment. Physical and structural components of the environment (habitat), as affected by weather and forest type, make up the elements affecting deer autecology. The difference between weather and climate is time: weather is conditions of the atmosphere are over a short period of time (years), and climate is how the atmosphere "behaves" over relatively long periods of time (decades to millennia) (NASA 2015—https://www.nasa.gov/mission_pages/noaa-n/climate/climate_weather.html). If landowners and managers are to manage deer and their impact on forest vegetation, and the wildlife species dependent on forest vegetation, they must have in-depth knowledge of deer autecological factors as they affect deer density and impact.

Deer relate to weather and the quality, quantity, and juxtaposition of habitat components over time with a number of physiological and behavioral adaptations. Habitats containing high amounts of quality forage and providing cover from weather extremes and predation will support more deer than habitats that are forage and cover poor. Mortality factors, such as starvation and exposure, can reduce deer welfare and abundance.

2.2 WEATHER AND FOREST TYPE

Environmental factors that influence and determine deer density and impact vary as affected by weather and forest type. Weather includes precipitation, wind, temperature, humidity, and duration and intensity of sunlight. Geology and latitude, in concert with climate, produce a number of forest types, each providing differing qualities and quantities of deer habitat requirements. Forest types include the boreal forests of the northeastern United States and Canada, northern and Appalachian hardwood forests of the northeastern United States, central broad-leafed and oak forests of the midwestern and southeastern United States, and southeastern oak and pine forests (Young and Giese 2003).

The 1984 compendium on ecology and management of white-tailed deer (*Odocoileus virginianus*) (Halls 1984) described deer management within 18 different forested habitats; the most recent version (Hewitt 2011) cut that number to three: eastern North America, midwestern North America, and western North America, and identified an additional category based on high human population— exurban, suburban, and urban. Whatever habitat classification forestland managers choose to develop plans for forest and deer management, the requirement is the same: make sure deer, forest science, and sources of information chosen for reference are as good a match with their forest type and climate as possible, especially concerning deer density and impacted forest resources. Using input from forest and deer specialists with specific experience and expertise in relevant forest types, as well as forest management related to even- and uneven-aged management, is critical. As identified in the preface, deer and forest science and relevant management expertise selected for illustration in this book were drawn from the northern hardwood forest type, as found in Michigan, Wisconsin, Minnesota, New York, Pennsylvania, West Virginia, Virginia, Massachusetts, Vermont, and New Hampshire.

2.3 HABITAT

The impact deer have on forest resources is related to how many of their habitat requirements are met within individual forestlands in concert with habitat components provided in the surrounding landscape. As quality, quantity, and completeness of habitat components within individual properties increase, so will the use of these properties by deer. This increases the potential for negative impact on vegetation. Habitat needs include composition and vertical structure of vegetation (for forage and protective cover), soil composition, and water. Forestlands larger than 200 ha may contain all habitat requirements for individual deer family groups (does, female yearling offspring, and fawns); lands encompassing 500 or more ha may provide habitat needs for multiple family groups as well as for nomadic bucks. Forestlands smaller than 200 or so ha may provide only some deer habitat requirements and may be used seasonally by deer rather than year-round. Whether deer use and impact these smaller properties is dependent on the degree to which the surrounding landscape provides complimentary habitat requirements. Hence, it's important to determine whether forestlands and surrounding landscapes provide some or all deer habitat requirements to gain a perspective on potential deer impacts.

2.3.1 FOOD

Like cattle, deer have a unique digestive system featuring four different stomachs. The first, the reticulum, is a small outpouching of the next and larger stomach, the rumen. Named the "hardware stomach" in cattle, the reticulum serves as a collector of large, ingested solid objects, like stones, to prevent them from passing on to subsequent stomachs where they could clog up the system. The reticulum also moves small, digested particles of forage for additional digestion to the other stomachs. The rumen, the largest stomach, is a fermentation vat where browsed material is mixed with water and converted by microbes into energy packets called free-fatty acids, and into microbial protein which is passed on to subsequent stomachs for absorption. The third stomach in line, the omasum, acts as a screen to prevent large undigested fibrous material from moving prematurely into the last stomach, the abomasum. In the abomasum, rumen microbes are digested to provide deer with protein, and fats (liberated from prior digestion of forage in the rumen) are broken down and absorbed.

Woody and herbaceous vegetation provides deer with energy, protein, and minerals for movement, maintenance of body temperature, growth, and reproduction. Woody vegetation (plants with above-ground woody vegetation that persists from year to year) includes trees, shrubs, vines, and some ferns. Herbaceous vegetation (plants with no above-ground woody parts that persist from year to year) includes some fern species, grasses, sedges, rushes, and forbs. Forbs are flowering nonwoody plants (including wildflowers) that are not grasses, rushes, or sedges.

Soils provide mineral elements, such as phosphorus, calcium, and magnesium required for bone (and antler) growth that are obtained from plants that take them up. Minerals may be obtained from other sources such as licks—places where minerals are concentrated in soils and deer can obtain them by licking the soil.

Deer use a wide variety of foods to meet their nutritional needs on a seasonal basis. They forage on high-energy foods such as nuts (e.g., acorns, beech nuts, and other high-energy seeds), berries, and agricultural crops (e.g., alfalfa, corn, and beans) in late summer and fall to store body fat. This fat is metabolized in winter to keep deer warm through cold winds, low temperatures, freezing precipitation, and food scarcity. Their unique digestive system allows deer to consume and digest woody stems from ferns, seedlings, and shrubs and turn these plants into energy for movement, temperature regulation, and reproduction, and into fat for winter survival. Deer browse on protein-rich woody growth (e.g., twigs and leaf buds) during winter and early spring for protein and on herbs during spring and summer for energy needs, and use both for fawn development during gestation.

2.3.2   FOOD PREFERENCES

Deer exhibit definite preferences for some forage species. As a general rule, they prefer plant species high in easily digestible carbohydrates and nitrogen, such as red maple (*Acer rubrum*) and white oak (*Quercus alba*), and shun species with high fiber content such as American beech (*Fagus grandifolia*) and striped maple (*Acer pensylvanicum*). These less-preferred plants often are heavily browsed in winter as "starvation" foods—possibly because more digestible species have already been nearly eliminated by deer browsing. Unfortunately, many preferred browse species are of high commercial value as mature trees for timber and as sources of soft mast (cherries) and hard mast (acorns) for wildlife. Deer replenish depleted energy reserves in spring by foraging on emerging herbs, which are high in digestible energy. In most years, summer is a season with high availability of nutritious foods that all deer use to store fat for winter, and does use for production of milk for their fawns. The extent to which forested lands provide these seasonal foods influences deer use and subsequent damage to woody, shrubby, and herbaceous vegetation.

2.3.3   WEATHER

Precipitation (rain and snow, primarily), temperature, and wind conditions are collectively called *weather*, as occurring on an hourly, daily, and seasonal basis, and can become limiting if extreme and lead to deer mortality. Winter snow depth and duration have a major impact on northern deer herds (Hewitt 2011). Deer accumulate fat reserves during fall, and use them for energy when forage is limited in winter. In a typical winter, deer may lose 20%–30% of their body mass (Mautz

1978). Parker et al. (1984) described the high energetic costs of deer moving through deep snow. Malnutrition, although probably always causing some deer deaths even during moderate winters, may cause massive mortality when deep snow cover exceeds 100 days (Saunders 1988). Deer voluntarily restrict food intake and activity during winter (Thompson et al. 1973), as the cost of searching for food may be greater than the energy obtained from low-quality winter browse.

In areas with harsh winters including snow, deer may concentrate in *deer yards* composed of blocks of conifer trees such as hemlock (*Tsuga canadensis*), balsam fir (*Abies balsamea*), or Eastern white cedar (*Thuja occidentalis*) that cut wind flow, reduce heat loss to the environment, intercept snow and reduce depth, and provide survival food. Deer may migrate more than 100 km from spring-fall home ranges to deer yards for winter survival. Deer yards may be wholly contained in landscapes smaller than typical home ranges. If a forest property contains a deer yard, the understory vegetation in the yard may be subjected to recurring and severe deer impacts, and it may be difficult to develop or maintain an abundance or diversity of tree seedlings, shrubs, or herbaceous vegetation within them.

2.3.4 THERMAL REGULATION

Precipitation, wind, and temperature, singly or in combination, may produce heat/cold stress on deer if they exceed threshold levels. Deer do not sweat to reduce internal temperature on hot days, but pant and seek out cool, moist environments such as riparian zones with an overstory of conifers/ leafy deciduous trees where there is less heat absorption from the sun and where they can lie upon cooled ground and drink water to lower body temperature.

The deer's coat of dense, hollow hair provides good insulation from cold temperatures, but when it is cold and windy enough, deer use additional ways to reduce heat loss. Bedding on the ground with their legs under them beneath dense conifer canopies reduces heat loss to wind flow. Dense overstory cover provided by conifer trees reduces radiant heat loss to the atmosphere, and some energy may be reflected downward to warm the microenvironment of bedded deer.

Deer will concentrate in areas that reduce heat uptake or loss, depending on the season. Consequently, additional time spent in these areas may result in increased foraging on understory vegetation with resulting negative impacts on vegetation and habitat. If forestlands provide coniferous thermal cover from heat and/or cold weather, they may experience additional browsing impact by deer.

2.3.5 WATER

Water is involved in all these nutritional functions for conversion of foods to muscle, bone, and other tissues; movement of nutrients and waste products within and out of the body; and regulation of body temperature (cooling in warm temperatures). Deer must consume considerable amounts of water daily (3–6 liters, some of which may be obtained from forage) to maintain the required volume in their stomachs for efficient digestion of woody browse and other vegetation. Most habitats provide readily available water from springs, bogs, streams, rivers, ponds, and lakes—but deer will travel considerable distances daily to obtain required water. Indeed, deer daily movements to sources of water (e.g., stream edges) are so predictable that mountain lions frequent riparian zones for ambushing them (Gagliuso, unpublished MS thesis, Oregon State University).

2.4 HABITAT AND FOREST SUCCESSION

Species composition, vertical structure of vegetation, and proportions of different habitat conditions determine the nature of habitat components available to deer. Forest vegetation is dynamic, changing as affected by environmental factors (e.g., wind, fire, ice storms, drought, insect defoliations, and soil composition) and human activities (e.g., timber harvest and thinning, land clearance for development such as oil/gas wellhead sites or residential/commercial development).

Once a disturbance clears most, if not all, trees from a site, a progression of stages of forest growth occurs, called *stand development stages* by foresters (Oliver and Larson 1996) and *seral stages* by ecologists (Odum 1983).

For clarification, the basic forestland unit manipulated by foresters is the stand, identified by Nyland (2002) as, "a contiguous community of trees sufficiently uniform in (species) composition, structure, age and class distribution, spatial arrangement, site quality, condition, or location to distinguish it from adjacent communities." The definition is extended to forestland units created by natural disturbances (e.g., fire, windthrow, ice storms).

Natural disturbances (windthrow, fire, disease, insect infestation) leading to death of individual trees open the forest overstory and create a succession of habitat configurations varying in size, timing, and frequency as affected by randomness of the disturbances. When death of single trees or small groups of trees occurs repeatedly within a landscape, the limited amount of light reaching the forest floor in the small openings favors growth and domination by tree seedlings tolerant of low lighting conditions (shade-tolerant). When these small openings occur at irregular intervals within a landscape, multiple-aged groups of shade-tolerant tree seedlings such as hemlock (*Tsuga canadensis*) and American beech (*Fagus grandifolia*) dominate, resulting in an uneven-aged structure of trees that are primarily shade tolerant. When the openings are of much larger size, ranging from single to hundreds of hectares, development of shade-intolerant tree species is favored, and trees of a single age class predominate—such landscapes are classified as even-aged.

Depending on the forest condition desired by forestland managers, sufficiently large patches of trees may be harvested to create an even-aged structure and species composition, or small groups of trees or single trees may be harvested to promulgate an uneven-aged structure and species composition. In their evaluation of 99 studies comparing stands managed for even- and uneven-aged species composition and structure, Nolet et al. (2017) found that diversity of resulting vegetation varied among even-aged and uneven-aged managed stands (23 studies indicated higher diversity in uneven-aged stands, 16 indicated higher diversity in even-aged stands, and 60 indicated no differences in diversity), with the overall result being no consistent difference in plant diversity between the two management practices, notwithstanding the (expected) difference in composition of shade-tolerant and shade-intolerant species.

2.4.1 EARLY SUCCESSION HABITAT

In the first stage (*stand initiation* to foresters, *early succession* to ecologists), after removal of overstory trees, increased sunlight on the forest floor, and less competition for water and nutrients, herbs (grasses, sedges, rushes, and flowering vascular plants called forbs), shrubs, and tree seedlings dominate in a layer generally under 1 m tall. Depending on opening size, dominant tree species are shade tolerant or shade intolerant. Resultant deer forage in the openings is abundant, high in nutritive quality, and highly digestible. Hiding cover for fawns may develop if the openings are large enough to produce patches of dense understory vegetation large enough to hide fawns from predators. Because even-aged stands provide a higher landscape proportion of area in openings within managed landscapes than uneven-aged stands, there will be more forage and dense understory vegetation in even-aged stands. From a deer management perspective, we will call this the *forage and fawn cover* stage.

2.4.2 FORAGE AND FAWN COVER HABITAT

The forage and fawn cover stage persists for ~10–15 years, after which tree seedlings begin to dominate the site, surpassing 2 m in height and blocking passage of sunlight to understory herbs, shrubs, and ferns. Understory vegetation begins to thin and tree seedlings advance to the sapling-pole stage, characterized by stem diameters of growing trees exceeding 5 cm in diameter at 1.4 m (called diameter at breast height or *dbh*) above ground level. Because of intense competition for sunlight, some of the sapling-pole trees die in what foresters call the *stem exclusion* stage and ecologists call the *mid-succession* stage. Deer are still able to obtain forage from this stage, but also heavily use it as protective cover, as the dense thickets of closely growing trees make it harder for predators (including hunters) to see and kill them. From a deer management perspective, we will call this the *hiding cover* stage.

2.4.3 HIDING COVER HABITAT

The hiding cover stage starts at about 15 years after the succession process starts and lasts for about 20 years. Dbh of sapling/pole trees begins to exceed 25 cm. Height of trees begins to exceed 7 m, and when it includes conifer trees, it provides deer with thermal cover as it reduces amounts of rain/snow falling on deer, cuts wind flow, and may reflect some escaping heat back downward to deer. Because a higher proportion of landscapes under even-aged management feature openings in the overstory, there will be higher proportions of hiding cover in these landscapes than in landscapes managed for uneven-aged structure and species composition.

As the overstory trees continue to self-thin (and some die), more sunlight reaches the ground. Seeds cast by overstory trees begin to germinate and grow, and herbs and shrubs become more abundant. Deer forage becomes more prevalent. Foresters call this stage *understory reinitiation* and ecologists call it *late succession*. From a deer perspective, we will call it the *thermal cover* stage.

2.4.4 THERMAL COVER HABITAT

The thermal cover stage persists beyond 35 years after the original disturbance as both understory and overstory continue to grow in abundance. Some trees fall out, having lost out in the competition for sunlight and nutrients, and the resulting openings in the overstory canopy allow more sunlight to reach the forest floor, stimulating germination and growth of tree seedlings. Dbh of maturing trees continues to increase and foresters begin to harvest some trees that have attained commercial-value diameter (>40 cm dbh). By this time, some decades have passed since the site was disturbed sufficient to initiate stand development. Deer continue to use this stage for foraging and thermal cover, especially under uneven-aged management where shade-tolerant hemlock trees provide year-round thermal cover. Increasing density of understory/midstory vegetation once again begins to provide deer with hiding cover.

On sites where foresters have begun to harvest trees, the amount of tree removal may cause the site to revert to the forage and fawn cover stage. Sites not harvested for trees progress in age and size (dbh of trees may exceed 60 cm). Overstory trees continue to die and fall out of the site, and emerging sapling/pole trees produce additional height/diameter classes of trees as the sites begin to develop an uneven-aged structure (multiple layers of trees of varying diameter, height, and species composition favoring tree species that are tolerant of shade). Ecologists call this the *climax* stage; biologists call it *old growth*.

2.4.5 OLD GROWTH HABITAT

Old growth is characterized by tree species and vertical structure typical of a "steady state." Deer use this stage for foraging, hiding, thermal cover, and may use it for fawning cover. From a deer perspective, this could be labeled the *multi-purpose cover* stage that can sustain deer herds when densities are low enough to prevent overuse of forage.

Forestlands containing all deer-related cover stages provide all deer requirements and will sustain deer populations indefinitely, especially if large enough (>2 ha) to include sufficient sizes of different deer habitat stages. When two or more deer cover stages are adjacent, deer concentrate along common boundaries in what is known as the "edge effect." Forestlands that lack a complete complement of deer habitat stages but are adjacent to forestlands providing missing stages will also sustain deer herds.

REFERENCES

Halls, L. K. (ed.) 1984. *White-tailed deer ecology and management.* Harrisburg PA: Stackpole Books.

Hewitt, D. G. (ed.) 2011. *Biology and management of white-tailed deer.* Boca Raton FL: CRC Press.

Mautz, W. W. 1978. Sledding on a bushy hillside: The fat cycle in deer. *Wildlife Society Bulletin* 6:88–90.

Nolet, P., D. Kneeshaw, C. Messdier et al. 2017. Comparing the effects of even- and uneven-aged silviculture on ecological diversity and processes: a review. *Ecology and Evolution* 8:1217–1226.

Nyland, R. D. 2002. *Silviculture, concepts and applications.* New York: McGraw Hill Co.

Odum, E. P. 1983. *Basic ecology.* San Diego CA: Harcourt Brace College.

Oliver, C. P. and B. C. Larson. 1996. *Forest stand development.* Hoboken NJ: John Wiley and Sons.

Parker, K. L., C. T. Robbins, and T. A. Hanley. 1984. Energy expenditures for locomotion by mule deer and elk. *Journal of Wildlife Management* 48:474–488.

Saunders, D. A. 1988. *Adirondack mammals.* Syracuse NY: State University of New York, College of Environmental Science and Forestry.

Thompson, C. B., J. B. Holter, H. H. Hayes et al. 1973. Nutrition of white-tailed deer. I. Energy requirement of fawns. *Journal of Wildlife Management* 37:301–311.

Young, R. A. and R. L. Giese (eds.) 2003. *Introduction to forest ecosystem science and management.* 3rd edition. Hoboken NJ: John Wiley and Sons.

3 Autecology
Landscape and Temporal Dimensions

David S. deCalesta

CONTENTS

Think landscape, act site scale.

Gregory Ruark (2010)

Think for tomorrow, but act for today.

Attributed to
Mahatma Ghandi

MANAGER SUMMARY

Arrangement in time and space of habitat features in surrounding landscapes affect deer density and impact and the manager's ability to manage both on forestlands. Managers must be able to assess how changes in habitat features through time and spatial arrangement influence deer density and impact so that they may direct and alter management activities that reflect influences of time and spatial factors.

3.1   LANDSCAPE (SPATIAL) EFFECT

Deer are landscape animals, and properties managed by landowners/stewards are embedded within the larger landscape that includes others' properties and deer thereon. Deer from these other properties may use habitat on managers' forestlands in ways that lead to impact on forest resources. In their chapter on spatial use of landscapes by deer in Hewitt's (2011) comprehensive book on white-tailed deer (*Odocoileus virginianus*), Stewart et al. (2011) described the relationship between deer and landscapes by focusing on landscape scale and deer movements (including migration and dispersal), home range, habitat selection, deer response to heterogeneity and human uses of landscapes, impact of deer on landscape vegetation components, and how landscapes are used differently by male and female deer. However, Steward et al. (2011) did not discuss or describe how spatial arrangements of habitat components affect impact of deer on forest resources, but did describe how deer overabundance can effect landscape shifts in vegetative species composition that affect dependent wildlife species. They also noted the role of deer in long-distance transport of seeds (ostensibly of tree, shrub, and herbaceous species).

Deer abundance on managed forestlands increases when deer move onto properties from adjacent lands and/or when fawns born during the previous summer are recruited (by surviving summer, fall, and winter mortality factors) into the population as yearlings. Deer movement onto managed forestlands from adjacent properties may occur when habitat resources are less available on adjacent properties than those on the managed forestland being managed. Examples would be an increase in forage production in a managed forestland resulting from timber harvest, and opening of the overstory canopy for increased sunlight on the understory. Or, the amount of forest habitat could decrease on an adjacent area resulting from commercial or residential development of the landscape.

Landowners can manage deer and vegetation only within the boundaries of their land but are at the mercy of how vegetation and deer are managed within the surrounding landscape, over which they have little if any influence/control. Therein lies a major deer management conundrum—deer and their impact are managed within individual properties actively through deer harvest/exclusion and vegetation manipulation, but regulations regarding how deer may be managed through hunting are applied at larger landscape levels by natural resource agencies—within deer management units numbering hundreds to thousands of hectares (see Chapter 9 and deCalesta 2017).

To the known impact of deer density on forest resources, researchers (Rutherford and Schmitz 2010, Hurley et al. 2012, Barrett and Schmitz 2013) added landscape configuration and level and characteristics of landscape vegetation (deer habitat, including forage and thermal cover, representing the degree to which deer habitat requirements are met within the Deer–Forest Management Area (DFMA) as opposed to presence in the surrounding landscape).

Other research related deer impact on agricultural crops (soybeans and corn) to characteristics of surrounding forestlands (e.g., deCalesta and Schwendeman 1978, Stewart et al. 2007, Rogerson et al. 2014). All reported that crop fields with the highest deer impact had the majority of the site boundaries bordered by forest and/or were short distances from forestlands harboring deer, and that a gradient of deer impact existed, with the greatest level on the edges of crop fields that lessened with distance into crop fields.

Deer from adjacent properties can impact resources within managed forestlands that owners/managers are unable to influence. Because there is no research on this topic (no data exist to provide guidelines or thresholds), managers are left to proceed on their own for responding to impact from deer on adjacent property. The problem is exacerbated when owners of adjacent properties are not sympathetic or do not control deer density. Adjacent landowners may also feed deer in winter, which may cause deer to use habitat on managed forestland as hiding cover when ruminating or seeking shelter from weather if such a habitat is not available on the adjacent property.

Several of the case histories in Section V identify problems with adjacent properties (stakeholders with management priorities including no hunting; movement of deer during hunting seasons; different carrying capacities; winter feeding of deer, food plots). To the degree that they can, landowners of managed forestlands should assess deer and habitat conditions within their own and adjacent properties as they affect deer density, impact, and the ability to manage both. Obviously, coordination, cooperation, and collaboration among owners of adjacent forestlands may be critical to successful deer management on managed forestlands.

3.2 HABITAT COMPONENTS: DIFFERENCES BETWEEN MANAGED AND ADJACENT FORESTLANDS

Forest property owners must be aware of the quality and extent of deer habitat components their property provides for deer. If they provide the full spectrum of successional stages deer require (forage and fawning, hiding, thermal, multipurpose) and the quantity and quality of deer forage are high, deer will make heavy use of their property, if it is within a forested landscape. If adjacent properties owned by others provide some, but not all, deer habitat components (e.g., surrounding crop fields that provide high-quality and abundant deer forage during crop growing seasons but not hiding

FIGURE 3.1   Deer feeding on waste corn meet cover needs in the adjacent woodlot. (Photo by Tom Ellis.)

or thermal cover), deer foraging within those properties (Figure 3.1) may use managed forestland for cover and browse incidentally on understory vegetation (Figure 3.2), eliminating native seedlings and forbs with resulting invasion of nonbrowsed exotic plants. Conversely, if managed forestland doesn't provide all deer habitat components, such as hiding cover during deer hunting seasons, deer may move off the managed forestland to adjacent properties that provide hiding cover but do not allow deer hunting.

Forestlands smaller than the typical deer family group home range (~200 ha) may provide some (or all) habitat requirements of these groups, but members of these groups may also use parts of adjacent forestlands as parts of their home ranges. More to the point, small managed forestlands surrounded by large, adjacent forested landscapes may be used by multiple matriarchal deer groups that include the managed forestland as an integral part of their home ranges.

FIGURE 3.2   Understory dominated by exotic garlic mustard in overbrowsed woodlot adjacent to agricultural field in Figure 3.1. (Photo by David S. deCalesta.)

Small managed forestlands (<200 ha) are highly susceptible to deer impacts from adjacent forestlands because deer may use the managed forestland for habitats/resources not found in the adjacent landscape. If control of deer density within adjacent forestlands is not practiced, the deer in these forestlands may drift onto managed forestlands and cause major impacts on habitats. Large managed forestlands (thousands of ha) will be less affected by deer from adjacent forestland, as an increasingly smaller proportion of the managed forestland will be within the home ranges of deer within adjacent forestland. Owners of small woodlots face a much greater potential for impact of deer from adjacent forestlands, whereas forest resources on large managed forestlands will be less impacted.

3.3 TEMPORAL FACTORS

Regarding time, management steps taken to abate deer impact sometimes result in rapid recovery and sometimes in prolonged recovery. Seedlings that have been heavily browsed but not eliminated may recover one growing season after deer density is dramatically reduced. In the case of plants (seedlings, shrubs, forbs, and grasses) eliminated by deer browsing, recovery may occur shortly (seeds shed by overstory trees and/or carried on-site by vectors such as squirrels, birds, and even deer germinate and grow under reduced deer impact), or may take years, decades, or even longer if seed sources are remote or absent from the landscape, as seed-producing plants may have been eliminated by chronic, long-term deer overbrowsing.

Deer may use distinct areas within forested landscapes differentially, based on season—during cold, snowy winters, they may concentrate on south-facing slopes for warmth and where forage may be exposed by snow melt, and during fawning may favor early succession sites where ground vegetation may be dense. Deer may also use established travel lanes through forested areas based on movement from one habitat type to another or based on seasonal migration from summer to winter ranges, including deer yards.

Managing deer density and impact is not a one-time fix. Rather, deer density and impact need to be monitored and managed on a recurring and often annual basis for the duration of forestland ownership. If forest resources are protected from deer impact by deer exclosure devices such as individual plant protectors (tubes and netting), group protection as afforded by small fenced areas, or area protection provided by erection and maintenance of larger fenced regeneration sites, the structures must be monitored and maintained frequently.

If reduction of deer density by public hunting is the chosen method for reducing deer impact, it must be practiced annually and in perpetuity. Because of their high reproductive potential, deer can virtually double their numbers in 3–4 years in the absence of predation, which in many cases may only be provided by hunters.

Because deer do not use forestlands uniformly, either spatially or temporally, landowners must be aware of these time and landscape differences and design monitoring for deer density and impact such that they representatively assess deer use of their forestlands within enclosing landscapes and over time. This may help landowners determine where and when deer reduction/exclusion efforts should be concentrated.

REFERENCES

Barrett, K. J. and O. J. Schmitz. 2013. Effects of deer settling stimulus and deer density on regeneration in a harvested southern New England forest. *International Journal of Forestry Research* https://www.hindawi.com/journals/ijfr/2013/690213/.

deCalesta, D. S. 2017. Bridging the disconnect between agencies and managers to manage deer impact. *Human Wildlife Interactions Journal* 11:112–115.

deCalesta, D. S. and D. B. Schwendeman. 1978. Characterization of deer damage to soybean plants. *Wildlife Society Bulletin* 19:46–52.

Hewitt, D. G. ed. 2011. *Biology and Management of White-Tailed Deer*. Boca Raton FL: CRC Press.

Hurley, P. M., C. R. Webster, D. J. Flaspohler et al. 2012. Untangling the landscape of deer overabundance: Reserve size versus landscape context in the agricultural Midwest. *Biological Conservation* 146:62–71.

Rogerson, J. E., E. L. Tymkiew, and V. L. Vasilas. 2014. Impacts of white-tailed deer browsing and distance from the forest edge on soybean yield. *Wildlife Society Bulletin* 38:473–479.

Ruark, G. 2010. Beyond the Fenceline. *Inside Agroforestry*, Volume 16, Issue One.

Rutherford, A. C. and O. J. Schmitz. 2010. Regional-scale assessment of deer impacts on vegetation within western Connecticut, USA. *Journal of Wildlife Management* 74:1257–1263.

Stewart, C. M., W. J. McShea, and B. P. Piccolo. 2007. The impact of white-tailed deer on agricultural landscapes in 3 national historical parks in Maryland. *Journal of Wildlife Management* 71:1525–1530.

Stewart, K. M., R. T. Bowyer, and P. J. Weisberg. 2011. Spatial use of landscapes. In *Biology and Management of White-Tailed Deer*. ed. D. G. Hewitt. Boca Raton FL: CRC Press.

4 Autecology
Home Range, Social Structure, and Landscape Use

David S. deCalesta

CONTENTS

Home range: the area to which an animal usually confines its daily activities.

Merriam-Webster

MANAGER SUMMARY

Habitat use, and impact by deer on forest resources, are related to deer use of landscapes. Specifically, how deer occupy habitat, including size and dimensions of areas (referred to as home ranges) they use to meet life cycle requirements, are determining factors in the level of impact they have on forest resources. Managers must understand how deer react to changes in habitat, including availability of forage, hiding, and thermal cover, to be able to predict and manage for not just how deer use of habitat may be affected by management activities, but also for how factors such as time and arrangement of habitat components affect how, where, and when deer respond to habitat changes.

4.1   INTRODUCTION

Deer conduct daily and seasonal activities within bounded areas called *home ranges*. Dimensions of home ranges and how deer use them are determined by social and seasonal factors, habitat features, and other influences such as predator presence. The impact of predation on deer habitat use is presented in Chapter 6. This chapter deals with autecological factors as they affect deer use of landscapes.

4.2   HOME RANGE

Numerous studies spanning the whitetail (*Odocoileus virginianus*) ranging from Texas to Quebec identify a winter home range size of 40–600 ha and a summer home range size of 20–280 ha (Stewart et al. 2011), representing wide variability in size and much smaller summer than winter home ranges. Walter et al. (2009) evaluated home range size of female deer in the Midwest (Illinois, Michigan, Nebraska, and Wisconsin) in agro-forested landscapes and discerned few differences among states (average home range size ranged from 0.99 to 1.47 km^2). Variables influencing size of home range were distance (from centroid of home range) to forest, roads, and urban development, and four landscape patterns related to cropland, edge density, nearest neighbor habitat, and shape and patch size of habitat components. Another landscape feature, connectivity of forest patches within agro-forestry landscapes, was a better determinant of home range size than "traditional measures of habitat."

McCoy (unpublished PhD thesis, Auburn University) studied home range size of 37 radio-collared male deer within a 2600-ha forestland managed for commercial timber production and wildlife and found the average adult buck (yearlings and older) home range was 142 ha. He attributed the fairly small home ranges to the close proximity of required habitats. There were no differences attributed to age: the two largest and two smallest home ranges were of yearling deer. The universal rule of thumb defines an average deer home range of about 200–250 ha. Seasonal home ranges include areas that provide deer with all their habitat requirements.

Deer are characterized as exhibiting fidelity to home ranges, more so in summer than in winter when there is less certainty of forage sources. Home range dogma asserts that home ranges of individual deer, or matriarchal family groups, overlap without territorial defense. Studies have shown that deer may leave defined home ranges in movements termed "excursions" to access one or more habitat components (e.g., salt licks) missing from identified home ranges or used on seasonal bases (Jacobsen 2017), including breeding season (Karns et al. 2011).

However, Powell and Mitchell (2012) took issue with the research focus on identifying the best method(s) for quantifying home range dimensions, preferring instead that the emphasis be on understanding why animals use various habitat components in their established home ranges as related to behavior and how they may change the dimensions of their home ranges based on updating mental maps they develop for their environment. Countless studies have established the multiple habitats used for deer for meeting daily and seasonal requirements (Chapter 2).

4.3 SOCIAL STRUCTURE AND HOME RANGE

Deer exist in family groups, composed of the matriarchal doe, fawns of the year, and yearling and older female offspring. These groups meet their habitat needs generally within home ranges of 150–200 ha (1.5–2 km^2), but sometimes must travel farther to meet their needs. Home range shapes are variable, dependent on location and arrangement of areas containing specific habitat needs.

In areas with extreme winter conditions (e.g., the upper peninsula of Michigan with deep snow, low temperatures) deer may travel varying distances (2–90 km) from home ranges used in spring, summer, and fall to wintering areas (deer yards) farther south (Van Deelen et al. 1998, Doepker et al. 2015) with milder temperatures, lower snow depth, and more contiguous overhead conifer cover. Deer traveled farthest from more northern areas and stayed longer in deer yards.

As Doepker et al. (2015) noted, attempting to reduce deer density by public hunting in areas with high impact such as deer yards may not be effective as a result of temporal and landscape differences in habitat use. Deer may not yet have arrived in deer yards during typical hunting seasons, so focusing hunting on these areas during regular hunting seasons may not achieve the desired reductions in deer density and impact.

4.4 HOME RANGE FIDELITY

Deer, especially mature does, exhibit home range fidelity when densities are similar to those prior to European colonization of North America (<5 deer/km^2). Porter et al. (1991), based on research in the Adirondack Park in northeastern New York, U.S., stated that white-tailed deer does and their female offspring are highly philopatric, and suggested that local deer populations expand slowly as a series of overlapping home ranges they likened to the petals on a rose. They formulated a "rose-petal concept" of deer management whereby removal of deer family units through hunting or other removals could reduce deer impacts on forest resources in localized areas for as many as 10–15 years.

Deer (*Odocoileus* sp.) are slow to move to additional forage created outside their home range areas. Research suggests that unless increases in forage availability occur within their home ranges, it takes deer more than a year to discover and use the new sources of forage. Doman and Rasmussen (1944) noted that starving mule deer (*Odocoileus hemionus*) in Utah would not leave traditional wintering areas to obtain forage provided to them in areas the next ridge over from their established winter home range. In north-central Pennsylvania, radio-collared white-tailed does did not leave home ranges to travel to areas of newly created forage in nearby clearcut sites in the year following tree harvest (R. Brenneman, personal communication). In West Virginia's Appalachian Mountains, home range, core areas, and movements of radio-collared does did not change during timber harvests, nor were they different from home range, core area, and movements of deer whose home ranges were composed of maturing forestlands without timber harvests (Campbell et al. 2004).

Lutz et al. (2016) determined that deer density levels above precolonial levels (>5 deer/km^2) were associated with movements of does outside established home ranges—they reported a gradient of increasing home range size with increasing deer density. Lutz et al. (2016) suggested that deer expand their home ranges under high deer density levels that compel them to seek out areas with less competition for resources for the welfare of their fawns.

However, providing corn and other crops in winter does lead to high deer densities on feeding grounds, but deer numbers grow over the years, as it takes them time to discover and use these new sources of additional forage outside their home ranges. Thus, deer can move outside established home ranges in response to high deer density and availability of forages—the former action possibly occurring over a shorter period than the latter, depending on the magnitude of deer density increase. Regardless, the take-away message is that large removals of deer from local areas may temporarily depress deer density and impact, but the effect is short-lived and management steps should be taken, probably on an annual basis, to monitor and maintain low deer density and impact as achieved by deer removals.

4.5 HOME RANGE MALLEABILITY

What is not well understood is the malleability of home ranges based on changing habitat components requiring changing landscape use and how rapidly deer can reshape their home ranges to respond to habitat changes, as exhibited by home range shifts by bighorn sheep (*Ovis canadensis*) in response to habitat alterations (Clapp and Beck 2015). Deer home range size, shape, location, and use are not static but rather shift in response to changes in availability of habitat components (Jacobsen 2017). Such shifts may take varying amounts of time to develop, dependent on the nature and importance of habitat components involved.

Deer movement onto managed forestlands from abutting properties may occur when resources are less available on abutting properties than those on the managed forestland. Examples would be an increase in forage production in a forestland resulting from timber harvest, and opening of the forest overstory canopy to increase levels of sunlight on the understory and stimulate plant growth, or the amount of forest habitat could decrease on an abutting area resulting from commercial or residential development of the landscape.

Enhancing understanding of where and when deer shift/adapt use of habitat components helps managers anticipate deer movements and habitat-associated changes in deer density and impacts on forest resources. Where, when, and how should methodology, including hunting, be altered to address changes in habitat use by deer?

4.6 SUMMARY

As proposed by Tierson et al. (1985), deer management based on projected home range location and size should conform to observed use by deer to avoid ineffective management (e.g., deer harvest strategy should target location[s] of deer impacting forest resources based on where they are during hunting season). Cooperation of and collaboration with neighboring landowners whose properties include home ranges of deer affecting individual forestlands is encouraged, with an emphasis on managing landscape-specific deer populations such as matriarchal groups.

REFERENCES

Campbell, T. A., B. R. Laseter, W. M. Ford et al. 2004. Movements of female white-tailed deer (*Odocoileus virginianus*) in relation to timber harvests in the central Appalachians. *Forest Ecology and Management* 19:371–378.

Clapp, J. C. and J. L Beck. 2015. Evaluating distributional shifts in home range estimates. *Ecology and Evolution* 5:3869–3878.

Doepker, R. V., C. A. Albright, M. A. MacKay et al. 2015. *Characteristics of white-tailed deer trapped and tagged in Michigan's Upper Peninsula.* East Lansing MI: Michigan Department of Natural Resources, Wildlife Division Report Number 3605.

Doman, E. R. and D. I. Rasmussen. 1944. Supplemental feeding of mule deer in northern Utah. *Journal of Wildlife Management* 8:317–338.

Jacobsen, T. C. 2017. Extra-home range and excursive movements by white-tailed deer. MS thesis, Auburn University.

Karns, G. R., R. A. Lancia, C. S. DePerno et al. 2011. Investigation of adult male white-tailed deer excursions outside their home range. *Southeastern Naturalist* 10:39–52.

Lutz, C. L., D. R. Diefenbach, and C. S. Rosenberry. 2016. Proximate influences on female dispersal in white-tailed deer. *Journal of Wildlife Management* 80:1218–1226.

Porter, W. F., N. E. Mathews, H. B. Underwood et al. 1991. Social organization in deer: Implications for localized management. *Environmental Management* 15:809–814.

Powell, R. A. and M. S. Mitchell. 2012. What is a home range? *Journal of Mammalogy* 93:948–958.

Stewart, K. M., R. T. Bowyer, and P. J. Weisberg. 2011. Spatial use of landscapes. In *Biology and management of white-tailed deer.* D. G. Hewitt (ed.) Boca Raton FL: Taylor & Francis.

Tierson, W. C., G. F. Mattfeld, R. W. Sage et al. 1985. Seasonal movements and home ranges of white-tailed deer in the Adirondacks. *Journal of Wildlife Management* 49:760–769.

Van Deelen, T. R., H. Campa III, M. Hamady et al. 1998. Migration and seasonal range dynamics of deer using adjacent deeryards in northern Michigan. *Journal of Wildlife Management* 62: 205–213.

Walter, W. D., K. C. VerCautheren, H. Campbell et al. 2009. Regional assessment on influence of landscape configuration and connectivity on range size of white-tailed deer. *Landscape Ecology* 24:1405–1420.

5 Autecology
Reproduction and Recruitment

David S. deCalesta

CONTENTS

As the number of females goes up, the fawns recruited per female goes down.

Dale McCullough (1979)

MANAGER SUMMARY

Deer density and impact are affected by deer reproductive behavior regarding how reproduction and recruitment contribute to annual increases in deer density. Managing deer density includes estimating annual recruitment so as to balance it with deer harvest—managers must have some appreciation for the size of herd increase from reproduction to balance it with mortality to retain deer density and impact at desired levels. Managers must be able to anticipate at least qualitatively how herds will increase annually so they can manage deer density and impact through hunter harvest. Production of fawns, and survival to adulthood, are affected by habitat quality and quantity, including forage and ratio of adult bucks to adult does—too few bucks means not all does will be bred and produce offspring, resulting in a reduced rate of recruitment. Managing forage production and recruitment are tools managers should at least be aware of as they plan for managing deer density and impact.

5.1 INTRODUCTION

Deer density and associated impacts on forest resources result from the sum of herd gains (reproduction and recruitment—the topics of this chapter) and immigration (Chapter 4), and herd losses (caused by starvation/winter kill, predation—including modern hunting—and parasites and diseases, as covered in Chapters 1, 6, and 7)

Managers need to address and incorporate provision for all the above factors including immigration, when relevant, in developing a deer management plan.

5.2 REPRODUCTION

Prior to the 1990s, the reproductive strategy of species (including wildlife such as deer) was classified as *r* or *K* (MacArthur and Wilson 1967). For *r*-selected species, adult body size was relatively small, life-span was relatively short (a few years), and reproductive strategy was characterized by production of many offspring per litter, each litter mate with a relatively low probability of surviving to adulthood, and rapid maturation of young (e.g., mice and rabbits). *K*-selected species were characterized as those of larger body size and longer life-span that produced single litters of small size and invested considerable effort and time in parenting, resulting in a higher probability of offspring surviving to adulthood than those of *r*-selected species. Unlike *r*-selected species, reproductive output of *K*-selected species was density dependent and influenced by resource availability (reproduction and recruitment were dependent on quality and quantity of forage relative to population density). In his treatise on the population ecology of white-tailed deer (*Odocoileus virginianus*), McCullough (1979) identified deer as being *K*-selected in the title of his book. Guynn (1981) described additional characteristics of deer as *K*-selected species as: adaptation to live in relatively stable habitats, with population density at or near carrying capacity (*K*), defined as that population size where births are exactly matched by deaths, and strong competitive ability. The *r-K* dichotomy ran into disfavor in the late 1980s (Reznick et al. 2002).

Reproduction is the most direct way that deer herd density increases. Whether does become pregnant and reproduce is directly related to the ratio of adult does to adult bucks. Research has indicated that bucks rarely breed more than three does during the rut, so when doe:buck ratios run higher than 4:1, fewer does may be bred. Data from the Kinzua Quality Deer Cooperative (Chapter 34) illustrates the inverse relationship between the ratio of does to bucks in fall prior to hunting season and ratio of fawns to adult does (Figure 5.1) the following fall after fawns have survived the gamut of fawn predators (bears and coyotes) and adverse spring/summer weather. Fawn fall abundance is highest when the doe:buck ratio the previous breeding season is <3:1 and lowest when the ratio is >3:1. Roadside counts of fawn predators (coyotes, bears, and bobcats) remained stable throughout the time span—changes in fawn abundance were not related to changes in predator numbers (deCalesta, unpublished information).

This phenomenon, coined "managers' dilemma, hunter's paradox" by Susan Stout, USDA Forest Service, represents a managerial quandary—should hunters be told that their attempts to increase fawn numbers by sparing does and harvesting only bucks (and likely boosting doe:buck ratio above 3:1) has the opposite effect and can lead to reduced deer abundance, or should managers use the misperception as a stealthy way to depress recruitment and deer density by emphasizing harvest of antlered deer?

Does require protein and energy from forage to support growth and development of fawns during gestation in winter and spring. When energy and protein intake by captive pregnant does were

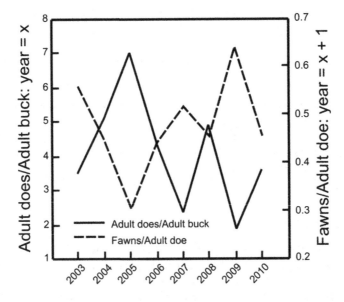

FIGURE 5.1 Adult does per adult buck vs. fawns per adult doe the following fall.

restricted, they exhibited higher pre- and postnatal losses than does for which nutrition was not restricted (Verme 1962, 1979). Nutritionally stressed does produced fawns of lower birth weight with reductions in survival (Verme 1963, Langenau and Lerg 1976), and scientists claimed that nutritionally stressed does may resorb or abort developing fetuses. Based on research with vaginal implant transmitters, if fawns are stillborn, this may happen in early May, prior to the main fawning season in late May and early June. Also, fawns born from nutritionally stressed does may not survive the rigors of early summer, especially if the weather is cool and wet. Whether managers are attempting to manage deer abundance, deer impacts, or understory vegetation, a universal goal is to increase abundance and diversity of understory vegetation. If managers are able to achieve this management objective, which includes control of local deer density, forage should not be a limiting factor in deer reproduction.

5.3 RECRUITMENT

Recruitment traditionally has been described as the difference between loss of deer to mortality factors and the increase in deer resulting from overwinter survival of fawns. Recruitment may be low when forage resources available to lactating does are insufficient to provide adequate nutrition to newborn fawns. Recruitment will also be low if adverse weather (cold, wet springs/summers) results in low fawn survival, or if predators kill young deer. Large, regional increases in bear and coyote densities and potential predation on newborn fawns may be a major cause in recently noted reductions in recruitment. And, as indicated above, when doe:buck ratios are >3:1 during the rut, the number of fawns produced per doe, and surviving summer mortality factors, declines and there will be fewer fawns available to survive winter mortalities, lower recruitment, and a slowing of rate of herd increase.

Female deer have recurring estrus cycles (they come back into heat if they are not bred during the first cycle), but this occurs after many of the available bucks have been removed during hunting season, so the does may again not be bred. Most deer that are going to become pregnant will do so by the end of December. However, photoperiod (increasing day length) will not end estrus cycling until late February or March. For does that are bred later, there is an unfortunate side effect. Fawns conceived during the second, third, or later estrus cycles are born in midsummer to early fall and may not have enough time to accumulate sufficient fat before severe northern winters create energy demands they cannot meet. These deer will not survive to be recruited.

Landscape features may contribute to reduced recruitment by preventing bucks from finding and impregnating does. In farm/forest areas with interspersion of forests and agricultural lands, local forested areas may support matriarchal groups but not provide sufficient security cover to hold bucks. If these areas also are primarily hunted by hunters who avoid harvesting antlerless deer, harvest focused on antlered deer may result in depletion of bucks of breeding age and lower impregnation rates of does, or later impregnation of does with resultant higher fawn mortality.

REFERENCES

Guynn, D. C., Jr. 1981. How to manage deer populations. *Proceedings International Ranchers' Roundup.* eds. L. D. White and L. Hoerman. Texas Agricultural Extension Service.

Langenau, E. M. and J. M. Lerg. 1976. The effects of winter nutritional stress on maternal and neonatal behavior in penned white-tailed deer. *Applied Animal Ecology* 2:207–223.

MacArthur, R. and E. O. Wilson 1967. *The theory of island biogeography.* Princeton NJ: Princeton University Press.

McCullough, D. R. 1979. *The George Reserve deer herd.* Ann Arbor MI: The University of Michigan Press.

Reznick, D. N., M. J. Bryant, and F. Bashey 2002. r- and K-selection revisited: The role of population regulation in life-history evolution. *Ecology* 83:1509–1520.

Verme, L. J. 1962. Mortality of white-tailed deer fawns in relation to nutrition. *Proceedings of the First National White-Tailed Deer Disease Symposium. Southeastern Section of the Wildlife Society.* Athens GA: University of Georgia.

Verme, L. J. 1963. Effect of nutrition on growth of white-tailed deer fawns. *Transactions of the North American Wildlife Conference* 28:431–443.

Verme, L. J. 1979. Influence of nutrition on fetal organ development in deer. *Journal of Wildlife Management* 41:791–796.

6 Synecology
Predation

David S. deCalesta

CONTENTS

> Remove the predators and the whole ecosystem begins to crash like a house of cards.
>
> **Brian Skerry (2019)**

MANAGER SUMMARY

Science teaches us that deer do not control their own numbers, at least not until density is so high that deer eliminate forage and contribute to their own starvation deaths. Science has also established that one of the natural brakes on deer density increase is persistent predation by a suite of large predators such as mountain lions, wolves, bears, and Native Americans. The onslaught of development of the country was partially enhanced by elimination of these predators, including Native Americans, by the turn of the twentieth century. The primary mortality factors since then have been coyotes, bears, and modern deer hunting. The majority of eliminated major predators cannot be recovered by reintroduction, so mortality by hunting remains increasingly important for keeping deer density at levels associated with sustainable production of forest resources.

6.1   INTRODUCTION

The initial position of the wildlife profession on the impact of predators on deer—that predation is a major factor in deer population regulation—was influenced by a massive die-off of mule deer (*Odocoileus hemionus*) on the Kaibab Plateau in Arizona in the 1920s after natural predators (mountain lions [*Felis concolor*], wolves [*Canis lupus*], coyotes [*Canis latrans*], and bobcats [*Lynx rufus*]) were greatly reduced and the deer population skyrocketed, depleted available forage, and collapsed (Leopold 1943). This dogma was challenged in the 1970s by wildlife theorists who argued that predators did not control deer abundance (Caughley 1970, Burke 1973, Colinvaux 1973, Binkley et al. 2006), postulating instead that reduced competition with livestock for forage, or other factors, were responsible for the deer irruption, and not reduction of predation caused by predator control efforts.

6.2   TROPHIC CASCADES

However, research by McCullough (1979) demonstrated that without external limiting factors, including predation, deer populations can rapidly overpopulate. Ripple and Beschta (2004, 2006) asserted that mountain lions and wolves, among other top predators such as bears, control deer density and prevent deer from rapidly overpopulating and greatly impacting vegetation in a phenomenon called trophic cascading (Eisenberg 2010, Terborg and Estes 2010). In "top-down" trophic cascades, where top predators such as wolves and mountain lions are eliminated, their primary prey species, such as deer, greatly increase in number and reduce species richness, abundance, and structure of vegetation, reducing habitat complexity and diversity of wildlife species dependent on the habitat. Ripple and Beschta (2004, 2006) noted that as wolf and mountain lion populations were greatly reduced in western national parks (Zion and Yellowstone), deer and elk populations soared, and major reductions in structure and species composition of vegetation occurred, resulting in simplified habitats.

Horsley and Marquis (1983) and Horsley et al. (2003) reported a similar trophic cascade syndrome involving white-tailed deer (*Odocoileus virginianus*), predators (e.g., wolves and mountain lions), and understory vegetation dynamics in Pennsylvania northern hardwood forests. Deer were nearly eliminated from Pennsylvania by market hunting by the end of the 1800s. Severe restrictions on deer hunting, elimination of wolves and mountain lions, reductions in bear populations, and lack of coyote predation (coyotes did not reach Pennsylvania in their eastward range expansion until about 1930 following the extirpation of wolves) were followed by a rapid and permanent expansion of deer density and subsequent reduction in species richness and abundance of understory vegetation. Overabundant deer caused a shift in the dynamics of understory vegetation, wherein tree, shrub, and forb species were dominated by ferns, grasses, and seedlings of tree species resistant to deer browsing.

6.3   DEER PREDATORS

Wolves and mountain lions prey on deer of all ages and sexes, including mature bucks. Other predators, primarily black bear (*Ursus americanus*), coyote, and bobcat, prey primarily on young fawns (Kunkle and Mech 1994, Carstensen et al. 2009, Grovenburg et al. 2012). Deer predators are territorial and generally maintain upper limits to their own population density (Wallach et al. 2015) when they are top, or apex, predators (no predators exist above them to control their numbers, such as lack of wolves to limit coyote density). They generally do not increase much in abundance, if at all, with increases in deer abundance. Consequently, there is an upper limit to the number of deer they can remove (e.g., there will be less predation impact at higher fawn density as occurs when deer densities are high).

Prior to their near elimination as deer predators in North America, indigenous peoples (American Indians) worked in concert with other predators to maintain deer densities at levels that did not negatively impact vegetation. Waller and Reo (2018) compared forest plant composition and structure and deer density between tribal lands of American Indians in Wisconsin and similar forest lands not controlled by these peoples. They reported that cultural practices and more liberal hunting on tribal lands (no restrictions on hunting seasons or bag limits for tribal members) resulted in far lower deer density (3.1–7 deer/km^2) on tribal lands, which was 22%–59% lower than on surrounding deer management units of similar forest type with existing forest management practices and deer hunting regulations typical of lands regulated by state natural resource agencies. They also reported higher overstory and understory plant diversity, higher rates of tree regeneration, and fewer invasive plant species on tribal lands than within deer management units surrounding the tribal lands.

6.4   EFFECT OF PREDATORS ON DEER

When deer numbers are sufficiently low, such as occurs when deer do not negatively impact understory vegetation, predators can suppress deer density and recruitment to the point where deer density remains low, in balance with the environment, and consistent over time.

Predators can affect deer impact on vegetation by changing their foraging behavior in response to the risk of being preyed upon. Studies (Ripple and Beschta 2004, Fortin et al. 2005, Mao et al. 2005, Cherry et al. 2015, Flagel et al. 2016) indicate that deer and other large ungulates change their foraging behavior in the presence of top predators (e.g., wolves and mountain lions). They avoid areas of high predator abundance, seek out areas where there is more protective cover, and spend less time foraging. Wolves can cause deer to spend more time in heavily forested areas, on steeper slopes, and at higher elevations.

Wolf packs avoid the home ranges of neighboring wolf packs, creating "buffer zones" in areas where their home ranges abut/overlap. Deer tend to avoid wolf pack home ranges and make disproportionate use of these buffer zones in local landscapes (DelGuidice 1998: http://www.wolf.org/wow/united-states/minnesota/prey-and-predation-3/). Conversely, mountain lions and wolves may avoid areas of high human habitation/use (e.g., roads and residential areas) (Theuerkauf et al. 2003, Kaartinen et al. 2005, Nicholson et al. 2014), thereby creating habitat voids where ungulate density may be high with little to no downward predator pressure.

Such altered deer habitat use and foraging behavior may result in diffusing deer browsing impact over larger areas and concentrating deer browsing in others, resulting in more or less impact on forbs, shrubs, and seedlings.

6.5   HUMAN DIMENSIONS, PREDATORS, AND DEER

Can these numerical and behavioral responses of deer to predators be used in a manner that can reduce deer impact on forest resources? LaRue and Nielsen (2016), modeling mountain lion movement patterns, suggested that in a few decades, cougars might repopulate areas where they were extirpated in selected large habitat patches in the midwestern United States that provide suitable cougar habitat. Licht et al. (2010) proposed introducing small populations of wolves (e.g., single packs) for the purpose of ecosystem restoration in large areas, such as national parks, where hunting of ungulates (to reduce their impact) is not permitted, or on large islands (e.g., within the Great Lakes) and other areas where negative impact with humans and human interests (growing livestock) is minimal. The idea is attractive but must be tempered with reality. Stakeholder groups (e.g., livestock associations and hunter associations) have far more members and more clout with agencies that might administer predator reintroductions or population expansions. Also, the highest deer densities are currently found in more suburban areas, and it is inevitable that predators will eventually find deer in these high-population areas, coming into direct conflicts with people and pets (Baron 2005).

After wolves repopulated parts of Wisconsin, public support for wolves began to decline (Browne-Nuñez et al. 2015), especially in places where wolves negatively impacted humans (e.g., areas where livestock are farmed). There was public support for reducing wolf density by legal lethal control, and a feeling that outside groups had influenced decisions regarding management of wolves. Contrarily, wolves are viewed with reverence and not hunted on Menominee tribal lands in Wisconsin, and the combination of wolf predation and unrestricted hunting by tribal members likely has kept deer at densities associated with retention of diversity of forest vegetation and success of seedlings to get past deer browsing to form future forests. Magle et al. (2014) studied deer-coyote interactions in the Chicago metropolitan area, finding that coyotes occupied areas farther from the urban center in areas with low road and housing density. Deer used habitat with high canopy cover and reduced human and dog presence. Deer did not totally avoid areas with high coyote density, but the propensity of coyotes to prey on fawns rather that adult deer may have been a factor in habitat sharing. The authors suggested reducing human foot traffic in green spaces as a way to increase coyote predation on deer.

6.6   USING PREDATION TO CONTROL DEER IMPACT

It is unlikely that managers could utilize reestablishment of apex predators to control deer density and impact. Possible exceptions could be in large national parks where deer hunting is not allowed, or in large and private commercial forests where the influence of outside stakeholder groups is much less than on public forest lands. Intentional reintroductions in state forests and parks, and national forests where deer hunting is allowed, would face high stakeholder resistance and likely would not pass stakeholder reviews generally required for such actions by state wildlife agencies. Additionally, deer density would first have to be reduced, likely by public hunting, before such introductions might work, to drive deer density down to levels where natural predation could be a controlling factor. Such deer density would be in the 5–10 deer/km^2 range in northern hardwood forests. Introductions of predators on small forest lands, such as small woodlot owners manage, or in forests adjacent to and/ or included in grazing lands for domestic livestock would never be acceptable to most stakeholders, including abutting landowners who might be negatively affected. Without overwhelming stakeholder support, state wildlife agencies are unlikely to authorize the permits and funding needed to support a predator reintroduction program.

If deer do indeed avoid predators by selective use or avoidance of areas with/without predators, it might be possible to establish frequent visits of deer predators (e.g., dogs and humans), which might deter deer use of at least some portions of a landowner's forested landscape. The problem is that in the natural setting, predators are present 24/7 most seasons (researchers noted that predators were more or less present in seasonal home ranges, but that ungulates changed their foraging behavior only in areas occupied by predators), and that the presence is accompanied by actual predation on deer. Thus, infusing deer habitat with human or canid (domestic dog) presence may only work with frequent occupation of the habitat by the "predators" and might have to also be accompanied by predation on deer. Deer temporarily avoid areas infused with hunters during hunting season, but the shortness of deer season precludes deer permanent/prolonged avoidance of habitats based on that kind of predatory pressure (unless of course, poaching occurs year-round).

6.7 BOTTOM LINE

The take-away message is that deer density is controlled by external factors, with predation being a primary cause of mortality. If predatory pressure is reduced, deer density quickly escalates and may result in a population crash caused by massive starvation loss. If annual deer removals by hunting are not maintained, deer density can rapidly escalate to prehunting levels, overabundance, negative impacts on forest resources, and eventual starvation deaths of deer.

REFERENCES

Baron, D. 2005. *The beast in the garden: The true story of a predator's deadly return to suburban America.* New York: W.W. Norton Company Inc.

Binkley, D. M., M. Moore, W. H. Romme et al. 2006. Was Aldo Leopold right about the Kaibab deer herd? *Ecosystems* 9:227–241.

Browne-nuñez, C. A., D. Treves, Z. MacFarland et al. 2015. Tolerance of wolves in Wisconsin: A mixed-methods examination of policy effects on attitudes and behavioral inclinations. *Biological Conservation* 189:59–71.

Burke, C. J. 1973. The Kaibab deer incident: A long-persisting myth. *BioScience* 23:113–114.

Carstensen, M., G. D. DelGuidice, B. A. Sampson et al. 2009. Survival, birth characteristics, and cause-specific mortality of white-tailed deer neonates. *Journal of Wildlife Management* 73:175–183.

Caughley G. 1970. Eruption of ungulate populations, with emphasis on Himalayan thar in New Zealand. *Ecology* 51:53–71.

Cherry, M. J., L. M. Conner, and R. J. Warren. 2015. Effects of predation risk and group dynamics on white-tailed deer foraging behavior in a longleaf pine savanna. *Behavioral Ecology* 26:1091–1099.

Colinvaux, P. 1973. *Introduction to ecology.* New York: John Wiley and Sons.

DelGuidice, G. D. 1998. *The ecological relationship of gray wolves and white-tailed deer in Minnesota.* St. Paul MN: Minnesota Department of Natural Resources.

Eisenberg, C. 2010. *The wolf's tooth: Keystone predators, trophic cascades, and biodiversity.* Washington DC: Island Press.

Flagel, D. G., G. G. Belovsky, and D. E. Beyer. 2016. Natural and experimental tests of trophic cascades: Gray wolves and white-tailed deer in a Great Lakes forest. *Oecologia* 180:1183–1194.

Fortin, D., H. L. Beyer, M. S. Boyce et al. 2005. Wolves influence elk movements: Behavior shapes a trophic cascade in Yellowstone National Park. *Ecology* 86:1320–1330.

Grovenburg, T. W., K. L. Monteith, R. W. Klaver et al. 2012. Predator evasion by white-tailed deer fawns. *Animal Behavior* 84:59–65.

Horsley, S. B. and D. A. Marquis. 1983. Interference by weeds and deer with Allegheny hardwood reproduction. *Canadian Journal of Forest Research* 13:61–69.

Horsley, S. D., S. L. Stout, and D. S. deCalesta. 2003. White-tailed deer impact on the vegetation dynamics of a northern hardwood forest. *Ecological Applications* 13:8–118.

Kaartinen, S., I. Kojola, and A. Colpaert. 2005. Finnish wolves avoid roads and settlements. *Annales Zoologici Fennic* 42:523–532.

Kunkle, K. E. and L. D. Mech. 1994. Wolf and bear predation on white-tailed deer fawns in Northeastern Minnesota. *Canadian Journal of Ecology* 72:1557–1565.

LaRue, M.A. and C.K. Nielsen. 2016. Population viability of recolonizing cougars in midwestern North America. *Ecological Modelling* 321:121–129.

Leopold, A. 1943. Deer irruptions. *Wisconsin Conservation Bulletin* 8:3–11.

Licht, D. S., J. J. Millspaugh, K. E. Kunkle et al. 2010. Using small populations of wolves for ecosystem restoration and stewardship. *Bioscience* 60:147–153.

Magle, S. B., L. S. Simoni, E. W. Lehrer et al. 2014. Urban-predator-prey association; coyote and deer distributions in the Chicago metropolitan area. *Urban Ecosystems* 17:875–891.

Mao, J. S., M. S. Boyce, D. W. Smith et al. 2005. Habitat selection by elk before and after wolf reintroduction in Yellowstone National Park. *Journal of Wildlife Management* 69:1691–1707.

McCullough, D. R. 1979. *The George Reserve deer herd*. Ann Arbor MI: The University of Michigan Press.

Nicholson, K. L., P. R. Krausman, T. Smith et al. 2014. Mountain lion habitat selection in Arizona. 2014. *The Southwestern Naturalist* 59:372–380.

Ripple, W. J. and R. L. Beschta. 2004. Wolves and the ecology of fear: Can predation risk structure ecosystems?. *Bioscience* 54:755–65.

Ripple, W. J. and R. L. Beschta. 2006. Linking cougar decline, trophic cascade, and catastrophic regime shift in Zion National Park. *Biological Conservation* 133:397–408.

Skerry, B. 2006. In Quotes. BrainyQuote.com, BrainyMedia Inc, 2019. https://www.brainyquote.com/quotes/brian_skerry_661742,

Terborg, J. and J. T. Estes, eds. 2010. *Trophic cascades: predators, prey, and the changing dynamics of nature*. 2nd edition. Washington DC: Island Press.

Theuerkauf, J., W. Jedrezejewski, K. Schmidt et al. 2003. Spatiotemporal segregation of wolves from humans in the Bialowieza Forest (Poland). *Journal of Wildlife Management* 67:706–716.

Wallach, A. D., I. Izhaki, J. D. Toms et al. 2015. What is an apex predator? *Oikos* 124:1453–1461.

Waller, D. M. and N. J. Reo. 2018. First stewards: ecological outcomes of forest and wildlife stewardship by indigenous peoples of Wisconsin, USA. *Ecology and Society* 23:45 (https://www.ecologyandsociety.org/vol23/iss1/art45/).

7 Synecology
Parasites and Diseases

David S. deCalesta

CONTENTS

Nowhere is it more true that "prevention is better than cure," than in the case of parasitic diseases.

Rudolf Leuckart (1886)

MANAGER SUMMARY

Parasites and diseases have always been minor factors in controlling deer density, primarily kicking in when deer density is too high and deer are crowded into conditions where spread of disease is high and deer resistance is low. The current exception is chronic wasting disease, which is less related to deer density than to spread from infected captive deer, such as in game farms and facilities designed to produce live deer for sale or to wild deer through escapes of penned deer or deer-deer contact through fences. Parasites and diseases are not factors in a manager's tool kits to manage deer density: density must be at levels destructive to other forest resources before they control deer density.

7.1   INTRODUCTION

Deer are afflicted by a multitude of internal parasites (tapeworms, liver flukes, abomasal worms, lungworms, stomach worms, arterial worms, nose bots, and brain worms) and external parasites (ticks and biting flies known as "deer keds"). Infestations of parasites normally are not fatal, unless deer are overabundant and/or concentrated when well-meaning people feed them corn and other high-energy food sources at deer winter feeding sites or during periods of water scarcity summer-fall when deer congregate at watering sources. Deer may also carry and transmit disease organisms (viruses and bacteria predominantly) to humans through insect vectors and to livestock by direct contact. This chapter's emphasis is on hazards to the welfare of humans and domestic livestock exposed to diseases and parasites vectored by deer and how to reduce these hazards through deer management. A more comprehensive treatment of deer parasites and diseases is provided by Campbell and VerCauteren (2011).

7.2 INTERNAL PARASITES

Deer rarely are debilitated by internal parasites except when they are subjected to repeat infestations at areas of high deer density, including winter supplemental feeding sites. Such feeding sites may attract and hold hundreds of deer within small areas for weeks. In these situations, parasite loads (e.g., abomasal round worms—*Haemonchus contortus*) may build up. Davidson (2006) recommended maintaining deer populations at densities not exceeding nutritional carrying capacity (see Chapter 15, "Deer Density, Carrying Capacity, and Impact of Forest Resources")—a density that is far exceeded on supplemental winter feeding grounds. This advice holds for reducing the potential for deer to transmit diseases and parasites to humans, domestic livestock, and wild ungulates such as elk (*Cervis sp.*) and moose (*Alces alces*).

Deer may be parasitized by the meningeal brain worm (*Parelaphostrongylus tenuis*), which, while not normally debilitating to deer (natural host), may fatally parasitize moose and elk (Lankester 2001). Brain worm larvae migrate to brain tissue in moose and elk, instead of meningeal tissue as in deer, and wander through brain tissue without encysting, destroying brain tissue instead of lodging in meninges. Infected moose and elk may exhibit slow movements, staggering, heads cocked at an angle, weakness, lameness, walking in circles, blindness, and loss of fear for humans. The frequency of infection increases when their ranges include high deer densities. Deer rarely exhibit symptoms of *P. tenuis* infection due to acquired resistance and different loci where the worms encyst (but they may exhibit typical symptoms when massively infected).

7.3 EXTERNAL PARASITES: TICKS AND KEDS

External parasites, primarily black-legged deer ticks (*Ixodes scapularis*) and less often Lone Star ticks (*Amblyomma americanum*) and dog ticks (*Dermacentor variabilis*), are vectors for bacterial, viral, protozoa-caused diseases (bacterial: Lyme disease—*Borrelia burgdorfi, Ehrlichiosis*, tularemia—*Francisella tularensis,* and Rocky Mountain spotted fever—*Rickettsia rickettsia*; viral: Powassan, *Flavivirus* sp.; and Bourbon, *Thogotovirus* sp.), and a few parasites (e.g., babiosis—*Babisia microti,* a malaria-like parasite, and *Theileria cervi*—a protozoan parasite spread by Lone Star ticks that infects white-tailed deer [*Odocoileus virginianus*] red blood cells). When tick and protozoa numbers are high, *Theileria* may cause death in deer and other ungulates in overpopulated herds (Kingston 1981).

Because ticks and the disease organisms they transmit to humans are found in nearly all states east of the Mississippi River, risks to humans are prevalent in all parts of the white-tailed deer range in the eastern United States. All three ticks are small: deer ticks are the smallest of the three stage-by-stage (Figure 7.1).

Ticks have four life stages. Females completing a blood meal on a host mate with adult males, drop off the host, and lay hundreds of eggs under ground litter. Larval ticks (with six legs rather than the eight of the succeeding two stages) hatch, climb up vegetation, and wait to grab onto passing hosts (mice, chipmunks, squirrels, deer, humans). These are the tiny ticks often infesting humans in batches and are often referred to as seed ticks. The disease organisms carried by adult ticks do not transfer from adult females to larvae, so the larval stage does not carry disease organisms. Larval ticks complete a blood meal on their first host, drop off, and metamorphose into the next stage (nymph). These ticks may have picked up disease organisms from their first blood meal and so pose danger to humans. Engorged nymphs drop off hosts and metamorphose into adult ticks, ready to find another host, get another blood meal, and reproduce. These are the largest-sized of the three tick stages, and they also can carry disease organisms.

The main concern for human health is exposure to ticks and the diseases they carry in forested areas such as parks and in wooded human residential developments. In these areas, human visitors/residents generally are not aware of ticks as disease vectors or of the precautions to take while traveling in wooded areas or even while working in their lawns and gardens. They may only realize

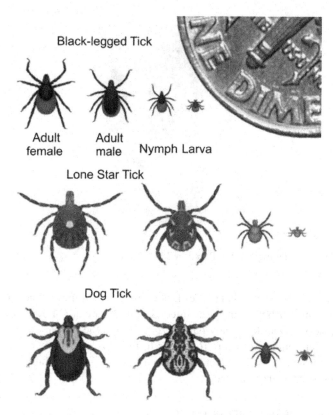

FIGURE 7.1 Relative sizes of black-legged, Lone Star, and dog ticks by stage. (From http://www.minipiginfo. com/first-aid-for-mini-pigs.html)

the danger after they have received tick bites, become infected with pathogens ticks transmit, and require medical treatment. In the rare cases where they become infected with the newly identified Powassan and Bourbon viruses, treatment is mostly palliative and death rates may be high, especially in the elderly and young.

Because of their relatively small home ranges (usually less than 5 ha), smaller hosts such as mice and squirrels do not carry ticks of various stages and infectiveness very far. Deer, on the other hand, with their vastly larger home ranges and daily movement distances, have the greatest potential for widespread dissemination of ticks and the disease organisms they host. This factor provides an argument for keeping deer out of areas of high human use such as woodland parks and forested residential communities by enclosing them within deer-proof fences. On other forestlands, reducing and maintaining deer density at the level recommended to enhance and maintain plant and animal diversity (≤ 5 deer/km^2: see Chapter 15, "Deer Density, Carrying Capacity, and Impact on Forest Resources") will minimize the number of deer carrying ticks onto properties and should reduce the overall density of ticks, reducing human exposure. Forest workers who work in forests during tick season (late spring to midsummer) are at risk: hunters who generally are not in the woods until after tick season are at lower risk. Generally, both these groups are aware of the potential for tick-transmitted disease and know they should conduct bodily inspections after being in the woods to find and remove ticks before they can transmit disease organisms.

Biting flies (deer keds—*Lipoptena cervi*) (Figure 7.2) introduced from Europe and China, fly to deer, lose their wings, and suck blood. They are often encountered by hunters while dressing out harvested deer. Their bite causes severe itching in humans, but, more worrisome, they have the potential to transmit *Anaplasma phagocytophilum*, a bacterium that causes anaplasmosis, and

FIGURE 7.2   The deer ked, a wingless fly. (Photo by Mark Evans.)

Borrelia burgdorferi, the bacterium that causes Lyme disease, to humans. There is also concern that keds may transmit prions (see below) that may infect deer and humans (Mark Edwards, entomologist, personal communication). Deer keds only parasitize deer, and the primary way to contract disease from them is by direct contact transfer from deer to humans as occurs when hunters dress out harvested deer, when scientists handle deer as research subjects, or when humans handle deer under other circumstances, such as deer farming.

7.4   BACTERIA, VIRUSES, AND PRIONS

7.4.1   Epizootic Hemorrhagic Disease and Blue Tongue

Occasionally, deer herds may be decimated by outbreaks of infectious viral diseases (blue tongue and epizootic hemorrhagic disease [EHD]) transmitted by biting midges of the *Culicoides* genus. These infestations may lead to deer die-offs under high deer densities in locations close to aquatic areas required by the midges for breeding. The midges transmit EHD and blue tongue from infected deer to noninfected deer. Female midges ingest the blood (and viruses) of an infected deer and transmit the viruses when they feed on an uninfected deer. The midges breed in muddy areas (e.g., puddles, ponds, streams) in summer and fall, so the viruses are spread among deer congregating at watering sites when midge activities peak. EHD and blue tongue symptoms include swelling of the face, lips, tongue and gums or neck, lethargy, weakness, lameness, respiratory distress, and excessive salivation. An additional symptom exhibited by victims of blue tongue is the swollen, purple-colored tongue. Deer may die after exhibiting signs of the disease: EHD has high mortality and is an important viral disease of deer. Because midges breed in water, management of breeding sites (treating areas with *Bacteroides thuringiensis* [BT] a bacterium that attacks and destroys midges) has the potential to reduce midge abundance and minimize the number of deer infected, though this may not be practical in areas where lakes or ponds are midge breeding sites. Maintaining deer density at the lower levels required for ecosystem diversity ($\leq$5 deer/km²) reduces the likelihood of crowded deer conditions at watering places and also minimizes number of deer infected—transmission of the viruses is through midge bites and not deer-deer contact. Because the midges and viruses are short-lived, usually dying off in winter in northern states and greatly reduced in southern states, EHD and blue tongue deer die-offs usually do not reoccur on an annual basis. For this reason, forestland managers may choose to simply let the disease take its course, noting the short-term impact on local deer herds of such diseases and that the resulting reduction in deer density may result in abundance levels more preferable for sustainable forest management.

7.4.2 BOVINE TUBERCULOSIS

Bovine tuberculosis (TB) is a bacterial disease of deer and domestic livestock (and sometime humans) that doesn't require a vector: it is spread from deer to livestock by direct contact or when livestock ingest forage contaminated by deer body fluids (saliva, urine, and feces) at watering and foraging sites. Bovine TB is caused by the bacteria (*Mycobacterium bovis*); symptoms include coughing and eventual death. Cattle were the initial source of infection, and deer served as a reservoir after infection by cattle. Humans can contract bovine TB when field-dressing deer and by consuming meat from infected deer. Preventing spread of this disease from deer to cattle focuses on restriction of supplemental feeding and baiting of deer, reducing deer density by harvest, and fencing to prevent contact between deer and cattle.

7.4.3 CHRONIC WASTING DISEASE

Chronic wasting disease (CWD) is a neurological disease of deer, elk, and moose that results in fatal degeneration of brain, spinal, and other nervous tissues. Concern exists regarding potential transmission to other species of wildlife and domestic livestock. Affected animals appear emaciated, drool excessively, lose their fear of humans, and usually carry their heads and ears low and walk with a wide-legged stance in circles. The causal agent is an infectious, proteinlike substance called a prion. Unlike other protein-based disease agents like viruses and bacteria, prions are not destroyed by heat (e.g., cooking), nor are they destroyed by protein-digesting enzymes. Diagnosis is confirmed after examination of brain and spinal tissue from infected animals.

It is believed that CWD is spread by animal-to-animal contact through saliva and other body fluids. Some evidence exists of transmission from pregnant does to fetuses. Deer may become infected through contact with vegetation contaminated by urine and feces of infected animals as well as by their carcasses. Soil may serve as a reservoir for prions where they can survive for at least 4 years. Prions appear to attach to clay particles and may survive longer in soils with high clay content.

Initially, CWD was identified in farm-raised deer exposed to CWD-infected deer, but it has also been identified in free-ranging deer harvested by hunters. CWD has been confirmed in free-ranging and/or farmed deer and elk from Colorado, Illinois, Kansas, Maryland, Michigan, Minnesota, Montana, Nebraska, New Mexico, New York, North and South Dakota, Oklahoma, Pennsylvania, Texas, Utah, Virginia, West Virginia, Wisconsin, Wyoming, and the Canadian provinces of Alberta and Saskatchewan. Once established in free-ranging deer, it is almost impossible to eliminate CWD. Most states have passed laws and regulations limiting interstate transport of live deer and carcasses.

CWD is a serious management problem with no known treatment. Attempts at localized reduction of deer herds, regulating transportation of farm-raised deer, and banning baiting and feeding may have slowed but not prevented spread of CWD in deer. Human exposure to CWD-infected deer in past decades is likely. However, studies of humans living in CWD-endemic areas have not shown occurrence of human infections, and researchers have not found convincing cases of CWD transmission to humans, supporting the conclusion that a species barrier protects humans from CWD. However, hunters harvesting deer in areas where CWD-infected deer are identified should consider having carcasses tested, and should not consume meat from harvested deer that exhibited CWD symptoms.

Current recommendations in Wisconsin for controlling distribution and intensity of CWD promote reducing deer density through hunting, especially by landowners allowing hunters to hunt their forestlands, by using agricultural damage control permits to harvest additional deer, by extending length of such hunting seasons, and by disposing of infected animals (Witkowski 2017).

Individual states have enacted regulations specific to management of captive deer that should prevent spread of CWD from pen-raised to free-ranging deer. CWD usually is identified within defined and isolated pockets in individual states. State and federal natural resource agencies work with landowners in affected areas to apprise them of potential infections and coordinate management activities to minimize impact. Current information on CWD can be obtained from the CWD Alliance website (www.cwd-info.com). The mission of the CWD Alliance is "to promote responsible and accurate communications regarding CWD and support strategies that effectively control CWD to minimize its impact on wild, free-ranging cervids, including deer, elk and moose."

7.5 MANAGING FOR PARASITES AND DISEASES

What should managers do about deer parasites and diseases? The feasible answer given by veterinarians (Gortazar et al. 2014) for reducing spread from deer to other ungulates—use deer-proof exclusion fencing if feasible and, if not, reduce deer density—is the same for reducing risk of spreading disease to humans. Gortazar et al. (2014) noted that most wildlife diseases afflicting humans and other ungulates result from unbalanced situations in which wildlife has increased in numbers, often as the result of anthropogenic factors. Control schemes should target re-establishing an ecological balance of deer populations for long-term disease control. Simply getting deer density to levels compatible with sustainable regeneration of desirable tree species and diversity of understory vegetation ($\leq$5 deer/km^2) and maintaining such density should minimize most parasite and disease problems. Winter feeding of deer, which concentrates large numbers of deer within small areas and provides the opportunity for spread of disease and parasites, should never be conducted.

Small woodlot owners, who own and manage forested properties smaller than typical deer home ranges ($\sim$200 ha), are limited in their ability to treat for diseases and parasites, short of enclosing their properties in deer-proof fencing. They have little control over density of deer using their property and that of abutting properties unless they can essentially eliminate by hunting deer that move onto their forestlands from abutting lands.

It is not feasible to immunize deer against disease organisms—in many cases, vaccinations do not exist, and if they were available, the management challenge of trying to immunize herds of free-ranging deer is insurmountable.

Some states have attempted to eliminate free-ranging deer herds identified as spreading disease organisms (e.g., foot and mouth disease transmitted by mule deer [*Odocoileus hemionus*] in California in 2014 wherein 22,214 deer and 109,000 cattle and hogs were slaughtered to contain the disease). Attempts to eradicate deer from identified locales with CWD-infected deer in Wisconsin failed partly due to intense public outcry over the wholesale attempt to eliminate the deer herd.

As a general rule, state natural resource agencies should promote managing for deer densities that enhance biodiversity that coincidentally will reduce the severity and spread of deer-borne parasites and diseases.

REFERENCES

Campbell, T. A. and K. C. VerCauteren. 2011. Parasites and diseases. In *Biology and management of white-tailed deer*. D. G. Hewitt ed. Boca Raton FL: CRC Press.

Davidson, W. R. 2006. *Field manual of wildlife diseases in the Southeastern United States* (3rd edition). Athens GA: Southeastern Cooperative Wildlife Disease Study.

Gortazar, C., I. Diez-Delgado, J. A. Barasona et al. 2014. The wild side of disease control at the wildlife-livestock-human interface: A review. *Frontiers in Veterinary Science* 1:1–27.

Kingston, N. 1981. Protozoan parasites. In *Diseases and parasites of white-tailed deer*. Davidson, W. R., F. A. Hayes, V. F. Nettles et al. eds. 193–236. Tallahassee FL: Tall Timbers Research Station Miscellaneous Publication No. 7.

Lankester, M. W. 2001. Extrapulmonary lungworms of cervids. In *Parasitic diseases of wild mammals* (2nd edition). Samuel, W. M., M. J. Pybus, and A. A. Kocan eds. Ames IA: Iowa State Press.

Leuckart, R. 1886. In, Hoyle W. E. *The parasites of man, and the diseases which proceed from them: A textbook for students and practioners*. Oxford, UK: Oxford University Museum.

Witkowski, K. 2017. *Recommendations for reducing the spread of chronic wasting disease (CWD)*. Madison WI: Wisconsin Department of Natural Resources.

8 Synecology
Deer and Plant and Animal Communities

David S. deCalesta

CONTENTS

No native vertebrate species in the eastern United States has a more direct effect on habitat integrity than the white-tailed deer. In many areas of the country deer have changed the composition and structure of forests by preferentially feeding on select plant species.

Allen Pursell (2013)

MANAGER SUMMARY

Deer densities are affected by habitat quality and quantity, predation, and hunting pressure as controlled by hunting regulations. Depending on how habitat and deer harvest are managed, deer density can expand to levels resulting in habitat degradation and negative impacts on forest vegetation and wildlife communities, or it can be brought to and remain at levels that sustain healthy deer populations, habitat, and other wildlife species. Attempts to increase deer density while reducing deer impacts by increasing amounts of forage may succeed under limited conditions (Chapter 15).

8.1   INTRODUCTION

Deer are identified as a *keystone species*—one that affects the structure of plant and animal communities by nature of its impacts on species composition and structure of vegetation (McShea and Rappole 1992). These impacts in turn affect numbers and abundance of species within the wildlife community dependent on that vegetation (Hunter 1990, Paine 1995, deCalesta 1997). Too many deer affect plant/animal relationships, with resulting impacts on ecosystems. A high abundance of large herbivores has major impacts on the composition, structure, and functioning of forest ecosystems, which can result in regeneration failures (Brousseau et al. 2017).

Deer welfare, abundance, and impact on forest resources are inextricably linked to dynamic interactions with the plant and animal communities coexistent with deer. If managers are to manage deer and their impacts on forest resources, they must understand and incorporate knowledge of these interactions into management plans and operations. By achieving this objective, managers can conduct informed and effective deer management.

8.2 PLANT COMMUNITIES

Deer have contributed to the differences between understory and overstory vegetation, with browse-tolerant species increasing in abundance at the expense of preferred browse species (Long et al 2007, Russell et al. 2017).

Deer affect forest vegetation (Rooney and Waller 2003, White 2012, Sullivan et al. 2017) at all height intervals: ground, shrub, and intermediate and overstory canopies. Although deer are selective feeders, their unique digestive system allows them to utilize all forms of forage, including succulent and low-fiber forbs; less palatable and more fibrous grasses and ferns; fruits, nuts, and berries from shrubs and trees; leaf and flower buds from shrubs and trees; and woody, less digestible stems from tree seedlings and saplings. Deer can dig in the ground for mushrooms, roots, and tubers of forbs; beech (*Fagus grandifolia*) and hickory (*Carya* sp.) nuts and acorns; eat the growing points of ferns (crowns) in winter; and strip bark off trees in extreme cases. When hard pressed, deer will even eat unthought-of foods, such as alewives washed up on shorelines of Lake Michigan islands.

When in balance with their environment under densities characteristic of pre-European colonization of North America (4–5 deer/km^2), deer browsing was spread out across many herb, shrub, and tree species. Because of predation risk, deer did not use landscapes randomly. Rather, they avoided areas where they were susceptible to predation by wolves and mountain lions, and spent less time foraging. Browsing intensity at this level did not eliminate categories of forage (e.g., forbs, shrubs, and highly preferred tree seedlings), as occurs today under browsing by overabundant deer herds in the absence of the complete suite of natural deer predators.

At the end of the 1800s, deer were released from predation risk by the elimination of major deer predators. At the same time, vast clearcutting across North American landscapes created enormous quantities of deer forage, allowing deer density to escalate rapidly because of the lack of predation pressure. This caused a *bottom-up trophic cascade*, wherein deer population density soared. As sources of deer forage declined when seedlings grew out of the reach of deer, overabundant herds began to eliminate highly palatable and easily reached forage in the ground and shrub levels.

Browsing by overabundant deer herds has "homogenized" forests in much of the eastern United States, creating depauperate understory communities (forbs, shrubs, and tree seedlings) dominated in some cases by ferns (Royo et al. 2010, Goetsch et al. 2011, Pendergast et al. 2016). Overabundant white-tailed deer (*Odocoileus virginianus*) are a significant management problem in North America that exerts unprecedented herbivory pressure on native understory forest communities (Knight et al. 2009).

Some forb species (e.g., *Trillium* sp., *Maianthemum canadense*), which depend on insects to disperse seeds (for short distances), may be locally extirpated as persistent removal of adult plants by browsing deer culminates in plant mortality with little chance of reintroduction by seeds from unbrowsed adult plants (Knight et al. 2009). Additionally, deer affect size, stage, and population dynamics of *Trillium grandiflorum* and other forbs by selectively browsing flowering and large nonflowering stages, reducing or eliminating seed production, and likely regressing in stage and size in the following growing season. Knight et al. (2009) further stated that levels of deer herbivory commonly experienced in states such as Pennsylvania are sufficient to cause the loss of *T. grandiflorum* and other forbs. Similarly, heavy browsing by deer on preferred tree seedlings may result in their elimination by deer browsing as mature trees fall out of the overstory (through natural disturbance or to timber harvest) and are not replaced by seedling offspring, which may have been eliminated by deer. Forests so browsed may be dominated by a reduced tree and forb species richness consisting of propagules resistant to deer browsing or not preferred by deer. *In extremis*, even nonpreferred or resistant tree seedlings of species such as American beech may be eliminated or perpetually suppressed by overabundant deer, resulting in a future landscape largely depauperate of trees.

Deer browsing can result in an extremely low diversity of the herbaceous understory dominated by ferns, local extirpation of shrubs, and extremely low forb abundance. Where deer density

FIGURE 8.1 Protected habitat for browse-sensitive herbaceous vegetation. (Photo by David S. deCalesta.)

and browsing pressure have been too high for long periods of time, favored herbaceous species may survive only in places protected from deer browsing, such as atop monolithic boulders (Figure 8.1).

Natural revegetation by understory plants reduced by deer browsing can only occur if individual plants survive deer browsing. Following intensified removals of overabundant deer herds, limited recovery of the herbaceous community can occur but diversity may not improve, suggesting that reducing deer populations alone may not result in restoration of plant diversity without strategies. Because forbs have low dispersal and reproductive rates, long-term legacy effects (delayed revegetation) should be expected even if deer numbers can be reduced (Sullivan et al. 2017).

Reduction of abundant forbs, shrubs, and seedlings by deer browsing gives a competitive advantage for resources to native ferns (New York [*Thelypteris noveboracensis*] and hay-scented [*Dennstaedtia punctilobula*]) and grasses (e.g., short husk grass [*Brachyelytrum erectrum*]) that deer do not eat (Horsley and Marquis 1983) (Figure 8.2). Overbrowsed sites encourage spread of exotic,

FIGURE 8.2 Dominance by fern (left panel) and garlic mustard (right panel) resulting from browsing removal of other species by overabundant deer. (Photos by David S. deCalesta.)

invading plants (Russell et al. 2017) such as Japanese stilt-grass (*Microstegium vimineum*) (Aronson and Handel 2011), Japanese barberry (*Berberis thunbergii*), and garlic mustard (*Alliaria petiolata*), which deer also do not eat. These changes in forest vegetation result in a much less diverse plant community, both species- and structure-wise, that persists to the present, especially in areas where deer remain overabundant (Figure 8.2).

Forb and shrub species preferred by deer are perpetually threatened by overabundant deer because their growth characteristics do not allow them to grow out of browsing height. Once gone, they are gone—elimination of forb and shrub species by decades of overabundant deer browsing is well documented (Rooney and Dress 1997). Restoration of extirpated forb and shrub species calls for extraordinary management actions, including re-establishment by planting adult plants. However, such restorative actions take years for success (Aronson and Handel 2011) and will fail if deer density is not concurrently reduced to and maintained at levels compatible with coexistence of deer and forb/shrub communities, or if restored areas are not protected indefinitely by enclosure within deerproof fencing.

Anderson (1994) recommended managing for deer densities of 4–6 individuals/km² for deciduous forests. Koh et al. (2010) documented a significant negative relationship between deer density (7–40/km²) and maximum *Trillium* height over a 15-year period. Pierson and deCalesta (2015), based on data from Stout et al. (2013), compared changes in height, flowering, and leaf size of three herbaceous species with changes in deer density over time, noting that impact was lowest when deer density was maintained at approximately 5 deer/km².

The natural solution to the negative impact of deer on forest vegetation—reduction of deer density to precolonial levels by enhanced predation and maintenance of those densities indefinitely—is best and likely cheapest. The precolonial period effect of native predators that once held deer numbers in check can be replaced by increasing and maintaining higher levels of deer mortality through required annual removals by humans. Combinations of enhanced public hunting (by increasing availability of permits for harvesting antlerless deer and increasing length of deer hunting seasons) and removals by contracted sharpshooters can be effective (see Chapter 37). Such removals may be conducted landscapewide, in contrast to the use of deerproof fencing that limits protection of forest vegetation to the relatively small areas enclosed by such fencing. Long-term restoration and survival of heavily impacted forest vegetation will require persistent, annually recurring reduction of deer density to environmentally friendly levels.

8.3 WILDLIFE COMMUNITIES

Reduced abundance of plant species and habitat structure produces an additional negative trophic cascade on dependent wildlife species. Reduced diversity and structure of ground and shrub layers, resulting from browsing by overabundant deer herds, and competition for food resources, including fruits, nuts, and berries, are associated with losses in species richness and abundance of forest songbirds and small mammals (Casey and Hein 1983, McShea and Rappole 1992, deCalesta 1994, Flowerdew and Ellwood 2001, Allombert et al. 2005, Tymkiw et al. 2013, Sullivan et al. 2017) (Figure 8.3).

Where other wild ungulates (elk [*Cervus* sp.] and moose [*Alces alces*]) are cohabitant with deer, interactions with deer may lead to infection of, and death by, parasites fatal to elk and moose.

As in the case with protection of forest vegetation from excessive browsing by overabundant deer, the best way to re-establish and maintain habitat components threatened and/or eliminated by overabundant deer is to conduct annually recurring landscapewide reductions of deer density to precolonial levels. Establishing and maintaining regular monitoring of deer density and impacts on forest vegetation will identify and direct needed levels of deer removal. Maintaining ecologically viable deer densities will also reduce the incidence of disease transmission from deer to domestic and other native ungulates.

FIGURE 8.3 Good breeding habitat for forest songbirds, top panels. Poor habitat, bottom panels. Black-throated blue warbler, left panels. Chestnut-sided warbler, right panels. (Bird photos courtesy Jake Dingle; habitat photos courtesy David S. deCalesta.)

REFERENCES

Allombert, S., A. J. Gason, and J. Martin. 2005. A natural experiment on the impact of overabundant deer on songbird populations. *Biological Conservation* 126:1–13.

Anderson, R.C. 1994. Height of white-flowered trillium (*Trillium grandiflorum*) as an index of deer browsing intensity. *Ecological Applications* 4:104–109.

Aronson, M. F. J., and S. N. Handel. 2011. Deer and invasive plant species suppress forest herbaceous communities and canopy tree regeneration. *Natural Areas Journal* 31:400–407.

Brousseau, M., N. Thiffault, J. Beguin et al. 2017. Deer browsing outweighs the effects of site preparation and mechanical release on balsam fir seedlings performance: Implications to forest management. *Forest Ecology and Management* 405:360–366.

Casey, D., and D. Hein. 1983. Effects of heavy deer browsing on a bird community in a deciduous forest. *Journal of Wildlife Management* 33:524–532.

deCalesta, D. S. 1994. Effects of white-tailed deer on songbirds within managed forests in Pennsylvania. *Journal of Wildlife Management* 58:711–718.

deCalesta, D. S. 1997. Deer density and ecosystems management. In *The science of overabundance: The ecology of unmanaged deer populations.* McShea, W. J., H. B. Underwood, and J. H. Rappole (eds). Washington DC: Smithsonian Press.

Flowerdew, J. R. and S. A. Ellwood. 2001. Impacts of woodland deer on small mammal ecology. *Forestry* 74:277–287.

Goetsch, C., J. Wigg, A. A. Royo et al. 2011. Chronic over browsing and biodiversity collapse in a forest understory in Pennsylvania: Results from a 60 year-old deer exclusion plot. *Torrey Botanical Society* 138:220–224.

Horsley, S. B. and D. A. Marquis. 1983. Interference by weeds and deer with Allegheny hardwood reproduction. *Canadian Journal of Forest Research* 13:61–69.

Hunter, M. L. 1990. *Wildlife, forests, and forestry.* Englewood Cliffs NJ: Regents-Prentice Hall.

Knight, T. M., H. Caswell, and S. Kalisz. 2009. Population growth rate of a common understory herb decreases non-linearly across a gradient of deer herbivory. *Forest Ecology and Management* 257:1095–1103.

Koh, S., D. R. Bazely, A. J. Tanentzap et al. 2010. Trillium grandiflorum height is an indicator of white-tailed deer density at local and regional scales. *Forest Ecology and Management* 25:1472–1479.

Long, Z. T., T. H. Pendergast IV, and W. P. Carson. 2007. The impact of deer on relationships between tree growth and mortality in an old-growth beech–maple forest. *Forest Ecology and Management* 252:230–238.

McShea, W. J. and J. H. Rappole. 1992. White-tailed deer as keystone species within forest habitats in Virginia. *Virginia Journal of Sciences* 43:177–186.

Paine, R. T. 1995. A conversation on refining the concept of keystone species. *Conservation Biology* 9:962–984.

Pendergast IV, T. H., S. H. Hanlon, Z. M. Long et al. 2016. The legacy of deer overabundance: Long-term delays in herbaceous understory recovery. *Canadian Journal of Forest Research* 46:362–369.

Pierson, T. G. and D. S. deCalesta. 2015. Methodology for estimating deer impact on forest resources. *Human Wildlife Interactions Journal* 9:67–77.

Pursell, A., T. Weldy, and M. White. 2013. Deer overabundance and ecosystem degradation: A call to action. *Science Chronicles* August 2013 The Nature Conservancy.

Rooney, T. P. and W. J. Dress. 1997. Species loss over sixty-six years in the ground-layer vegetation of Heart's Content, an old-growth forest in Pennsylvania, USA. *Natural Areas Journal* 17:297–305.

Rooney, T. P. and D. M. Waller. 2003. Direct and indirect effects of white-tailed deer in forest ecosystems. *Forest Ecology and Management* 181:165–176.

Royo, A. A., S. L. Stout, D. S. deCalesta et al. 2010. Restoring forest herb communities through landscape-level deer herd reductions: Is recovery limited by legacy effects? *Biological Conservation* 143:2425–2434.

Russell, M. B., C. W. Woodall, K. M. Potter et al. 2017. Interactions between white-tailed deer density and the composition of forest understories in the northern United States. *Forest Ecology and Management* 384:26–33.

Stout, S. L., A. A Royo, D. S. deCalesta et al. 2013. The Kinzua Quality Deer Cooperative: Can adaptive management and local stakeholder engagement sustain reduced impact of ungulate browsers in forest systems? *Boreal Environment Research* 18:50–64.

Sullivan, K. L., P. J. Smallidge, and P. D. Curtis. 2017. *AVID—Assessing vegetation impacts from deer: A rapid assessment method for evaluating deer impacts to forest vegetation (Draft).* Ithaca NY: Cornell University Cooperative Extension.

Tymkiw, E. L., J. L. Bowman, and W. G. Shriver. 2013. The effect of white-tailed deer density on breeding songbirds in Delaware. *Wildlife Society Bulletin* 37:714–724.

White, M. A. 2012. Long-term effects of deer browsing: Composition, structure, and productivity in a northeastern Minnesota old-growth forest. *Forest Ecology and Management* 269:222–228.

9 Human Factors
Deer–Forest Management Areas vs. Deer Administrative Units

David S. deCalesta

CONTENTS

Wildlife management can be defined as the manipulation of wildlife populations and habitat to achieve a goal.

Mark Sargent and Kelly Carter (1999)

MANAGER SUMMARY

Managers of individual forestlands need and benefit from the flexibility and specificity to control site-specific deer density and impact afforded by deer density reduction programs designed for these lands. State natural resource agencies attempt to affect deer density and impact on large deer administrative units (DAUs) that enclose individual forestlands (deer–forest management areas or DFMAs) by issuing permits to harvest antlerless deer on these large administrative units. Some recognized the need to provide permits to reduce deer density on DFMAs and provided programs entitled deer management assistance programs (DMAPs). Managers in states that offer such programs should use them as an effective tool to reduce deer density and impacts at effective and appropriate landscape levels. Managers in states that do not offer the programs way wish to lobby their respective state agencies/legislatures to enact and make available DMAP programs.

9.1   INTRODUCTION

The wildlife management profession identifies, recommends, and practices manipulation of habitats and wildlife populations. Volume 2 (Management) of the 7th edition of the *Wildlife Techniques Manual* (Silvy 2012) addresses and describes management categories, including human dimensions; communications and outreach; adaptive management; managing forest, rangeland, wetland, farmland,

and urban habitats; harvest management; managing animal damage; ecology and management of small populations; captive propagation and translocation; and habitat conservation planning. Deer management should incorporate the above-identified components, as required, to meet landowner goals. However, the profession does not differentiate between individual forestlands managed by the owner and/or by managers (of private lands) or by stewards (of public lands) and large blocks of forestland managed by multiple managers/owners/stewards loosely lumped into deer administrative units (DAUs).

9.2 DEER ADMINISTRATIVE UNITS

Season and bag limits regulating deer harvest are established by state natural resource agencies for large, discrete landscapes commonly referred to as deer management units, but also as deer season zones, deer hunting zones, deer zones, deer management zones, deer hunting units, deer reduction zones, deer areas, deer management regions, deer gun hunting units, wildlife management areas, wildlife management units, and wildlife management zones (collectively referred to as DAUs in this book). Some states have as few as four; others identify almost 100 such units. Some are composed of individual counties, others encompass more than one county, and some have more than one unit within individual counties. DAUs are composed of multiple deer–forest management areas (DFMAs), which leads to a management disconnect.

9.3 DEER–FOREST MANAGEMENT AREAS

Deer–forest management areas are discrete parcels of private or public forestlands where deer density, deer impact, and forest vegetation are actively manipulated/managed. Many are <200 ha owned by private landowners, but others are quite large, exceeding 500 ha as in national forests, national parks, state forests and parks, and county parks. Management of deer density on DFMAs by hunting is regulated by state agencies at state and DAU levels. Management activities on individual DFMAs include (1) creation of forage, fawning, and hiding deer habitat by thinning and/or harvesting timber stands; (2) creation and maintenance of within-site roads used for forest management and by hunters for access; (3) protection of tree seedlings from deer browsing with individual tree shelters or by enclosing tree regeneration areas with deer fencing; (4) use of herbicides and/or mechanical weed reduction to revert understory composition to tree seedlings, shrubs, and forbs; (5) activities associated with managing deer harvest (posting boundaries, repairing road damage); and (6) monitoring of habitat conditions, deer density, and deer impact to determine whether management activities result in progress toward achieving goals.

The above management activities are conducted to create optimal conditions for tree seedlings and/or other forest resources, which may also result in enhancing deer habitat and hunter access to forestlands. Costs of managing forest vegetation, deer density, and hunting access are borne by managers in the case of private lands and by public funding of managing agencies on public lands. Hunters granted hunting access to DFMAs generally do not contribute to deer management expenses unless they pay for exclusive hunting rights within private forestlands through lease hunting fees.

9.4 THE DEER ADMINISTRATIVE UNIT–DEER–FOREST MANAGEMENT AREA DISCONNECT

Forest vegetation and deer are not managed at the DAU level because no single managing entity has jurisdiction over the collection of disparate properties within DAUs. Rather, individual DFMAs within DAUs are managed by their respective owners/managers/stewards, which range in size from individual private managers to large and comprehensive staffs at state and national forests composed of supervisory staffs that oversee the activities of dozens of forest resource specialists for forest-related fields. State natural resource agencies may occasionally assess deer density within DAUs using aerial censuses or deriving them from sex and age ratios of deer harvested within DAUs, but they do not assess deer density, deer impact, or habitat for individual DFMAs within DAUs. Therefore, any deer harvest regulations designed to adjust deer density are based on variable DAU-wide estimates of deer density and nonstandardized anecdotal reports of deer impact and forest vegetation conditions provided by managers of DFMAs within DAUs. Deer harvest regulations developed for DAUs are rarely, if ever, derived from collaborative integration of deer and forest vegetation conditions among DFMAs.

A further and important distinction between deer management on DFMAs and DAUs is in how different silvicultural systems (see Chapter 13) are employed. Some systems (uneven-aged) feature small openings with an emphasis on establishing and maintaining species typical of uneven-aged management and do not emphasize creation of deer forage. Others (even-aged) may feature creating larger and more frequent openings in the forest overstory and focus on tree regeneration, resulting in more abundant, high-quality deer forage. Managers of DFMAs probably employ a consistent silvicultural system including management across entire properties. Managers of differing properties within DAUs, however, may employ differing silvicultural systems resulting in different proportions and arrangements of deer habitat and forage and creating differing deer habitat conditions and deer densities among DFMAs within DAUs. A likely outcome of landscape deer forage and habitat arrangements and availabilities is differences in deer abundance, movements, and impact on habitat/vegetation within DFMAs within such DAUs.

Deer hunting is governed by deer harvest regulations developed for DAUs by state natural resource agencies. Many of these agencies have no jurisdiction to manipulate forest vegetation or deer populations within the DAUs they define, relying instead on embedded landowners to manage forest vegetation and deer density (via public hunting). The exceptions are state natural resource agencies that have management authority and responsibility ceded them on identified public land DFMAs, such as state forests, state parks, or, as in Pennsylvania, State Gamelands. And, in states like Pennsylvania where the Pennsylvania Game Commission is responsible for management of forest vegetation and wildlife species on public Gamelands, harvest regulations on these lands are the same as those assigned to the larger DAUs the lands are embedded within.

Deer harvest regulations controlling deer hunting (seasons and bag limits) as developed by agencies for DAUs are not sufficiently area-specific to meet the management needs of DFMAs for deer management (deCalesta 2017). The ability of landowners to reduce deer density and impact within their DFMAs by public hunting is constrained by restrictions on deer harvest specific to encompassing DAUs.

9.5 REGULATIONS DESIGNED TO REDUCE DEER DENSITY AND IMPACT ON DEER–FOREST MANAGEMENT AREAS

9.5.1 DEER ASSISTANCE MANAGEMENT PROGRAM PERMITS

As some state natural resource agencies recognized the costs and negative impacts incurred by overabundant deer herds on understory forest vegetation (and associated wildlife communities), they attempted to remedy the situation by increasing the number of DAU permits for harvesting antlerless deer to reduce deer density and impact. In some cases, DAU hunting season lengths were increased when agency biologists recognized that hunting conditions were affected by poor weather and extended seasons to increase deer harvest.

Ensuing efforts by natural resource agencies to reduce deer density and impact focused on increasing number of permits for harvesting antlerless deer within specified DAUs. In some states, such as New York, state natural resource agencies held DAU-level stakeholder meetings to arrive at consensus desired deer abundance to determine the number of antlerless permits required to remedy the overabundant deer situation. Unfortunately, DAU-specific antlerless permits do not identify

DFMAs needing reductions in deer density: rarely are desired DFMA-specific reductions achieved with permits, as hunters may use DAU permits elsewhere.

State natural resource agencies that recognized the DAU/DFMA disconnect when attempting to increase hunting pressure and reduce deer density on DFMAs began to provide DFMA-specific allocations of antlerless permits by developing deer management assistance programs (DMAPs). DMAPs provided additional permits for harvesting antlerless deer usable only for specific DFMAs. Generally, DFMA managers provide evidence that deer are causing unacceptable damage to forest resources and maps of their properties displaying boundaries and access roads. DMAP permits are used during regular hunting seasons only for harvesting antlerless deer on DFMAs. Some states require hunters using DMAP permits to send in reports detailing whether they harvested a deer. Depending on the state, DMAP permits may be issued by DFMA managers or by state natural resource agencies.

9.5.2 Deer Damage Permits

In some states (e.g., New York, Indiana) deer damage permits (DDPs), usable only within defined deer damage areas/zones, may be provided to private landowners and managers of municipalities and public forestlands for harvest of antlerless deer. Generally, DDPs are provided in situations where regular season hunting (including properties within DAUs wherein antlerless permits are available and for properties identified as DMAP areas) does not reduce deer populations sufficiently to alleviate deer impacts on specified properties (deer damage areas). DDPs are designed to increase removal of deer above numbers harvested during regular season regulations (including DAU and DMAP permits) on identified deer damage areas. DDPs are generally for deer removals outside dates for regular deer hunting seasons. As with DMAP permits, applications for DDPs must include documentation of impacts, use of regular season antlerless permits (DAU and DMAP), failure to achieve desired reductions in deer density and impact, plans for monitoring effectiveness, and reporting of numbers of deer harvested.

DDPs can be especially effective when available after regular deer hunting seasons close because they represent last opportunities for hunters to harvest a deer. Research demonstrated that providing extra hunting days for antlerless deer within special permit areas after regular hunting seasons are over results in additional harvest of antlerless deer on forestlands where existing season length and bag limit did not achieve desired reductions in deer density and impact (Roseberry et al. 1969, deCalesta 1985).

The proof of the effectiveness of DFMA-directed deer management is in the pudding—case history sites (Section V) that employed one or more antlerless permit programs for increasing deer harvest to reduce deer density and impact succeeded in reducing deer density and impact. The few (smaller) sites that chose methodologies other than reducing deer density (fencing and other exclusion devices, providing sources of alternate forage) did not reduce deer density but did reduce deer impact. The one case history (Chapter 40) where landowners chose neither enhanced reduction of deer density nor exclusion of deer failed to achieve goals for their deer herd and forest vegetation.

Most states have one or more kinds of DAU, DMAP, or DDP permits for reducing deer density and impact: a few have all three, most have one or two, and a few (e.g., Maine and Connecticut) have none. The most effective are DMAP permits and DDPs because they are designed to reduce deer density on specific properties.

REFERENCES

deCalesta, D. S. 1985. Influence of regulation on deer harvest. In *Symposium on game harvest management.* eds. Beasom, S. L. and S. Roberson. Kingsville TX: Texas A & I Univ.

deCalesta, D. S. 2017. Bridging the disconnect between agencies and managers to manage deer impact. *Human-Wildlife Interactions* 11:112–115.

Roseberry, J. L., D. C. Autry, W. D. Klimstra et al. 1969. A controlled deer hunt on the Crab Orchard National Wildlife Refuge. *Journal of Wildlife Management* 33:791–795.

Sargent, M. and K. S. Carter (eds). 1999. Managing Michigan's Wildlife: A Landowners Guide. Michigan Department of Natural Resources (MDNR). East Lansing. Michigan United Conservation Clubs (MUCC), and Michigan State University. http://www.dnr.state.mi.us/wildlife/Landowners_Guide/Introduction/TOC.htm).

Silvy, N. J. (ed.) 2012. *The wildlife techniques manual: Volume 2: Management.* Baltimore MD: The Johns Hopkins University Press.

10 Human Factors
Hunters and Hunting

David S. deCalesta

CONTENTS

Hunters are the [Pennsylvania Forestry] bureau's strongest ally in its bid to achieve deer populations in balance with state forest habitat.

Cindy Dunn (2015)

MANAGER SUMMARY

Reducing deer density and impact by public hunting is by far the cheapest and most effective way to manage deer impact on forest resources. Some hunters, termed alpha and beta hunters, and novice hunters hunting for subsistence rather than for trophy deer (identified as locavore hunters), are more accepting of science-based deer management and support the need to control deer density to reduce impact on forest resources, including deer. These are the hunters most effective in reducing deer density and impact and need to be identified, recruited, and rewarded for reducing deer density through their hunting efforts. Omega hunters are older and less accepting of science-based deer management, get their philosophy of deer management from peers, and want maximum deer density. Managers need to identify needs of alpha, beta, and locavore hunters, and develop timely and effective communications with them to make hunting work.

10.1 INTRODUCTION

The Chinese proverb, "know thy enemy" has a corollary for deer management: "know thy ally"—in this case, the deer hunter. Hunters are a keystone component in programs integrating forest and deer management. Success in managing deer density and impact is largely dependent on recruiting and retaining hunters and facilitating their success in harvesting deer. And, just as managers need to understand deer ecology to design effective deer management programs, they also need to understand hunter "ecology" (needs/interactions related to their physical surroundings and to other humans), especially as it relates to hunter type. If managers are to enlist hunters to reduce deer density and impact, they must understand their knowledge level, needs, skills, capabilities, and motivations regarding deer hunting so they can appeal to them to help reduce deer density and impact. Once managers are informed on these issues they are better prepared to provide hunters with what they need to help manage deer density and impact. Needs, and how they are addressed, vary by hunter type. Meeting hunter needs requires an enduring, ongoing commitment by managers to become directly involved with hunters in off-site (communications/hunter solicitation) and on-site (meeting, greeting, and working with hunters at access points; checking stations and other opportunities) interactions. Last, forest managers must embrace and adjust to the change in the deer hunting paradigm from "deer camp" to "day hunt."

10.2 HUNTER TYPES

Declining numbers of deer hunters make deer management more difficult for natural resource managers. Ward et al. (2008) suggested that if managers could identify and target hunters more attuned to the concept of reducing deer density to reduce their impact on natural resources, they might be more effective in managing deer impact. They conducted a survey that classified deer hunters into four groups. Group A ("no-damage traditionalists") was composed of two subgroups: (1) those that favored high deer density, did not recognize deer damage, opposed changes in hunting regulations designed to reduce deer density, and hunted on lands of others, and (2) those that perceived deer impact as minimal, opposed changes in hunting to affect deer impact, favored high deer density, and did not like posting of private lands that prohibited them from hunting posted properties. Group 2 ("damage control managers") was also composed of two subgroups: (1) those that recognized deer impact as significant, supported changes in hunting regulations to reduce deer density and impact and reduction in deer density, considered themselves deer managers, and opposed posting of private lands, and (2) those that were similar to the preceding subgroup but hunted on available lands and did not oppose posting.

The problem with the hunter classification posed by Ward et al. (2008) is that it is academic rather than practical: managers would have no way to tailor strategies for managing hunters in these categories with management activities (other than educational). I much prefer the three hunter categories (alpha, beta, and omega) identified by Alt (quoted in Frye 2006) and add a fourth category: locavore. Alpha hunters are capable of consistently harvesting deer. They are well informed on the science of deer biology and management and on deer hunting strategies and techniques, and understand that for habitat to produce healthy and trophy animals, deer must be in balance with their habitat. Alpha hunters support deer density that preserves and sustains healthy forests and healthy deer. Because alpha hunters are the most effective in harvesting deer, they must be recruited and retained if landowners rely on hunting to maintain deer density and impact at desired levels. For alpha hunters, the message that deer density must be managed at the level that consistently produces high-quality deer will resonate with their values, as long as there is evidence to support it.

The same may be said of beta hunters (Figure 10.1), but this group needs more than recruitment and maintenance. Beta hunters are alpha hunters in the making: open to scientific evidence regarding the bases for managing deer, developing hunting skills, and aware of the relationship between deer density and deer and habitat health. As with alpha hunters, the values to stress for beta hunters are quality deer and habitat, and quality hunting experiences. Efforts must be made to furnish beta hunters with information that provides better understanding of the relationship between deer health and deer density and how to be more successful in harvesting deer. Pairing these hunters with mentors (experienced alpha deer hunters) will help them gain the skills and knowledge to become successful alpha hunters.

FIGURE 10.1 Beta hunters benefit from pairing with mentors who can help them harvest deer.

Omega hunters, Alt's third group, are not as skilled at hunting, are dependent on culture and hunting lore for their understanding of deer management and deer density, and are less successful than alpha and beta hunters in harvesting deer. Omega hunters are resistant to scientific evidence that equates high deer density with poor habitat and deer health and may hunt in ways (e.g., road hunting) wherein success is dependent on high deer density and high visibility (as seen in depauperate forest understories caused by high deer density). Omega hunters want high deer density and will not accept information that high deer density is neither sustainable nor results in optimal deer and habitat health.

The fourth hunter category is locavore (or localvore, meaning persons who obtain and eat foods produced locally). Persons in this group have little to no previous hunting experience but want to hunt wild game (e.g., deer, rabbits, turkeys) to obtain local (and ostensibly chemical-enhancement free) sources of protein. As their objective is obtaining wild meat rather than trophy animals, hunters in this group are not affected by arguments concerning whether there are too many or too few deer—they simply want a place where they can hunt with the expectation of harvesting a deer. Because they are inexperienced, they require education and training on safe and effective use of arms used to harvest deer, methods for hunting and harvesting deer, ethics of hunting, and techniques for processing (field-dressing, skinning, boning-out, and cooking) deer they have harvested. Two states, Minnesota and Wisconsin, have offered workshops that provide information and training for locavore hunters. Locavore hunters, like beta hunters, would benefit from pairing with mentors who can teach them how to hunt safely and effectively. Also needed is a system by which locavore hunters can be matched with managers seeking reduction in deer density through hunting.

Because alpha, beta, and locavore hunters are most likely to support reducing deer density to produce quality deer and quality habitat, managers should solicit these hunter types and familiarize themselves with, and help provide for, the needs specific to each type.

10.3 CHANGE IN THE DEER HUNTING PARADIGM

As deer populations climbed in the early 1900s and harvest regulations became more liberalized, deer hunting became increasingly popular. Although deer hunting in the early days was practiced

FIGURE 10.2 Early twentieth-century deer camp (left); late twentieth-century deer camp (right). Photos by Wikimedia Commons (left) and David S. deCalesta (right).

by some solitary hunters, the advent and popularity of "deer camp" altered how deer were hunted. Deer hunting became a treasured and much-anticipated tradition as hunters annually made extensive preparations for lodging, feeding, and equipping themselves to enjoy the camaraderie of a week or so in hunting camps (Figure 10.2) away from the pressures and realities of work. Deer camp expanded over time to include family members as well as friends and became a primary recruiting tool for young hunters and a way of ensuring the continuation of the deer hunting tradition. An additional benefit of deer camp was that deer drives, an efficient way of harvesting deer, became a standard practice in addition to still and stand hunting.

In the deer camp heyday in the 1940s–1970s, local business experienced a surge of spending in the weeks leading up to and including deer season. Grocery stores stocked up ahead of the season to meet the demand, reservations at hotels and motels had to be made a year or more in advance, and restaurants had to hire extra help to serve all the hunters.

However, as the hunting population became less rural and more urban/suburban, and electronic technology expanded (television, computers, the internet, and cell phones), potential hunters, especially preteens and teenagers, became captured by the popularity and pervasiveness of instant and personal communications. There was less time for hunting, adult hunters had increasing conflicts for time, and hunting received less priority and time. Deer camp began to wane, replaced by "day hunting," wherein hunters would drive to their hunting spot, hunt opening day, and maybe stay overnight in a motel/hotel to hunt the next day. For many hunters, deer hunting shrank to a 1 or 2-day exercise, possibly extended an additional day on the Saturday following opening day. Hunters hunted out of their vehicles, and deer drives became more difficult to organize and conduct with few hunters to participate. Overall, deer hunting declined and deer populations began to increase, abetted by a second wave of timber harvest, which increased amounts and quality of deer forage.

One consequence of the shift from deer camp to day hunting was that most of the deer harvest was limited to 3 or fewer days by hunters strapped for time. As deer harvest declined, deer population density (and impact on forest resources) increased. Coincidentally, local businesses catering to deer hunters during deer season experienced a major shrinkage in business in the decades following the decline of deer camp.

The realities of a deer season limited to 2–3 days by hunters with less time for recreation and the resulting potential for reduced deer harvest require that forestland owners and managers focus on attracting hunters and retaining them. They need to make their forestlands and deer management programs more attractive to hunters by anticipating and responding to hunter needs and by aggressively seeking out (advertising) hunters and enhancing hunting on their properties.

10.4   NEEDS COMMON TO ALL HUNTER TYPES

Needs for all solicited hunter groups are: (1) forested landscapes of sufficient size and habitat characteristics to support deer populations and produce sufficient recruitment to allow for a deer harvest of sufficient size and quality to retain hunters year to- year; (2) information and instructions on how to locate and travel to forest landowner properties; (3) vehicular access to interior portions of forested landscapes provided by improved all-weather roads and all-weather driving conditions, including during periods of snow and ice; (4) a controlled hunting environment where there are neither too many nor too few hunters; (5) effective and ongoing communications before, during, and after hunting season(s) between managers and deer hunters; (6) seasonal and nearby availability of sources of hunting equipment, lodging, and meals; (7) detailed topographic maps, including road systems and descriptions of habitat types, including deer habitat components (see Chapter 2) and productive hunting locations; (8) opportunities for, and encouragement of, preseason scouting; and (9) providing for and enhancing different hunting styles.

Hunters may request modifications/enhancements of deer management, as voiced by Pennsylvania deer hunters during hearings prior to restructuring the Pennsylvania Game Commission's deer management program in the late 1990s. Those requests may include providing estimates of local deer density, characterizing deer health as derived from harvest data collected at deer checking stations, and downsizing or restructuring deer management units to provide more homogeneous habitat landscapes at manageable scales (hundreds to thousands of hectares rather than the typically enormous management units encompassing tens of thousands—or more—of hectares of landscapes of highly variable deer habitat often encompassing multiple ownerships).

10.4.1   COMPREHENSIVE FORESTED LANDSCAPES

Quality deer are produced by quality forests that contain required deer habitat components. Early succession habitat, created by timber harvest/thinning and/or natural disturbances, provides high-quality deer forage and fawning cover. Sapling/pole stands provide hiding/security cover that will retain deer within local landscapes during hunting seasons. Maturing forest stands, especially those containing conifers, provide forage and thermal cover. The larger and more diverse such complete landscapes are, the more capacity there is for an increasing number of hunters. Smaller landscapes lacking one or more essential deer habitat components benefit from juxtaposition with neighboring lands that supply missing habitat(s). Managers of small properties lacking one or more of the essential deer habitats may wish to consider combining their forestlands with other like-minded managers in cooperatives to enhance deer habitat and facilitate hunter harvests (see Chapter 31).

10.4.2   HUNTER ACCESS

Thomas et al. (1976), studying the effect of access on hunter use of forestlands in West Virginia, found that hunter density decreases as distance from trails, parking sites, and roads increases and recommended management of trails as a way to increase hunter distribution. Hunters rarely traveled more than 600 m from forest edges/roads in a Pennsylvania study (Keenan 2010), raising the likelihood of higher deer harvests in forestlands providing high hunter access. Sparsely roaded forestlands (e.g., Pennsylvania Bureau of Forestry state forests) have lower deer harvest rates than forestlands with high access (e.g., the Kinzua Quality Deer Cooperative; Chapter 34). An average density of 10 hunters/km^2/day harvested an average of 1 black-tailed deer (*Odocoileus hemionus columbianus*)/km^2/day on a 47 km^2 special hunt area (open three consecutive weekends/year) in Oregon over a 23-year period (deCalesta 1985)—no area within the special hunt area was more 500 m from one or more well-maintained, graveled roads. Such roads provide hunters good access to hunting areas; plowing the roads after snowfalls during hunting season enhances hunter access and success. Providing hunters with maps detailing locations of access roads and foraging and hiding cover for deer increases the probability that hunters will hunt a higher proportion of forestlands, with likely increases in number of deer harvested.

10.5 CONTROL OF HUNTER DENSITY

Managing public hunting on forestlands is a balancing act between having enough hunters to move and harvest deer and having so many hunters that conflicts arise when hunters encroach on the hunting spaces of other hunters. On relatively small properties (e.g., <200 ha), landowners and managers may cap hunter density by allowing hunting by permission only and by denying access to nonsolicited hunters by posting property boundaries against hunting/trespass and fitting access roads with lockable gates. The problem is the opposite with larger forestlands ($\geq$200 ha)—it may be difficult to obtain sufficient landscapewide hunting pressure to reduce deer density to the goal level uniformly across forestlands. Landowners may have to solicit hunters via advertising through traditional advertising media such as newspapers and magazines and through digital media (internet access via websites, blogs, and Facebook). They can increase hunter participation by obtaining and distributing antlerless permits, if available, specific to their forestlands to hunters. Distributing harvest across large forested landscapes is enhanced by providing roaded access to all portions of properties: improving logging roads and trails can aggregate and distribute access when added to existing access roads. Adding additional attractions such as checking stations, warming huts, and information kiosks with maps and other information will help attract hunters and distribute hunting pressure more uniformly.

10.5.1 COMMUNICATION WITH AND EDUCATION OF HUNTERS

It will do the forest landowner little good if they provide comprehensive habitat components and access for deer hunting, but the hunters are unaware of the hunting opportunities available within such lands. Hunters who are informed of hunting opportunities and hunting-related amenities, are provided maps detailing location and access points of forestlands open to hunting, and are solicited and informed by landowners are much more apt to find and hunt on described forestlands and more likely to return on a recurring basis. Chapter 22 describes how managers can establish and maintain effective communications with hunters to optimize their hunting experience and success.

10.5.2 MEETING INFORMATION NEEDS: DEER DENSITY

State natural resource agencies downplay or decline to estimate deer density within management units. Collecting deer density data for management units of typical sizes (thousands of ha) is costly for state agencies, requiring large economic outlays for training and logistically supporting seasonal field personnel for collecting and analyzing data. Excepting the use of deer density estimates derived from expensive aerial surveys, state resource agencies rarely recognize or use published protocols or techniques for estimating landscape-level estimates of deer density or impact. Additionally, due to the large variation inherent in deer habitat components within traditional management units, confidence intervals around deer density and impact estimates would be so large as to make the estimates meaningless. However, as described in Chapter 37, one cannot manage deer density, and associated impact levels, if one cannot produce valid density estimates. The answer to providing information on deer density (and impact) lies in collecting and analyzing these data from individual deer–forest management areas, rather than from deer management units (Chapter 9).

10.5.3 MEETING HUNTER NEEDS: CREATION OF PRACTICAL MANAGEMENT UNITS

Some hunters complain about the unevenness of habitats, hunting access, availability, utility and applicability of antlerless permits, and level of deer and habitat management as encountered across large deer management units, calling for smaller, more homogeneous, and antlerless permit–friendly deer–forest management areas. Chapter 9 ("Human Factors: Deer–Forest Management Areas vs. Deer Administrative Units") distinguishes between deer management units and deer–forest management areas. Briefly, deer management units are large administrative units identified by state natural resource agencies that include ownerships neither managed nor influenced by the agencies defining them. Deer–forest management areas are forestlands administered by a single private owner or single public management agency (as in state forests and state parks) wherein monitoring deer density and impact are/can be practiced by the managing entity, antlerless permits are often available separately for the entirety of such areas (e.g., deer management assistance permits), and forest/habitat management is consistent and uniformly applied. These deer–forest management areas are much smaller than deer management units, generally encompassing less than 10,000 ha.

10.5.4 MEETING HUNTER NEEDS: QUALITY DEER

Factors identified by hunters as influencing whether they will hunt specific properties (deer–forest management areas) include quality of deer harvested. Hunters sought by forestland owners are usually alpha, beta, or locavore, and these hunters are interested in hunting and harvesting quality deer (deer in quality body condition, including, but not limited to, bucks with quality racks). Managers can use social media (Chapter 22) to post trail-cam photographs of quality deer taken on their property prior to hunting season as evidence that they have quality deer. A more comprehensive way to provide evidence of quality deer as well as a good venue to productive hunter-landowner interactions is to conduct deer checking stations during hunting season. Checking stations (Chapter 18) should be located at highly trafficked areas within forestlands and are places where hunters bring harvested deer to be weighed, sexed, and measured for trophy characteristics (antler measurements). Checking stations may be operated from permanent or temporary shelters or even the beds of pickup trucks to provide information sought by hunters (weight, condition, and age of deer they harvest) and can be used to offer written materials useful to hunters, including maps of the deer–forest management area.

10.5.5 SUPPORTING DIFFERENT HUNTING STYLES

Experienced deer hunters employ three basic hunting styles, which landowners can enhance to improve hunter success in harvesting deer. The first two, hunting from tree or ground stands and "still hunting," are done by solitary hunters. Hunters place tree or ground stands to intercept deer traveling at dusk to feeding grounds and at dawn from feeding grounds to bedding grounds. Such locations are usually on the edges of forested areas and openings with forage, such as farm fields or recent timber harvest sites. During other daylight hours, hunters may "still hunt," which is a misnomer—what they do is slowly move a short distance, stop, observe for a few minutes, and then move again, looking for deer in interior parts of forests where they may be bedding or feeding, such as on hillsides or along streams. The last hunting style, "drive hunting," requires teams of 10 or more hunters, about half of which move as a line where they are spaced 20 or so meters apart, slowly moving to drive deer to a second line of hunters similarly spaced apart waiting for deer moved by the drive line to move into their line of fire. Drives are conducted by groups of hunters who know each other well and are conscious of safety measures to employ to avoid shooting at each other. Hunters using all or any of these techniques benefit from preseason scouting and detailed topographic maps that identify feeding, bedding, and travel areas used by deer.

10.5.6 NEEDS OF BETA AND LOCAVORE HUNTERS

Beta and locavore hunters lack the experience and training to make them effective deer hunters. Such lack of experience often results in unsuccessful hunts and a lack of interest in hunting in subsequent seasons/years. The best way to provide essential training and experience is exemplified by the Quality Deer Management mentoring program (Chapter 27). Managers should emulate the mentoring program developed by the Quality Deer Management Association (QDMA)—contacting a local QDMA chapter would be an excellent start to developing and maintaining a mentoring program that attracts, trains, and retains effective deer hunters.

10.5.7 Is There a Place for Omega Hunters?

As noted above, omega hunters require high deer density as a prerequisite for hunting a forestland. When additional numbers of hunters are solicited to hunt a forestland, the rationalization is often high deer density accompanied by high deer impact. Under these conditions, omega hunters will respond to reports of high deer density, and their participation will be helpful in moving deer around and contributing to higher deer harvest, as some will harvest deer, and alpha and beta hunters will benefit from increased movement of deer.

On the Kinzua Quality Deer Cooperative (Chapter 34), the initially high density of $\sim$11.5 deer/ km^2 in the spring of 2003 was advertised in the run-up to the 2003 deer hunting season. Hunter density, as judged by number of cars/trucks parked along access roads, was high, and the harvest was sufficiently high to reduce deer density to 9.5 deer/km^2 by the following spring. In the following hunting season, a similarly high hunter turnout and deer harvest resulted in a spring 2005 density 6 deer/km^2, which represented target density. Deer density was maintained at approximately this level (4–6 deer/km^2) to the present. As deer density declined, checking station operators began to receive numerous complaints about low deer density (from what were perceived as omega hunters). Hunter participation, as judged by car/truck counts, and harvest were significantly lower. By that time, however, deer density was sufficiently low that a much-reduced harvest was all that was required to maintain desired deer density. Coincidentally, a large drop-off in visits to checking stations by hunters complaining of too few deer occurred. DeCalesta (2017) surmised that the number of alpha and beta hunters returning every year harvested enough deer to offset recruitment and maintain deer density at target level. Seemingly, omega hunters were no longer contributing to nor needed for balancing recruitment with harvest. Thus, omega hunters can contribute to heightened deer harvest in driving deer density down to the target level and may self-eliminate when deer density reaches that level.

REFERENCES

deCalesta, D. S. 1985. Influence of regulation on deer harvest. In *Symposium on game harvest management.* eds. S. L. Beasom and S. F. Roberson. 131–138. Kingsville TX, Texas A & I University.

deCalesta, D. S. 2017. Achieving and maintaining sustainable white-tailed deer density with adaptive management. *Human-Wildlife Interactions* 11:99–111.

Dunn, C. 2015. Resource newsletter. Harrisburg PA. Pennsylvania Department of Conservation and Natural Resources. December 2015.

Frye, B. 2006. *Deer wars: Science, tradition, and the battle over managing whitetails in Pennsylvania.* State College PA, Pennsylvania State University Press.

Keenan, M. T. 2010. Hunter distribution and harvest of female white-tailed deer in Pennsylvania. *MS thesis,* State College, Pennsylvania State University.

Thomas, J. W., J. D. Gill, J. C. Pack et al. 1976. Influence of forestland characteristics on spatial distribution of hunters. *Journal of Wildlife Management* 40:500–506.

Ward, K. J., R. C. Stedman, A. E. Luloff et al. 2008. Categorizing deer hunters by typologies useful to game managers: A latent-class model. *Society and Natural Resources* 21:215–229.

11 Human Factors
Science, Values, and Stakeholders

David S. deCalesta and Michael C. Eckley

CONTENTS

Scientists do live in ivory towers. They believe that facts win debates and speak for themselves.

Joseph Romm (Friedman and Mandelbaum 2011)

If history and science have taught us anything, it is that passion and desire are not the same as truth.
The human mind evolved to believe in the gods. It did not evolve to believe in biology.

E. O. Wilson (1999)

Logic will never change emotion or perception.

Edward de Bono (2016)

MANAGER SUMMARY

For managers to make hunting a successful tool for reducing deer density and impact, they must be aware of and incorporate the science behind deer management and the values and culture of hunters and other stakeholders whose actions will determine the success of public hunting to manage deer density and impact. This entails tailoring and developing management actions and communication efforts that resonate with stakeholders; reflect their values; and integrate culture, values, and science in effective deer management. Such management requires compromises among stakeholder groups to find common grounds for managing deer acceptable to stakeholder groups.

11.1   INTRODUCTION

The authors of every chapter in this book possess college degrees, some advanced, emphasizing science as the basis for management decisions. We were taught that management of natural resources should be based on quantitative science, wherein management practices were the product of rigorous scientific analysis. With science behind our management, we had an airtight case for managing natural resources. Until, that is, we tried to make science prevail in natural resource decisions based more on the faith, values, and beliefs of stakeholders, and the reality that these factors have a major impact on management decisions and actions enacted by administrators. We forgot the old axiom, "resource management is people management."

11.2   SCIENCE VS. VALUES

What differentiates science from beliefs is the application of a process (the Scientific Method) that seeks truth, advances understanding, and challenges beliefs based on values and faith (Katz 2014). In science, a belief can create hypotheses, but hypotheses must be proven by rigorous testing before being accepted. However, it is commonplace in the internet age to make claims based entirely on what we wish were true. Detractors of science routinely make the comment, "There is no evidence," which is code for systematically ignoring, avoiding, or rejecting any evidence that does not support existing beliefs. To complicate matters for landowners and managers further, there is a pecking order to values and beliefs.

11.3   INTEGRATING VALUES OF STAKEHOLDER GROUPS

In the case of deer management, where issues and discussions get hot and heavy, values and beliefs of managers are outweighed by those of the vastly more numerous and vociferous hunting public. Years of interactions with natural resource administrators acceding to values based on culture-based belief have taught us that we must first address stakeholder beliefs and values, particularly those of hunters, related to deer management before we can apply science.

We also need to recognize that even though we as scientists, educators, and managers tout science as the basis for making management decisions, at root, our proposed management decisions are designed to meet desired future conditions which in turn are based on values that *we* have for forest resources, including deer. Managers wishing to manage deer using public hunting as the primary tool to control deer density and impact must first identify and appeal to hunter beliefs and values. If they can do that and integrate them with their own values, they can then apply relevant science to formulate and execute successful deer management.

11.4   APPEALING TO STAKEHOLDER BELIEFS

How to appeal to the beliefs and values of hunters? In his paper relating to decisions based on science vs. values, Sundlof (2000) made a number of statements that apply to the beliefs/values vs. science deer management situation.

> A basic premise of all free societies is that decisions are based on a shared set of values among their members.
>
> Science and values provide completely different guides to decision-making. Values are emotional connections between individuals, whereas science is value neutral. The scientific process attempts to minimize the influence of values, because they introduce biases into decisions. Scientists strive to be dispassionate observers to prevent personal values from influencing the decision-making process. There is a dichotomy, therefore, between science and value-driven decision-making.
>
> Most societal controversies that take place are based on differing values among individuals within the society. Science, on the other hand, is a deliberate, rational process. Science is value neutral which is problematic if it is used as a sole guide to decision-making.

In order to resolve political conflicts that may arise from ... conflicting values, stakeholders must come together, discuss the issues, and find common ground.

However, as Seidman (2007) noted, two kinds of values are at play: short-term (situational values) and long-term (sustainable values). Situational values deal with what is available in the here and now and relate to what can and cannot be done in short-term situations. Sustainable values are about what we should and should not do in situations over the long term. Situational values relate to the size and scale of what you do now; sustainable values are about relating to the environment and future generations.

11.5 INTEGRATING TIME FRAMES OF STAKEHOLDER VALUES

Landowners managing forest resources are forced to think, act, and manage in terms of decades and long-term sustainability of the forest resources they manage (it takes 80 years to grow a tree to harvest size, 300 years to develop old-growth characteristics, decades to centuries to recolonize extirpated herbaceous plants vulnerable to deer overbrowsing). Time frames of hunters usually involve daily (what's the weather doing relative to the day's hunting plans?) and year to year (more or less deer than last year?) time frames. The same time frame applies to farmers, who grow crops on an annual cycle, and homeowners, who maintain their vulnerable-to-deer landscaping in seasonal cycles. All are related to deer density, but over varying time frames and with varying interpretations of what the "right" deer density is.

11.6 COGNITIVE DISSONANCE, MOTIVATED REASONING, AND GROUP THINK

Finally, there is the factor of "group-think." When groups of people coalesce around a given topic, such as deer management, and established science relevant to that topic conflicts with their values and views, the group collectively suppresses disagreement and prevents acceptance of alternatives in management decisions (Janas 1982). The associated phenomena of cognitive dissonance (Festinger 1962) and motivated reasoning (Kunda 1990) provide the bases for group-think. Social scientists tell us that when people confront information that disagrees with their value-based perception of reality (e.g., seeing 100 deer on opening day of deer season means there are too many), it causes dissonance—they don't want to change the way they think about an issue because it implies their thinking was wrong. In these cases, people resort to motivated reasoning: they only accept facts (science) that support and agree with their perception and discount science that disagrees with their perceptions. In short, people reject science if it contradicts their values.

Characteristics of group thinkers include: rejection of opposing views, failure to question the assumptions their group uses, belief in the rightness of their cause and disregard for the ethical or moral consequences of their decisions, negative views of those with opposing views, group pressure to not express arguments against any of the group's views, and distancing the group from information that is problematic or contradictory (Janas 1982).

Kahan et al. (2010) summarized the impact of group think as a shaper of individuals' beliefs and actions about the science behind management actions. Group thinking motivates individuals to place credibility on information that reinforces their cultural disposition. Their view of the trustworthiness of experts is conditional on the fit between their position, as promoted by their culture, and the one the expert is promoting. Individuals reject information inconsistent with their predispositions when they perceive that it is being advocated by experts whose values they do not share. In short, they process scientific information in a selective pattern. When shown information (e.g., too many deer means there needs to be higher harvest of deer) that threatens their cultural values, individuals tend to react dismissively toward that information.

11.7 BRIDGING VALUE DIFFERENCES WITH COMMUNICATIONS

Gharis (no date) stated, "Unsuccessful communication can be detrimental not only to science but to society. In a model world, citizens would come together to discuss and to decide on science issues, but often this type of engagement does not occur. How the media and scientists present information still matters. Studies show that not only how material is framed but who delivers the message affects an individual's willingness to take action."

To address these hurdles, Kahan et al. (2010) recommended that communicators pay attention to the cultural meaning as well as the scientific content of information. When presented with information in a way that reflects their cultural values, individuals are more likely to consider it. By crafting messages that are culturally acceptable to target audiences, there is a better chance that scientific information will receive considered attention. Kahan (personal communication) recommended providing groups with scientific evidence in a way that would favor their acceptance of the information by: (1) packaging the information in a way that appeals to and supports the identified group's culture as it relates to the management issue and (2) having a messenger who is credible to the target audience deliver the message.

11.8 EXAMPLES OF VALUE BRIDGING, MESSAGE DEVELOPMENT, AND SELECTION OF MESSENGER

Two examples from Pennsylvania illustrate how such messages can be crafted and delivered. In the 1990s, foresters' and wildlife biologists' science-based testimonies at annual season and bag limits hearings presented to sway Pennsylvania Game Commissioners to alter hunting regulations that would lead to reduced deer density and impact on forest resources failed. They were negated by hunter claims that reduced deer density would destroy their culture and values and lead to failure to recruit new hunters. Wrong message, wrong messengers, and failure to liberalize deer hunting and reduce deer density. An aggressive information program developed in by the Pennsylvania Game Commission in the late 1990s was pitched at hunters by a biologist (bear biologist Gary Alt) who hunters believed in, with the key message that producing quality deer requires better forage and better deer habitat, and that those conditions required reducing deer density. Enough hunters supported the concept that the Pennsylvania Game Commission enacted hunting regulations (concurrent buck-doe seasons and a deer management assistance program that allowed additional antlerless deer to be harvested on properties that claimed reductions in deer density were needed to improve habitat). The regulations led to reduced deer density and improvements in understory vegetation, including tree seedlings vulnerable to deer browsing. Why? A different message (that resonated with deer hunters), delivered by a professional credible to deer hunters.

In 2015, spokespersons for the Pennsylvania Game Commission began to lobby the commissioners to allow Sunday deer hunting on the grounds that hunters mostly hunted on weekends and needed more time to practice their sport and increase their chances of harvesting a deer. Should these efforts be successful, hunting regulations would be altered in a way that increased hunting pressure and resulted in reduced deer density and impact, although not advertised as byproducts of giving hunters more time to hunt deer. Formula for (potential) success: development of a message that resonates with deer hunters and delivery by a state game agency that has more credibility with deer hunters than professional foresters or ecologists.

11.8.1 Identifying Recipients of the Message

It is apparent that whereas forest managers (including park managers and farmers) have similar values and goals, natural resource agencies responsible for issuing regulations that affect hunting and ultimately deer impact have to deal with varying values and goals of different stakeholder groups, calling for variable deer densities. Therein lies the problem for managers and other non–deer hunter stakeholders: hunters have variable values, goals, and desired deer densities that may not agree with those of other stakeholders. Furthermore, the agency responsible for managing deer may not be sympathetic to the nonhunting stakeholders, or at least may be more responsive to the omega hunters, and thus promulgate regulations favoring the values and goals of omega hunters who outnumber alpha and beta hunters. The locavore movement is too new for most natural resources agencies to be aware of or deal with.

Managers desiring to manage deer by public hunting must appeal to those segments of the hunting public (alpha, beta, and locavore hunters) that do not include high deer density with their values. Conversely, omega hunters whose values require deer density ≥ 15 deer/km^2 must seek out landowners who can live with high deer densities; such landowners would be those that cater primarily to deer hunters wishing high deer densities. An example landowner could be the Pennsylvania Game Commission, which owns and manages over 1 million acres of forested gamelands in Pennsylvania primarily for its hunting public.

Table 11.1 identifies stakeholder values, goals, and desirable deer associated with categories of stakeholders based on conditions occurring in northern hardwood forests of northwestern Pennsylvania and supportive deer harvest regulations. Forest managers in other parts of the country may wish to construct their own tables based on region-specific information.

11.8.2 Addressing and Integrating Messages for Stakeholders

Using a model of: values → goals → deer density → hunting regulations → education/communication to achieve goals, we can construct a table that associates types of hunters whose values and goals are met by deer densities compatible with the values and goals of forest managers. These are the hunters that managers must attract if they are to achieve deer densities compatible with reaching and maintaining their goals for natural resource management.

11.8.3 Developing the Message

Landowners bear the costs of creating and maintaining deer habitat, providing and maintaining road access for hunters to hunt the deer, and implementing practices (deer fencing and spraying of undesired vegetation) caused by overabundant deer herds. However, few hunters are cognizant of, or sympathetic to, the concerns of foresters and ecologists regarding deer impact on forest resources. Landowners and managers must reframe their message to appeal to the values of hunters if they want those hunters to hunt their property and help reduce deer density and impact.

For alpha and beta hunters, the message that resonates is that deer health/trophy status and habitat health (for deer and other game species such as grouse and turkey) are optimal when deer density is in balance with diverse and abundant understory vegetation (seedlings, herbs, and shrubs). However, there is no message that will appeal to omega hunters that promotes sustainable deer density and sustainable and diverse forest vegetation. Locavore hunters simply need information on how to locate forestlands where they can obtain permission to hunt deer. Landowners should spend their limited time and dollars working with alpha, beta, and locavore hunters, as omega hunters harvest few deer and contribute proportionately less to proper deer management than alpha and beta hunters.

11.8.4 Selecting the Messenger

The messenger delivering the information to alpha, beta, or omega hunters cannot be a government or agency administrator or a college professor or government scientist wearing a coat and tie. Conversely, locavore hunters are usually well educated and seek out and respect information provided by scientists and other data-equipped experts. The most believable source of information for experienced deer hunters is another deer hunter. There are some hunters who are also recognized

TABLE 11.1

Representative Stakeholder Values, Goals, Associated Deer Densities, and Supportive Hunting Regulations for Deer Management

Stakeholder	Values	Goals	Requisite Density	Supportive Hunting Regulations[c]
Alpha and beta hunters	Quality hunt[a]	Harvest quality deer from quality habitat	$\leq$8 deer/km^2	Antlerless permits, antler point restriction
Omega hunters	Good-old-days hunting[b]	See many deer, harvest only antlered deer	>15 deer/km^2	No antlerless harvest, all antlered deer legal
Locavore hunters	Obtain source of wild protein	Harvest any deer	Not a factor	Antlerless permits, extra seasons for locavore hunters
Timber growers	High-quality timber, low production cost	Regeneration and stocking of desirable tree species	$\leq$8 deer/km^2	Antlerless permits, antler point restriction, special hunts
Park managers	Optimize forest benefits for all stakeholders	Diverse, abundant plant and animal communities	$\leq$6 deer/km^2	Antlerless permits, antler point restriction, special hunts/hunters[d]
Residents in forested communities	Safe, enjoyable environment, low impact on landscaping	Minimal deer for minimal impact	As few deer as possible	Special hunts[d] and/or fencing to exclude deer
Farmers adjacent to forests	Produce variety of crops at minimal cost	Minimal deer for minimal impact	<6 deer/km^2	Antlerless permits, antler point restriction, special hunts
Environmentalists	Balanced, sustainable ecosystems	Abundant, diverse plant and animal communities	<6 deer/km^2	Antlerless permits, antler point restriction, special hunts
Natural resource administrators	Professional, noncontroversial management of natural resources	Satisfy all stakeholders	Variable, depending on stakeholder group	Variable, depending on stakeholder group

[a] Opportunity to hunt in habitat supporting trophy/quality deer with low hunter density.

[b] Deer hunting typical of 1930–1970 era—high deer density, shared camaraderie in hunting camps with friends and family, excitement for young hunters by seeing many deer, primarily hunting for antlered deer.

[c] Hunting regulations related to sex and antler characteristics of deer legal for harvest.

[d] Special hunts designated for identified properties, at specified times, and using public hunting and/or deer removal by contracted marksmen.

as research/management experts by other deer hunters, and these are credible deliverers of the message about the need to manage for deer at sustainable levels. An option for landowners without access to such experts is to identify spokesperson hunters within the group(s) of alpha and beta hunters hunting on their land who are open to scientific information on deer management and rely upon them to promote sustainable deer density to their peers.

Hunters need landowners for the habitat to support deer and for a place to hunt deer. Landowners need hunters to keep deer density within limits that allow them to meet goals for sustainable management of selected forest resources. Once the values and goals of hunters and landowners are integrated and made compatible, the path to application of management based on science is cleared.

11.9 EXPANDING THE MESSAGE

Additionally, landowners must educate hunters on the need for the long-term time frame required to achieve their values and the associated need to manage for and support annually adjusted regulations that maintain deer densities at or below target densities indefinitely. State and governmental agencies (USDA Forest Service, the National Park Service, and the U.S. Fish and Wildlife Service) are large enough to include information and education staff for providing this information to hunters and other stakeholders.

Landowners should be able to use the information provided by this book to duplicate the Pennsylvania Game Commission's success in helping landowners manage deer density on the smaller landscapes they oversee to identify, recruit, and maintain a hunter base to manage their deer herd in perpetuity, if they appeal to hunter values.

In the following chapters, we discuss how forest managers can identify, attract, and educate and communicate with stakeholder groups, including hunters, whose values are compatible with deer densities landowners must achieve if they are to reach their goals for natural resource management. We also describe how the managers can develop and apply adaptive management strategies to ensure they are adapting and responding to the values of the hunters who are helping manage their deer through regulated harvests.

REFERENCES

deBono, E. 2016. *I am right, you are wrong*. London: Penguin Books.
Festinger, L. 1962. Cognitive dissonance. *Scientific American* 207:92–102.
Gharis, L. A. no date. *Communications in natural resources management—A research proposal submitted to the U. S. National Institute of Food and Agriculture*. Fayetteville AR: University of Arkansas.
Janas, I. L. 1982. *Groupthink: Psychological studies of policy decisions and fiascoes*. 2nd Edition. New York. Houghton Mifflin.
Kahan, D. M., H. Jenkins-Smith, and D. Braman. 2010. Cultural cognition of scientific consensus. *Journal of Risk Research* 14:147–174.
Katz, D. L. 2014. What 'There is No Evidence' Really Means. *US News & World Report*, July 8, 2014.
Kunda, Z. 1990. The case for motivated reasoning. *Psychological Bulletin* 108:480–498.
Romm, J. as quoted in Friedman, T. L. and M. Mandelbaum. 2011. *That used to be us: How America fell behind in the world it created and how we can come back*. New York: Farrar, Straus and Giroux.
Seidman, D. 2007. *How we do anything means everything*. New York: John Wiley and Sons.
Sundlof, S. 2000. The role of science in regulation and decision making. *Journal of Agrobiotechnical Management and Economics* 3:137–140.
Wilson, E. O. 1999. *Consilience: The unity of knowledge*. New York: Vintage Books.

12 Human Factors
Landscape, Politics, and Regulation

David S. deCalesta

CONTENTS

Deer management, like other political issues, isn't always guided by pure science. Often it's the constituents who complain the loudest that end up calling the shots.

Tony Hansen, Mark Kenyon, and Alex Robinson (2015)

MANAGER SUMMARY

Deer management exists on two landscape levels: that involving regulation of deer hunting within large administrative units identified herein as deer administrative units, and active management on individual forestlands herein identified as deer–forest management areas. The latter are individual forestlands managed by single entities wherein factors related to deer ecology and management, including habitat, are actively manipulated. In the past, state natural resource agencies attempted to manage deer by regulating hunting within deer administrative units erroneously identified as deer management units, where the only thing managed was control of hunting. State natural resource agencies regulate deer management by determining season and bag limits for harvesting deer. Unfortunately, these regulations are often manipulated in response to the political power of deer hunters. True management involves more, including control of access, manipulation of forest vegetation, and monitoring of deer density and impact. Some state agencies have realized the difference, and acknowledge as much by developing deer management assistance programs wherein permits for reducing deer density are made available for individual deer–forest management areas. This is the future and true dimension of deer management.

12.1 INTRODUCTION

Ideally, deer management should be directed by science and biology. In reality, these two absolutes often take a back seat to the elusive and malleable nature of politics. When politicians get involved with deer management, their decisions often are based on the political calculation that laws can be made and changed to the advantage of politicians and their base(s). In the case of deer–forest management, politicians who try to regulate wildlife management to suit their purposes fail to consult with the laws of nature before attempting to change those laws. When management regulations are made and changed with little regard or reference to the laws of nature, the resulting impacts on the welfare of managed forest resources and on human interests are often contrary to the politicians' intentions, to the sorrow of natural resources and the people affected by them.

Unfortunately, public confidence in wildlife professionals has been eroding: decades ago wildlife professionals "… carried more authority in decision-making than they did 10 years ago, and far more than they do today. The vast majority of decision-making authority has been usurped by legislators. Barely a fraction of our legislators are trained wildlife professionals, so when they disregard the input of agency staff, their policies often produce unintended results" (Adams 2016).

Managers of deer–forest management areas (DFMAs) must be aware of the disconnect between management and reality and address the regulatory/scientific dichotomy affecting deer management if they are to be successful in managing deer and other forest resources. In short, they must think like biologists *and* politicians and find ways to make them work together. This chapter describes the landscape, political, and regulatory frameworks managers must become familiar with and work within to help them deal effectively with political and regulatory influences on deer management. This chapter describes how landowners can do this by working with politicians and other stakeholders to meld science and politics to influence deer and forest management in their favor.

As preceding chapters noted, deer impact on forest resources is a direct function of deer density, which is affected by deer mortality factors. Because of the elimination of all but a few historic predators, predation can be an effective mortality factor only when it includes hunting by humans. Increasing deer mortality by hunting requires increasing opportunities for hunters to harvest more deer. This can be achieved by providing more liberal hunting seasons on lands where public hunting is allowed, and should be considered on lands where public hunting is not allowed (e.g., local, state, and national parks). Increasing opportunities for hunters to harvest more deer requires enacting/changing regulations to liberalize hunting as described below. Regulations for increasing deer harvest are many, and address different circumstances that affect the potential for hunting to reduce deer density and impact.

Liberalizing hunting regulations requires that politicians, be they elected to local, state, or federal offices, or appointed to relevant commissions or boards by local, state, or federal executives, alter existing regulations and/or enact new ones. Elected officials and appointees are subject to recall by stakeholders whose support may have gotten them elected/appointed in the first place. So the challenge is to motivate legislators/appointees to change regulations that control deer harvest in ways beneficial to good deer–forest management. Because legislators/appointees are sensitive to pressures from stakeholders, presenting them with facts supporting reduction of deer density to reduce impact on forest resources will not effect change in regulations *if* the majority of stakeholders do not support reducing deer density or impact through changes in hunting regulations. In this unfortunate reality, facts presented by scientists and managers who support reduction of deer density may be ignored by legislators/appointees who are more swayed by information/testimony presented by stakeholders based on values and culture. Expert testimony may be negated by "alternative facts" provided by sources favored by the stakeholders.

Legislators/appointees will only enact regulations designed to reduce deer density if they are supported by a majority of the stakeholders who interact with those legislators. In other words, convincing legislators/appointees to liberalize deer hunting regulations to benefit managers will occur only if a majority of (voting) stakeholders can be convinced that the liberalized regulations will be beneficial to their culture and values. This amounts to a mission impossible for individual deer–forest managers because deer hunters far outnumber managers and politicians will invariably side with the deer hunters and act in ways to enact/alter regulations supported by the hunters. To be effective, managers must organize within their ranks, but also join forces with other stakeholder groups to form sufficiently large and politically active blocs (including hunters who want quality over quantity of deer) to offset political pressures brought by those hunters who want more deer.

12.2 REGULATIONS SPECIFIC TO DEER ADMINISTRATIVE UNITS

Most hunting regulations designed to reduce deer density and impact have in the past been directed mostly at large landscape deer administrative unit (DAU) levels encompassing dozens to hundreds of DFMAs. These regulations may help reduce deer density and impact on DFMAs if hunters use them on individual DFMAs, but this effect is diluted by hunters being able to use them on other lands besides DFMAs within designated DAUs.

12.2.1 PERMITS FOR HARVESTING ANTLERLESS DEER

Many states issue permits for harvesting antlerless deer within identified DAUs where deer overabundance is recognized as endangering forest resources. Some states may offer multiple permits per hunter for harvesting antlerless deer or may allow hunters to obtain additional permits if not all permits designated for DAUs are taken by other hunters. However, as noted above, permits issued at the DAU level may not effect reductions in deer density at the DFMA level if recipients of the permits do not use them on DFMAs where needed. In order for permits to be effective in reducing deer density and impact, they must be specified for use on identified DFMAs as noted below under "Deer Management Assistance Programs."

12.2.2 CONCURRENT BUCK/DOE SEASONS

Many states had/have separate buck and doe gun seasons (in many states, archery seasons, which come earlier than gun seasons, allow harvest of antlerless as well as antlered deer). Hunters hunt for antlered deer the first part of the gun season, generally about 2 weeks, and then return for a later, separate, and shorter doe season. However, many hunters only hunt the first few days of deer season, perhaps the opening weekend (or opening 2 days not on a weekend, as in Pennsylvania) and may not return for a later doe season. In response to hunter-truncated deer seasons, some states offer concurrent antlered/antlerless seasons to increase the harvest of antlerless deer.

12.2.3 SUNDAY DEER HUNTING

Because of the limited time hunters spend hunting deer, and because fewer hunt during weekdays, some states (e.g., Maryland, North Carolina, and Virginia) that formerly did not allow Sunday deer hunting have relaxed the rule to increase deer harvest (e.g., deer harvest increased by 19% in Maryland from 2015 to 2016 when Sunday hunting was allowed; Maryland Department of Natural Resources 2017). However, other states (e.g., Connecticut, Delaware, Maine, Massachusetts, New Jersey, Pennsylvania, South Carolina, and West Virginia) continue not to allow Sunday deer hunting.

12.2.4 Increasing Season Length

When natural resource agency administrators are advised by their biologists that the level of desired deer harvest may be negatively impacted by weather not conducive to hunting deer or other factors affecting deer harvest, they may extend the length of regular deer hunting seasons. It seems intuitive that the longer the deer hunting season, the more days hunters have to hunt, and the more deer will be harvested. However, studies (Sunde and Asferg 2014; Chapter 34) have shown that harvest is highest on the first few days of hunting season and falls off dramatically as season length increases. To be sure, there is a small uptick in number of deer harvested the first weekend following the opening season date, but the additional deer harvested are too few to reduce density meaningfully. For one thing, deer become skittish and reclusive as hunting season length increases. McCoy (unpublished Ph.D. thesis, Auburn University) radio-collared 37 male deer and found that these deer left hunted areas at the beginning of hunting season and did not return until >3 days postseason.

Conversely, if hunting season is open only on weekends and extends over a sizeable timeframe, like a month or more, deer harvest remains fairly high, due partly to relaxation of hunting pressure and resultant less reclusive deer and partly to higher hunter participation (hunters are more apt to hunt on weekends without the need to take time off from work, and are more prone to conduct single-day hunts on properties closer to home) (Sturgis and deCalesta 1981).

12.3 PROMOTING REGULATIONS SPECIFIC TO DEER–FOREST MANAGEMENT AREAS

Natural resource agency administrators requesting deer hunting regulations designed to reduce deer impact within DAUs do so under considerable pressure from hundreds to thousands of multiple stakeholders within large areas. Obtaining similar regulations for individual DFMAs is considerably more difficult, as pressures exerted by individual stakeholders/managers are relatively insignificant unless supported by organizations with clout such as farm and forestry organizations. When managers attempt to influence legislators to enact regulations or other actions to affect deer management on DFMAs, they fail—they lack the pressure of numbers. They are, in essence, "vox clamantis in deserto," individual voices crying out in the wilderness. They need the strength and organization that comes with numbers, and the best way to do that at state and regional levels is to join and add their support and voices to those of organizations that represent their interests, such as small woodlot owner organizations, Society of American Foresters, the Wildlife Society and, specifically, the Quality Deer Management Association (QDMA). According to Kip Adams (Adams 2016), the QDMA engaged in over 100 legislative or regulatory deer issues in 20 of 37 states with major white-tailed deer (*Odocoileus virginianus*) populations in 2016 alone. The over 60,000 QDM members, combined with the QDMA organizational structure designed to press deer management issues at the individual state level, represent the kind of clout managers need to overcome the pressure of deer hunters clamoring for increased deer density. The QDMA is respected by state natural resource agencies, and the information it provides in pressing for quality deer/quality habitat is given credence.

12.4 OBTAINING REGULATIONS FOR DEER–FOREST MANAGEMENT AREAS

12.4.1 Educating Legislators, Agency Administrators, and Deer Hunters

12.4.1.1 Landscape

The first and most important step in seeking and securing regulations to increase deer harvest to reduce deer density and impact on DFMAs is to define the landscape for which enhanced hunting regulations are sought. A previous chapter (Chapter 9) identifies the landscape unit upon which deer management is actually practiced and for which change in hunting regulations can be meaningful—the deer forest management area. Although varying greatly in size, DFMAs are forestlands owned/

managed as single units (e.g., private forest lands, including small woodlots and commercial forests, and public forest lands such as state parks and state and national forests). In some cases, abutting landowners may combine their individual holdings into unified forestlands (cooperatives—see Chapter 31) that they manage jointly under a single management plan. Managers of DFMAs are aware of the level of deer impact because they measure it as a starting point to gauge successes of applied management actions. On some but not all DFMAs, managers may also monitor deer density.

Deer hunters do not hunt on DAUs but rather on DFMAs within DAUs. It is on DFMAs where they obtain access, where access roads are maintained, where deer habitat is managed, and where impact of too many deer reduces the ability of DFMA managers to provide quality deer habitat and deer hunting. Legislators/regulators must understand the difference between DAUs and DFMAs and support development of deer hunting regulations specific to DFMA units in addition to those provided for DAUs. When natural resource administrators/legislators provide regulations that help DFMA managers control deer density and impact, those managers will be more apt to allow public hunting on their forestlands and will be able to afford to make habitat improvements favorable to producing quality deer.

All states produce annual deer hunting regulations at the DAU level. The handful of states that adopted the deer management assistance program concept for controlling deer density at the DFMA level acknowledge that enhancing deer management at the DFMA level can be affected by issuing permits for harvesting deer specifically on forestlands managed by private and public managers.

12.4.1.2 Deer and Habitat-Specific Information on Deer–Forest Management Areas

The second step in obtaining hunting regulations leading to reduced deer density and impact on DFMAs is to develop and deliver educational programs designed to convince legislators/regulators and their stakeholders of the benefits of deer management at the DFMA level. The major emphasis of educational programs should be on promoting the health of forest resources, including timber, wildlife, deer, and other game animals for the benefit of managers who absorb the costs of providing access to, and improving the health of, deer herds. Politicians will only enact regulations designed to reduce deer density if a majority of their constituents support that action. Although their constituents include more than deer hunters, members of this group are most likely to contact and pressure their representatives in significant numbers. Because deer hunters far outnumber managers, politicians will invariably side with the deer hunters and act in ways to enact/alter regulations supported by the hunters.

12.4.1.3 Integrating Landscape and Education

The third step for obtaining regulations designed to reduce deer density and impact on DFMAs entails packaging education about the difference between deer management at DAU and DFMA landscape-level deer management with the benefits to hunters of reducing deer density to produce quality deer and deer habitat. Educational programs should be designed to help legislators/regulators understand the dichotomy between DAUs and DFMAs and to support development of deer hunting regulations specific to DFMA units for the benefits they provide to hunters and landowners. Pairing educational messaging with field trips that contrast the negative impact of too many deer on forest vegetation on some sites with sites where lowered deer density is associated with improved wildlife habitat and diversity and density of understory forest vegetation reinforces the message(s).

All states produce annual deer hunting regulations at the DAU level. The handful of states that have adopted the deer management assistance program concept for controlling deer density at the DFMA landscape level acknowledge, albeit implicitly, that enhancing deer management at the DFMA level can be affected by issuing permits for harvesting deer specifically on forestlands managed by private and public managers.

A model (developed in Pennsylvania) for development and delivery of educational programs is described in Chapter 22. Impetus for the program was a decades-long battle between hunters who wanted high deer density and groups negatively impacted by high deer density—managers of impacted forestlands, farmers whose crops deer decimated, ecologists lamenting the loss of forest plant diversity, and the professional staff from the Pennsylvania Game Commission that recognized the negative impacts of high deer density and the need to control the overabundant deer herd. The program was developed by Gary Alt, a renowned and credible bear biologist tasked with developing a deer program that addressed the growing negative impact by deer on forest resources. Alt developed an effective program by: (1) demonstrating how reducing deer density produces healthier deer and improved health of other forest resources, including game mammals and birds, with an educational program including field tours featuring comparison of forest vegetation inside and outside deer-fenced plots; (2) proposing a hunting regulation that made it easier to harvest antlerless deer (concurrent buck and doe seasons); and (3) proposing that regulations designed to reduce deer density with deer management assistance program (DMAP) permits be made available at the DFMA level. This was the genius of Alt's program: for the first time, regulations (DMAP) designed to reduce deer density and impact were directed at the DFMA level where deer management activities, including manipulation of forest vegetation, actually take place, and target reduction of deer density to specified (quantitative) levels on specific forestlands can occur. An additional benefit of this DMAP program was that managers could monitor the effectiveness of the program to reduce deer density and impact and make annual adjustments in number of DMAP permits requested to avoid under- or overharvesting deer. Chapter 34 provides a case history example, and the fact that permanent and sustained reduction of deer density and impact has been maintained since 2003 demonstrates the effectiveness of such programs if maintained annually.

Before commissioners/legislators vote to make deer hunting regulations more liberal to reduce deer density and impact, they determine whether they are responding to the wishes of a majority of their voting constituents. Democracy demands that they do this, as does their desire to retain enough votes in coming elections to ensure their re-election/reappointment. Doing the right thing for the environment is only right in their eyes if a majority of their constituents feel it is right. In some states (e.g., Pennsylvania) hunting license fees (needed to fully fund the wildlife agency's operating costs) can only be increased by vote by state legislators and not by appointed natural resource commissioners. Hunters wishing to force commissioners to restrict antlerless permits so that deer density will increase through reduced harvest may get their state representatives to deny requests for license fee increases if the commissioners do not, as a *quid pro quo*, reduce availability of antlerless permits or shorten antlerless seasons.

12.5 DEER–FOREST MANAGEMENT AREA-SPECIFIC REGULATIONS

12.5.1 Deer Management Assistance Programs

The deer management assistance program concept was developed decades ago by Guynn et al. (1983). Basically, the DMAP program is a system whereby private/public landowners of defined forestlands can apply for permits from state natural resource agencies for harvesting antlerless deer on specified DFMAs. State natural resource agencies issue permits to the managers (or directly to hunters) for a specified number of permits for harvesting antlerless deer. Generally, the managers must claim unacceptable deer impact on forest resources and in some cases must present quantitative data in support of their application. Usually, there is a minimum size requirement ($\sim$20 ha) and a maximum number of permits, generally in multiples related to size of property. End-of-season reports are generally required of landowners detailing how many permits were used and how many deer were harvested.

12.5.2 Deer Depredation Hunts

In some states, geographic areas with known high deer impact on forest and range resources are identified as depredation areas and established as special deer hunting areas. Limited numbers of antlerless deer permits are issued on a lottery, first-come, first-served basis. To increase hunter participation, depredation hunts generally are held after the close of regular seasons, providing a last opportunity for hunters to harvest deer. For example, in 2015, Indiana hunters could harvest additional deer beyond the statewide bag limits in designated Deer Reduction Zones. Beginning with an antlerless deer, they were allowed to harvest up to 10 additional deer under the Deer Reduction Zone bag limit, for a total of either 10 antlerless or 1 antlered ("earn-a-buck") and 9 antlerless deer. Harvest of these additional deer required the possession of a Deer Reduction Zone license for each deer harvested. An antlered deer harvested under the Deer Reduction Zone license did not count toward a hunter's statewide bag limit of one antlered deer. However, deer harvested in designated Deer Reduction Zones with other license types (e.g., archery, bonus antlerless, bundle) did count toward statewide bag limits. The Deer Reduction Zone season opened 2 weeks prior to the beginning of archery season and continued through the end of the year, well after the regular season closed.

12.5.3 SPECIAL DEER DAMAGE PERMITS ISSUED TO LANDOWNERS

Currently, these permits are available to persons with deer damage only to agricultural crops—in some cases, only the landowners may harvest the deer; in other states, landowners are permitted to have hunters they know personally harvest the deer. Some allow for harvest of antlered as well as antlerless deer. A major objection to this practice voiced by hunters—that these permits are used more to fill hunters' freezers than to reduce deer density and impact—can be avoided by requiring that deer harvested with these permits be taken to meat processing centers and the meat donated to the needy. For such permits to be of use to forestland managers, they must be made available for use on forestlands, in addition to use on agricultural lands.

12.5.4 ADDRESSING THE ENFORCEMENT OBJECTION

A commonly asserted downside to the above recommended legislative actions designed to reduce deer density, as voiced by persons not supportive of reductions in deer density, is that liberalizing deer hunting seasons will place greater demands on enforcement of game laws. However, discussions with wildlife conservation officers tasked with enforcing the enhanced hunting regulations revealed that their enforcement workload did not, in fact, increase. Actually, because of the trend of increasingly reduced participation of hunters in hunting, the officers stated that overall they had been spending less time enforcing hunting regulations and that the addition of regulations for harvesting antlerless deer had not increased their workload. They added that they were supportive of regulations designed to reduce deer impact on forest resources (other tasks for wildlife conservation officers include improving wildlife habitat and managing for wildlife species that would be impacted by overabundant deer herds).

12.6 RETAINING REGULATIONS FOR DEER–FOREST MANAGEMENT AREAS

Obtaining regulations is not "one and done." Rather, regulations directed at reducing deer density and impact on DFMAs are subject to revision/recall by politicians/administrators pressured by hunters responding to reduced deer density on DFMAs. Much work goes into nudging legislators/administrators to enact/improve regulations that favor reduction of deer density and impact; often, years of coordinated effort are required. However, these regulations are vulnerable to being rescinded or watered down by hunter groups wanting higher deer density—annual vigilance is required to identify and head off such attempts.

Second, providing annual evidence of the efficacy of the regulations in deer management in the form of reports/updates promotes the success of regulation and can head off claims that the regulations are "exterminating" the deer. Describing how successful adjustments to developed regulations made them sufficiently flexible to respond to changes in deer density and impact are additional proof of the effectiveness of the regulations. Presenting evidence of the intent, success, and flexibility of the regulations to legislators and administrators should be high on the list of associated activities designed to support and maintain those regulations. Providing field tours demonstrating how DFMAs have responded positively to the regulations to legislators and administrators should be designed, promoted, and conducted frequently.

Unfortunately, stakeholders rarely congratulate or thank politicians/administrators when things are going well as a result of their actions (e.g., reduced deer density and reduced impacts resulting from regulations designed to do both), so legislators/administrators can only respond to the (mostly negative) feedback they get. Publicly thanking politicians/administrators and supporting their re-election/reappointment campaigns for enacting regulations that reduce deer impacts should be an annual first order of business for managers. Repeated and annual recognition of, and appreciation for, legislators/administrators who enacted regulations that reduced deer density and impact should be a priority for managers.

REFERENCES

Adams, K. 2016. The dangers of blending politics and deer management. https://www.qdma.com/dangers-blending-politics-deer-management/.

Guynn, D. C., Jr., S. P. Mott, W. D. Cotton et al. 1983. Cooperative management of white-tailed deer on private lands in Mississippi. *Wildlife Society Bulletin* 11(3):211–214.

Hansen, T., M. Kenyon, and A. Robinson. 2015. The 6 new rules of white-tailed deer hunting. *Outdoor Life*, March.

Maryland Department of Natural Resources. 2017. http://news.maryland.gov/dnr/2017/02/16/85193-deer-harvested-during-2016-2017-hunting-season/

Sturgis, H. and D. S. deCalesta. 1981. The MacDonald Forest deer hunt: a second look. *Oregon Wildlife* 36:3–8.

Sunde, P. and T. Asferg. 2014. How does harvest size vary with hunting season length? *Wildlife Biology* 20:176–184.

13 Deer and Silviculture

David S. deCalesta

CONTENTS

We must learn more about how the components of a forested ecosystem interact and change when silviculture alters the character and development of the community it supports.

Ralph Nyland (2016)

MANAGER SUMMARY

Silviculture is a branch of forest science developed to manipulate forest vegetation to achieve goals for forest resources. Deer can negatively affect outcomes of silvicultural practices by their foraging on forest understory vegetation if deer density is too high. Negative impacts include elimination of tree seedlings needed to reforest areas wherein timber has been removed, changes in the dynamics of understory growth, including alteration of species composition and elimination of the structure of understory forest vegetation. Negative impacts by deer may necessitate additional, expensive silvicultural practices, such as the use of herbicides to eliminate unwanted vegetation, building of fences to protect forest vegetation from overbrowsing by deer, or abandonment of management to produce desired forest resources as being too expensive or not available. The flip side of silviculture is that it can be used to improve deer habitat, influence deer quality, and affect forest resources (as detailed in Chapter 21).

13.1 INTRODUCTION

Unless managers of deer–forest management areas (DFMAs) have an entirely "hands-off" attitude toward forest vegetation and are content to "let nature take its course," they employ a common set of practices to achieve their goals for forest vegetation and dependent wildlife species—a veritable tool box of vegetation manipulation practices. These practices are identified collectively under the broad heading of "silviculture." By their numbers and behavior, deer may dictate the direction and intensity by which silviculture is applied to affect the outcomes of management designed to manage species composition and structure of forest vegetation. Unfortunately, at high population densities, deer can negate the intended results of silvicultural practices, which may then require additional, and expensive, silvicultural practices to offset deer impact on forest vegetation.

Addressing this component of deer management is the focus of this chapter. Conversely, silvicultural practices may be employed to improve deer health and habitat by manipulating forest vegetation—understory, midstory, and overstory canopy—to change its composition and structure in ways that positively affect welfare and abundance of deer and other wildlife species (see Chapter 21). Silvicultural practices are applied to distinct areas (usually more than one) within forestlands identified as stands.

13.2 STANDS: SILVICULTURE MANAGEMENT UNITS

The basic management unit in silviculture is the stand, identified by Nyland et al. (2016) as, "a contiguous community of trees sufficiently uniform in (species) composition, structure, age and class distribution, spatial arrangement, site quality, condition, or location to distinguish it from adjacent communities." Class distribution refers to tree diameter classes based on trunk diameter 1.4 meters above ground. Classes represent different growth periods of trees identified by inclusive diameter (stem diameter at breast height above ground—1.4 m, referred to as dbh) ranges (e.g., seedlings [<5 cm], saplings [5–10 cm], poles [11–18 cm], pulpwood [19–24 cm], small sawtimber [25–35 cm], medium sawtimber [36–45 cm], and large sawtimber [>45 cm]).

Stands range in size from less than 1 hectare to several hundreds of hectares. They progress through a number of distinct vegetation stages as they move through time, each stage providing different components (forage and cover as provided by vertical structure and species composition of herbs, shrubs, and trees) of deer habitat requirements, as described in Chapter 2.

Goals and practices among DFMAs vary considerably within deer administrative units (DAUs) and are not coordinated or organized within DAUs. The landscape context of DFMAs is important: most DFMAs contain more than one stand—usually they are collections of stands of varying ages, size classes, size, landscape position, and, sometimes, of differing natural resource goals. DAUs rarely are composed of single DFMAs: rather they contain multiple DFMAs of differing ownership and often of different goals for forest resources, making uniform forest management using silviculture within DAUs impossible. The greater the number of different stands in DFMAs, the higher the habitat diversity and the more attractive the DFMA will be to deer for meeting all of their habitat requirements. Unfortunately, the presence of DFMAs of differing forest resource goals renders uniform deer management of DAUs impossible, a reality rarely acknowledged by state natural resource agencies. Failure of state natural resource agencies to relate to this dichotomy of differing forest management goals within DAUs hinders deer and forest management on DFMAs.

Silviculturalists manipulate the path and individual forest vegetation successional stages to meet the objectives of landowners for sustained production of forest amenities, including lumber, nontraditional forest products such as mushrooms, wildlife habitat, water quality, biodiversity, and aesthetics. Generally, this is accomplished by identifying limiting factors and designing manipulation of vegetation and other factors to negate effects of limiting factors.

13.3 WHAT IS SILVICULTURE?

As defined by the Society of American Foresters (Nyland et al. 2016), silviculture is:

- The art of producing and tending forest stands by applying scientifically-acquired knowledge to control forest stand establishment, composition, growth, health, and quality
- Applying different treatments to make forests more productive and useful to a landowner and society on a sustainable basis
- Integrating biologic and economic concepts to devise and carry out treatments most appropriate in satisfying the objectives of an owner

Implied in all of this is emphasis on more than just timber production In addition to timber production, silviculture is used for creation and maintenance of diversity and abundance of forest vegetation, including habitat for wildlife species, game and nongame.

13.4 SILVICULTURAL TREATMENTS

Silvicultural treatments are devised and implemented on stands within DFMAs by individual forest managers for the purposes of growing and harvesting timber, enhancing regeneration of seedlings that will form species composition and structure of subsequent stands, and improving habitat for wildlife and understory plant species. These practices vary among DFMAs and are specific to individual DFMAs as driven by forest resource goals and characteristics of individual stands.

13.4.1 SILVICULTURAL TREATMENTS: IMPROVEMENT

After overstory trees have been removed during timber harvests and stands progress to the stem exclusion stage, the majority of trees are saplings/poles of small diameter and rapidly grow into the overstory, seeking light. Some fall out naturally, having been outcompeted for light and nutrients by their peers. Foresters may selectively use herbicides and manual cutting to weed out undesirable tree species of remaining saplings.

13.4.2 SILVICULTURAL TREATMENTS: INTERMEDIATE HARVESTS

As stands mature into the understory reinitiation phase and trees increase in diameter and height, foresters may apply limited removals of some trees to enhance/improve growth characterizes of remaining trees (improvement thinnings). They may selectively remove trees of poor quality (termed unacceptable growing stock or UGS) to improve quality of trees for harvest, although this practice may remove trees prone to becoming valuable wildlife trees (snags) by providing nesting/roosting sites for other wildlife species in cavities. They may remove designated numbers and distribution of overstory trees (shelterwood seed cuts) to increase light reaching the forest floor, which stimulates germination and growth of seedlings that will become the next forest stand. And, they may select superior trees for producing seedlings in the forest following timber harvest and enhance their growth by removing adjacent, competing trees (crop tree release).

Because all these treatments are designed to improve the quality of desired tree species (for timber) and are followed by final removals (with some trees left to become snags and as sources of seeds for following trees) on areas generally greater than 5 ha, they tend to produce more deer forage (and hiding cover) than might occur on stands managed for multiple age classes favoring shade-tolerant species (uneven-aged management). The flip side to intermediate harvests is that if deer density is too high, deer will eat most of the regenerating seedlings, and generally select for the more nutritious and easily digestible, which also are more commercially desirable, defeating the purpose of silvicultural treatments.

13.4.3   Silvicultural Treatments: Final Removals/Regeneration Initiation

When overstory tress reach optimal size and quality for lumber and understory reinitiation has created adequate stocking of seedlings of desired tree species, most overstory trees will be removed in what used to be called "clear-cuts" but are now euphemistically called "final harvest removals," usually on stands >5 ha. Such removals produce large quantities of quality deer forage, as well as producing species composition and vertical structure of vegetation required by "early succession" wildlife species—unless, of course, deer density has been sufficiently high to prevent development of advance regeneration of seedlings prior to harvest or to remove by browsing enough seedlings to prevent full tree stocking of future forest stands.

For uneven-aged management, mature trees are removed singly or in small groups, ensuring continuance of an uneven age structure and dominance by shade-tolerant trees. This management style does not create nearly the amount of deer forage as even-aged management, and may fail to regenerate after limited tree removals because deer may, if overabundant, decimate the resulting advance regeneration of seedlings and other understory vegetation.

Silviculturalists estimate what is called the annual allowable-cut (AAC) (Leak 2011) as an important step in "placing a forest property under management and ensuring a continued supply of timber over time. Regular harvests also provide for the maintenance of needed wildlife habitat." Managers only remove a prescribed proportion of overstory trees (or other forest vegetation) that guarantees the successional path following vegetation removal is maintained for all stages. Area/volume control that is applied to forests under even-age management (the majority of DFMAs producing timber in the eastern United States) is designed to harvest a prescribed portion of forestlands on an annual basis. It also entails efforts to efficiently/optimally produce a forest with equal area in all age classes. Generally, the proportion is calculated by making sure that the duration in time of successional stages succeeding the mature stage is such that the progression from removal to mature stand includes the typical duration of individual successional stages.

For example, if early succession stands resulting from overstory removal persist for ~20 years before moving to the sapling/pole stage that persists for ~20 years and then moves to maturing stands of small sawtimber (~20 years) thence to medium sawtimber (~20 years) and finally to maturing timber (~20 years), the cycle of timber harvest should be approximately 100 years. For equal representation of all five successional stages, each successional stage should make up about 20% of the DFMA. If early succession stages last 20 years, then an annual allowable cut of about 4%–5% of mature stands should ensure the proper progression of seral stages in a sustainable proportion.

Exceeding the AAC in hopes of creating enough forage to support high deer density results in reduction of species composition and abundance of seedlings valuable for future timber harvests and future wildlife habitat. And, managers run out of timber to harvest sustainably when they exceed AAC and exhaust the potential for creating more early succession stands with diverse and abundant understory vegetation. What results is regeneration of only those species deer avoid eating, reduced carrying capacity for deer, and depleted diversity and abundance of understory vegetation. Of the nine case histories presented in this book, only one (Chapter 40) exceeded AAC. In disregarding the advice of a revolving-door succession of deer biologists to reduce deer density, and the advice of a forester regarding exceeding AAC, managers created a situation where deer density and quality were low, forest understory vegetation (and wildlife habitat) was depleted, and the potential for success of regeneration to produce commercially valuable timber was eliminated.

Removal of competing undesirable trees and trees of poor quality (called thinning or stand improvement) from maturing stands (small, medium, and large sawtimber) adds to the amount of harvestable timber, but of lower quality that may or may not be marketable. These tree removals do open up the overstory and stimulate growth of additional deer forage in the understory. Annual timber removals should be scheduled by professional (consulting) foresters familiar with the forest type represented by individual DFMAs.

There is a minimal size below which timber harvests are not economical or biologically feasible. Small woodlot managers facing this dilemma may scale down the size of final harvest removals (clearcuts), resulting in similarly smaller sizes for succeeding seral stages. However, if the size of such removals is smaller than a hectare, the resulting composition of the ensuing stand may revert to shade-tolerant tree species characteristic of uneven-age management. Additionally, if deer density is high, the meager amount of understory vegetation surviving browsing may have little if any advance regeneration of desirable tree species on the harvested sites.

Managers of larger forestlands (>100 ha) usually can harvest trees from sufficiently large areas to provide an annual source of income to forest landowners/stewards and adequate advance regeneration to provide for the next stand while providing desirable succession of wildlife (including deer) habitat.

As noted in the preface, deer range spans different forest types, including the boreal forest of the northeastern United States and Canada, northern and Appalachian hardwood forests of the northeastern United States and Canada, central broad-leaved and oak forests of the midwestern and southeastern United States, and southern oak and pine forests in Florida and other southern coastal states. Manipulation of vegetation by silvicultural prescriptions is dictated by species composition of plant communities associated with each forest type, as influenced by soils, weather, and vegetation specific to each type. Deer are found in each forest type and their density, welfare, and impacts on forest resources are shaped by weather (temperature, precipitation, wind over short periods of time—days and weeks), climate (temperature, precipitation, wind over long periods of time—years, decades, millennia), and plant and animal communities characteristic of each forest type.

Management practices used to manipulate deer density, deer impact, and forest vegetation to achieve landowner goals are specific to each forest type. This requirement for specificity in managing vegetation and deer by forest type is the reason that forest/deer management plans must be developed by silviculturalists and deer biologists with knowledge and experience in the forest type within which forest landowner properties lie.

Chapters 2–5 describe how deer relate to animals and plants they share forests with, and how deer density and impacts affect these relationships and interactions. The composition and structure of forest vegetation—understory, midstory, and overstory canopy—can be manipulated by silviculture to change stand composition and structure in ways that affect deer welfare and numbers. Conversely, by their numbers and behavior, deer may dictate the direction and intensity by which silviculture is applied to affect the outcomes of management designed to manage species composition and structure of forest vegetation.

13.5 THE NEED FOR SCIENCE-DRIVEN FOREST MANAGEMENT: AN EXAMPLE

When extraction of natural resources proves unsustainable, science is called upon to rescue the situation, generally after traditional measures have been tried and failed. Lacking an established and proven methodology for managing a natural resource, a new branch of science emerges, proceeding along a path of fits and starts, as research compiles and clarifies responses of natural resources to systematic and directed manipulation of assumed limiting factors. Such was the state of forestry in northern hardwood forests of northern Pennsylvania in the 1960s and 1970s when foresters tried to reestablish black cherry in the understory following harvest of abundant and highly valuable black cherry trees in the overstory. The desire was to perpetuate the establishment, growth, and successive harvests of valuable black cherry lumber.

The repetitive, successive clearcutting of Pennsylvania's northern hardwood forests in the 1800s and early 1900s was described in Chapter 14 as an example of the combined impacts of timber harvest and unregulated deer herds on composition and structure of resulting forest vegetation. The multiaged and complex multilayered forest dominated by sugar maple (*Acer saccharum*), beech (*Fagus grandifolia*), and hemlock (*Tsuga canadensis*) typical of the northern hardwood forest type was converted to an even-aged, much less structured forest comprised of a rapidly growing, less shade-tolerant tree species mix dominated by black cherry (*Prunus serotina*), birch (*Betula* sp.), red maple (*Acer rubrum*), and blackberry (*Rubus* sp.) brambles. Because black cherry is not a preferred deer forage species, black cherry seedlings proliferated while seedlings of more deer-preferred species that normally thrive in full sunlight, such as red maple (*Acer rubrum*), white ash (*Fraxinus americana*), and tulip tree (*Liriodendron tulipfera*), were heavily browsed, reducing competition with black cherry for resources including sunlight, moisture, and soil nutrients. As a result, black cherry dominated the overstory of the ensuing forest.

However, the highly abundant deer herd in the 1930–1960 span severely reduced species richness and abundance of tree seedlings that would form the overstory of the succeeding forest. After suppressing more preferred seedling species, deer increasingly browsed on black cherry seedlings more than on less-preferred species such as beech and striped maple. As shrubs and tree seedlings of preferred tree species declined in abundance, other plants, such as hay-scented fern (*Dennstaedtia punctilobula*) and New York fern (*Thelypteris noveboracensis*) proliferated in the understory because of the lack of competition. In conjunction with light browsing pressure on beech and striped maple, ferns and grasses began to dominate the understory, crowding out seedlings of tree species such as black cherry. When foresters harvested the lucrative overstory of black cherry trees in the 1960s, they discovered that because of the predominance of beech and striped maple in the seedling class, the succeeding forest was populated by low-value beech and striped maple rather than black cherry and other commercially valuable species. The need for scientifically based timber management was apparent and stimulated the promulgation of a USDA Forest Service silviculture project in the research laboratory in Warren, Pennsylvania, in the late 1960s (Stout and Brose 2014).

The ensuing research indicated that the problem was caused by multiple factors, including failure to establish adequate advance seedling regeneration of black cherry (Marquis et al. 1975), due partly to browsing by white-tailed deer (*Odocoileus virginianus*) (Marquis 1981 and Tilghman 1989) and partly to interference from competing ferns and hardwood seedlings (Horsley and Bjorkbaum 1983, Horsley and Marquis 1983). Marquis. (1979) developed inventory methodology for determining a successful stocking level of black cherry seedlings and produced strategies for developing understories dominated by black cherry prior to removing overstory black cherry trees. Silvicultural treatments designed to mitigate the impacts of identified limiting factors were researched and developed. Treatments included selective application of herbicides to reduce competition by ferns and seedlings of undesirable hardwood species, preharvest treatments to enhance development and growth of seedlings of desirable hardwood species, and reduction in deer density.

As the art and science of silviculture developed and expanded, interactions among deer, forest vegetation, and silviculture were discovered and led to better understanding of management requirements for forest vegetation and deer. In some instances, deer impact necessitated reducing deer density and adjusting and/or adding silvicultural treatments to offset the impacts of browsing. On the flip side of the coin, silvicultural treatments were developed/adjusted to enhance habitat for deer and other wildlife species.

13.6 SILVICULTURAL TREATMENTS AFFECTED BY DEER

13.6.1 Tree Regeneration

Replacing harvested overstory trees in northern hardwood forests generally is accomplished by stimulating germination and growth of seedlings from seeds shed by overstory trees (stand reinitiation). After overstory trees are harvested, the increased light and soil disturbance caused by harvesting operations foster rapid development of existing understory tree seedlings (stand initiation). In some cases, such as where the overstory does not contain the desired tree species, or seeds shed by the trees are inadequate or missing, as with red oaks (*Quercus rubra*), nursery stock may be planted on regeneration sites and protected from deer browsing with solid translucent or open mesh tubes.

If deer density is high enough prior to overstory removal, deer browsing eliminates or severely impacts seedlings of preferred species such that a competitive advantage is given to interfering ferns and seedlings of tree species not preferred as forage by deer, especially under uneven-aged management, which produces far less forage than even-aged management. To prevent this, managers may fence regeneration sites prior to overstory removal. However, an additional step may be required: to provide sufficient stocking of preferred seedlings under even-aged management. Foresters may need to thin overstory and midstory tree density in a treatment called *shelterwood seed cuts* to stimulate germination and growth of sufficient numbers of preferred tree seedlings. The problem is, sites treated with shelterwood seed cuts must be fenced prior to or shortly after the thinning is effected to eliminate deer browsing that would defeat the attempt to provide full stocking of seedlings.

Additionally, if the understory prior to overstory harvest has been affected by deer such that it is dominated by ferns, grasses, and seedling species competitive with desired species, stands must be treated with expensive herbicides to eliminate competition with preferred seedling species, and such sites must be further protected from deer by fencing until a new crop of seedlings has been produced from overstory trees and grows beyond browsing reach of deer, at which point the fencing may be removed.

13.6.2 TREE REGENERATION: CONVERTING STANDS TO UNEVEN-AGED SILVICULTURE

Managers wishing to maintain or restore stands to the species composition and structure typical of the uneven-aged, multicanopied stands, including the complex or old-growth successional stage, dominated by shade-tolerant tree species, do so by creating multiple small (less than 1/4 ha) openings in maturing forests by removing single trees (single tree selection) or small groups of trees (group selection) on a recurring basis, or removing immediately surrounding overstory trees to reduce competition for sunlight and nutrients by commercially desirable trees (crop tree release). If deer density is greater than that occurring under natural uneven-aged development (<5 deer/km^2 in northern hardwood forests), deer will rapidly forage on and degrade seedling regeneration stimulated by the small forest openings, as there will be far less forage available in the surrounding understory. Landowners wishing to develop and maintain uneven-aged forests under high deer densities will need to achieve and maintain low deer densities permanently or fence all created openings to prevent deer from destroying regeneration until seedlings attain sapling/pole size and grow above browsing height of deer (about 2 m).

13.7 SILVICULTURAL TREATMENTS FOR REDUCING DEER IMPACT

13.7.1 OVERWHELM DEER WITH OVERABUNDANCE OF FORAGE

Foresters have attempted to overwhelm deer with so much forage, by harvesting overstory timber from large (>50 ha) areas, that desirable seedlings escape deer browsing in sufficient numbers/ condition that they will dominate species composition in the ensuing maturing forest. This attempt capitalizes on the fact that deer will not leave traditional home ranges to forage on new sources of food outside their home ranges—at least not until the seedlings may have grown out of browsing height of deer.

Results vary with size of harvest sites. In West Virginia, Akins and Michael (1993) found that deer browsing impact on forest vegetation 1 year after timber harvest on sites ranging from 0.2 to 0.8 ha was higher on the smaller harvest sites, as was percent ground cover of interfering fern vegetation. Conversely, foresters on a private forestland in north-central Pennsylvania attempted to overwhelm local deer by creating a large clearcut (~160 ha) with so much forage creation that stocking levels of desired seedling species of commercially valuable trees would reach acceptable threshold levels. Unfortunately, on-site pin cherry (*Prunus pensylvanica*) seed germinated and rapidly dominated the understory, reducing stocking of black cherry below required levels (J. Kochel, lands manager for Forest Industry Associates, personal communication): pin cherry suppressed black cherry seedlings. The unintended dominance of pin cherry necessitated its removal, which was nearly impossible, as application of herbicides or other traditional methods would not differentiate between the two cherry species. Had the clearcut been smaller, the resident deer herd might have exerted sufficient browsing pressure on pin cherry to allow black cherry to establish itself and grow into the sapling stage.

13.7.2 Increasing Forage Production within Landscapes in Smaller-Sized Harvest Units

In the example comparing deer impact on seedling regeneration between an area with reduced timber harvest and an area with sustainable level of timber harvest (Chapter 15), optimizing the level of timber harvest can increase the amount of forage availability and result in lower deer impact, even with higher deer density. In the example, area of harvested stands generally was less than 20 ha. The critical factor is determining the level of deer density compatible with regeneration carrying capacity specific to forest type and associated environmental factors. Additionally, type of forest management (even-aged vs. uneven-aged, with more forage available in even-aged stands) dictates the amount and distribution of forage creation and associated deer carrying capacity for advance regeneration. It is also understood that designating a recommended deer density is keyed to the resources managed for and will be lower if diversity and abundance of forest understory vegetation, especially forbs, are management objectives.

13.7.3 Treatments to Alter Composition of Understory Vegetation: Herbicides

As indicated above, when long-term overbrowsing by deer leads to understories dominated by plants that outcompete seedlings of desired tree species, foresters may use herbicide mixes to selectively remove unwanted vegetation. Prohibition of applying herbicides in or near to riparian zones (e.g., streams, seeps, springs, bogs) containing sensitive vegetation may restrict their use, reducing their effectiveness. Also, costs of herbicide application are high.

13.7.4 Treatments to Alter Composition of Understory Vegetation: Mechanical Removals

Managers may elect to remove undesirable woody vegetation from forest understories by mechanical removals rather than with herbicides. Chiefly, this practice entails cutting of individual stems of unwanted tree species by technicians with hand-held trimming devices or by crushing vegetation with mechanized equipment. The latter practice eliminates all understory regeneration, whereas the more directed cutting by individuals allows for discriminating between desirable and undesirable seedling species. Both are time consuming and expensive. In oak-dominated stands, foresters may apply controlled understory burns of low intensity to remove vegetation competing with oak seedlings which are resistant to low intensity ground fires.

REFERENCES

Akins, J. W. and E. D. Michael. 1993. Impact of clearcut size on white-tailed deer use and tree regeneration. In *Proceedings Sixth Eastern Wildlife Damage Control Conference* 6:185–195.

Horsley, S. B. and J. C. Bjorkbaum. 1983. Herbicide treatment of striped maple and beech in Allegheny hardwood stands. *Forest Science* 29:103–112.

Horsley, S. B. and D. A. Marquis. 1983. Interference by weeds and deer with Allegheny hardwood reproduction. *Canadian Journal of Forest Research* 13:61–69.

Leak, W. B. 2011. *Estimating allowable–cut by area-scheduling.* Research Note NRS-115. Newtown Square PA. United States Department of Agriculture Forest Service, Northeastern Research Station.

Marquis, D. A. 1979. Ecological aspects of shelterwood cutting. In *Proceedings, National Silvicultural Workshop*. Charleston SC: U.S. Department of Agriculture Forest Service.

Marquis, D. A. 1981. Removal or retention of merchantable seedlings in Allegheny hardwoods: Effect on regeneration after clearcutting. *Journal of Forestry* 79:280–283.

Marquis, D. A., T. J. Grisez, J. C. Bjorkbaum et al. 1975. *Interim guide to regeneration of Allegheny hardwoods*. General Technical Report NE-19. Broomall PA: Northeastern Forest Experiment Station, U.S. Department of Agriculture Forest Service.

Nyland, R. D., L. S. Kennefic, K. K. Bohn et al. 2016. *Silviculture: Concepts and applications* (3rd edition). Long Grove IL: Waveland Press.

Stout, S. L. and P. H. Brose. 2014. The SILVAH saga: 40+ years of collaborative hardwood research and management highlight silviculture. *Journal of Forestry* 112:434–439.

Tilghman, N. G. 1989. Impacts of white-tailed deer on forest regeneration in northwestern Pennsylvania. *Journal of Wildlife Management* 53:524–532.

14 National and Regional Perspectives on Deer Management

David S. deCalesta

CONTENTS

Those who do not learn history are condemned to repeat it.

George Santayana (1905)

I personally believed, at least in 1914 when predator control began, that there could not be too much horned game, and that the extirpation of predators was a reasonable price to pay for better big game hunting.

Aldo Leopold (1944)

14.1 INTRODUCTION

About 100 years ago, deer were brought back from the brink of extinction by a consortium of concerned hunters and newly established state wildlife agencies acting to eliminate market hunting and heavily restrict deer hunting, including prohibition of doe hunting. To the great joy of hunters, these actions, in serendipitous concert with vast timber harvest operations creating vast quantities of quality deer forage, succeeded in vastly increasing deer abundance. To the sorrow of foresters, farmers, and ecologists, the resulting deer management overcorrections in the ensuing decades resulted in deer of such abundance that forest understories, the browse and cover they provided, and obligate wildlife species nose-dived and complaints by farmers of excessive deer impact increased. Corrective deer management steps, including encouragement of harvesting antlerless deer, occurred in roughly 30-year cycles of boom-and-bust deer abundance and managed to limit somewhat deer population explosions and related habitat degradation. Management efforts to reduce deer density represented responses to decades of complaints about impacts of overabundant deer herds by farmers, foresters, and biologists. These were followed by decades of complaints by hunters of too few deer and reductions in efforts by state game agencies to reduce deer density and impact, reigniting the rollercoaster ride of deer abundance and associated deer impacts. The repeated 30-year cycles of boom-and-bust deer density and impact occurred at intervals too short to permit recovery of defenseless understory seedlings, shrubs, and herbaceous plants in the down abundance cycles, and to this day, these plants are scarce or totally lacking in deer-impacted forests. The institutional memory of natural resource agencies responsible for regulating deer harvests should be long enough to factor in the environmentally devastating effects of yo-yoing deer density and avoid them. It is critically important for natural resource agencies to adopt the decades-long perspective required for sustainable management of deer and other forest resources and to resist pressures by hunters whose shorter-term memories may be influenced by values and culture that exclude consideration for sustainability of forest resources other than deer.

14.2 NATIONAL HISTORICAL PERSPECTIVE ON FLUCTUATIONS IN DEER ABUNDANCE

Prior to European colonization of North America, predation pressure by indigenous peoples, wolves, bears, and mountain lions, and irregular cycles of forage production by overstory openings created by natural events, maintained deer densities sufficiently low to prevent them from negatively affecting understory vegetation, including tree seedlings and eventual composition of overstory trees.

Adams and Hamilton (2011) described the history of white-tailed deer abundance and impact in North America as affected by humans beginning with Native Americans. The characterization is generic for North America and divided into five periods of alternating decrease/increase in deer abundance: (1) prior to 1800 (Native American Exploitation—moderate control of deer abundance), (2) 1800–1850 (Moderate Recovery—increase in deer abundance), (3) 1850–1900 (European Exploitation—decrease in deer abundance), (4) 1900–1975 (Protection and Recovery—increase in abundance), and (5) 1975-present (A Changing Paradigm—with failed attempts to reduce deer abundance and impacts).

Depending on how humans modified forest vegetation, reduced populations of deer predators, and exploited deer, deer abundance waxed and waned during these periods. The depiction of the period of moderate increase in deer abundance (1800–1850) did not mention elimination of major deer predators (e.g., mountain lion, bears, and wolves) that preyed on domestic livestock. Also ignored was the near-eradication of American Indians through introduction of epidemic diseases for which American Indians had no defenses, slaughter in the Indian Wars, and subsequent removals to reservations. The almost complete elimination of these predators greatly reduced their impact on deer by the beginning of the twentieth century. Indeed, a sixth period (near-elimination of historic predators) from 1850–1900 could be included as a defining time that affected deer abundance prior to the great expansion of deer abundance in the early twentieth century.

Prior to European colonization, eastern North America was covered in forests. Openings providing abundant and nutritious deer forage were sporadic and of great variety in size and location as created by various overstory-clearing natural events, including tornadoes, microbursts, ice storms, hurricanes, and fire (by natural events and by Native Americans as a way to increase forage for deer and visibility to hunt them and for growing crops). Tree species were dominated by shade-tolerant trees that germinated and grew under diffused lighting conditions in response to irregular overstory openings that produced multiple-aged groups of seedlings characteristic of uneven age structure and species composition. As eastern North America was developed by colonists, much land was cleared of trees for farming and grazing, and timber began to be harvested for lumber products—both creating abundances of deer forage likely much greater than provided by natural events. By the end of the nineteenth century, a second wave of extensive timber harvesting produced a second wave of deer forage and fawning cover. These waves of timber harvesting created a different species composition and structure of trees. Resulting forests were dominated by less shade-tolerant trees, and large groups of seedlings and trees of a single age, referred to as even-aged species composition and structure. These changes in forest cover and forage were not put in context with the boom-and-bust periods of deer abundance nor recognized for their contribution to the cycles.

By the late 1800s, market hunting of deer was outlawed and restrictions on deer hunting were put in place, at a time when timber harvesting had progressed from limited harvests for feeding tanning mills and clearing of settler lands to wholesale removal of trees during the timber exploitation period of 1850–1900. Large blocks of eastern, southern, and midwestern states were harvested multiple times, resulting in an unprecedented amount of understory forage for deer, which coincided with removal of major predators and heavy restrictions on deer hunting. Reduction/elimination of natural predators enhanced top-down trophic cascading by deer simultaneous with production of enormous quantities of nutritious deer forage that enhanced bottom-up trophic cascading created a perfect storm for deer irruptions followed by collapses of structure and species composition of understory vegetation, deer forage, wildlife habitat, and deer abundance.

14.3 A REGIONAL EXAMPLE OF FLUCTUATIONS IN DEER ABUNDANCE AND IMPACT

Deer population fluctuations in Pennsylvania prior to the 1900s mirrored those across America as described by Adams and Hamilton (2011). Deer density was 4–6 deer/km^2 in mature Pennsylvania hardwood forests (Dahlberg and Guettinger 1956) prior to the increasing impact of European development of America before the twentieth century. Market hunting and liberal deer seasons combined to steadily reduce deer populations until near elimination in the late 1890s (Kosak 1995). The Pennsylvania Game Commission (PGC) was created in 1895, primarily to restore lost wildlife populations, especially deer. Deer were trapped from Michigan and other states and transplanted into Pennsylvania. Short hunting seasons were established, bag limits introduced, and doe hunting was prohibited. Simultaneously, enormous quantities of deer forage had been created by clearcutting the entire state multiple times, and deer populations began to skyrocket statewide.

Data collected sporadically from the Allegheny National Forest in Pennsylvania since 1900 concerning white-tailed deer numbers and coincident annual levels of timber harvest provide a localized documentation of the history of deer management. Redding (1995) characterized the fluctuations in deer density and timber harvest 1907–1992 that he drew from individual accountings of deer density and timber harvest (Figure 14.1). Data collected for the case history of deer management on the Allegheny National Forest 2000–present (Chapter 34) complete the picture.

By 1900, with the elimination of major predators, severe limitations on deer harvest, and multiple and extensive timber harvests, deer densities skyrocketed, approaching 20 deer/km^2 by the 1930s. Biologists (Leopold et al. 1947) identified the northwestern part of Pennsylvania as being severely overpopulated, with 50% depletion of deer carrying capacity by 1938. Forests were maturing after timber harvests in the late 1800s/early 1900s, and with the closed overstory canopy and minimal

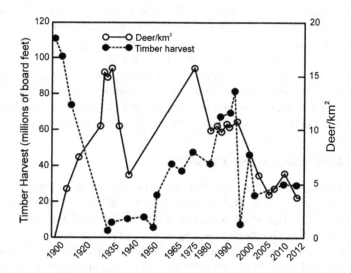

FIGURE 14.1 Variations in deer density and timber harvest 1900–present, Allegheny National Forest, Pennsylvania, USA.

tree harvest, there was little deer forage available. Deer density plummeted following an antlerless-deer-only hunting season in 1938, and a cold and snowy winter was followed by large-scale deer starvation mortality.

A new round of timber harvest in the 1940s and 1950s, coupled with restrictions on harvest of antlerless deer, resulted in the deer herd rebounding and, by 1978, reaching even higher abundance levels than previously. After three decades of browsing by an increasing deer population, composition of understory vegetation changed, reflecting the impact of overabundant deer on vegetation dynamics. Preferred deer foods (e.g., seedlings of sugar and red maple, black cherry, white ash and oaks, shrubs, wildflowers) were depleted, and species resistant to deer browsing (e.g., ferns, beech, and striped maple) began to dominate the forest understory. Foresters noticed that when they harvested overstory trees, they got only fern, striped maple, and beech growing on the forest floor. Deer browsing had removed shrubs, forbs, and seedlings of more desirable tree species. To return the understory to a desirable composition of tree and shrub species, foresters had to apply herbicides to remove ferns, beech, and striped maple. Also, timber harvest sites were fenced to keep deer from eliminating tree seedlings of desirable species following harvest of overstory trees.

In the mid-1980s, the PGC increased the number of antlerless permits in an attempt to reduce the deer herd to a goal density of 8 deer/km². The PGC also allocated "bonus" antlerless permits, wherein hunters could obtain a second antlerless permit by applying for permits initially offered that were not used, providing some hunters the opportunity to harvest more than one antlerless deer. However, the deer density remained above target level the following 15 years.

The PGC, under the guidance of a new deer program manager, made three modifications to deer harvest regulations to get deer to target density. In 2001, hunters were permitted to harvest antlered and antlerless deer concurrently during the general rifle season for the entire 12-day rifle season. In 2002, harvest of antlered deer was restricted to bucks with at least three antler points on at least one side to reduce harvest of yearling bucks and increase the proportion of mature bucks in the population. This measure resulted in an increased harvest of antlerless deer by hunters who hunted for meat in addition to trophy hunting. In 2003, a deer management assistance program (DMAP) was established to increase harvest of antlerless deer by allowing forest landowners experiencing high negative deer impacts to obtain permits and distribute them to hunters to harvest antlerless deer.

Following implementation of these three changes in deer harvest regulations, deer density was halved within 4 years in the Allegheny National Forest. Foresters were able to remove fencing to protect regeneration, and did not have to fence new regeneration sites, resulting in more forage available for deer. Timber harvest remained relatively low, due to poor market conditions for harvested lumber and lawsuits brought by environmental groups to restrict timber harvest. Reductions in numbers of DMAP permits issued in subsequent years resulted in a slight recovery of deer density. Body weights and antler characteristics of deer harvested during this period increased significantly from the years when deer density was much higher. Deer impact on regeneration decreased significantly, to the point where impact levels were identified as light-moderate. However, legions of hunters complained about seeing few deer during hunting season, and hunter discontent began to swell, leading to pressure on the PGC to scale back its DMAP program and increase deer density statewide.

14.4   THE DEER MANAGEMENT CONFLICT

Cycles of deer and forage abundance created the basic conflict among hunters, foresters, and ecologists that persists and confounds deer management. Present-day hunters, their fathers, and grandfathers remember the days when deer density was so high, and the understory so depleted, that dozens of deer would be seen during hunting season, especially opening day. They remember that the highs of deer density were preceded by restrictions in antlerless or doe hunting, and that the population crashes followed liberal antlerless seasons as the game agencies attempted to reduce deer density and avoid such crashes and associated depletion of understory vegetation. Not being foresters or ecologists, most hunters did not notice the depleted state of understory vegetation, lack of regenerating seedlings necessary for future forests, or lack of diversity of understory vegetation. Nor did they notice reductions in abundance and diversity of wildlife species associated with habitat degradation. Hunters also did not see that when deer density was high, deer quality (measured by body condition and size and quality of antlers) was much less than when deer were in balance with the forest vegetation. Many hunters still want deer densities at least as high as 15 deer/km^2 and are unaware that such densities are not sustainable and inevitably lead to winter die-offs. Their response in the past to deer die-offs was to demand that game agencies conduct and support emergency feeding operations for the deer and prohibit harvest of female deer.

On the other hand, foresters and ecologists are well aware of the effects of a century of overabundant deer herds on regeneration of seedlings and diversity and abundance of other understory vegetation. They have struggled to protect and enhance understory vegetation, but without reduction of deer density to sustainable levels, their efforts have been mainly palliative and limited to excluding deer from the vegetation with costly fencing. Vegetation outside fenced areas had no protection from overabundant deer herds and remained sparse and lacking in understory species richness and abundance. Scientists have documented severe reductions in diversity and abundance of herbaceous vegetation that never grew out of the reach of deer, and the effects on associated plant and animal communities, with some species being eliminated entirely.

REFERENCES

Adams, K. P. and R. J. Hamilton. 2011. Management history. In *Biology and management of white-tailed deer.* D. G. Hewitt ed. Boca Raton FL: CRC Press.

Dahlberg, B. L. and R. C. Guettinger. 1956. *The White-tailed Deer in Wisconsin.* Technical Wildlife Bulletin 14, Madison WI. Wisconsin Conservation Department.

Kosak, J. 1995. *The Pennsylvania Game Commission, 1895–1995: 100 years of wildlife conservation.* Harrisburg PA: The Pennsylvania Game Commission.

Leopold, A. 1944. Review of the wolves of North America. *Journal of Forestry* 42:928–929.

Leopold, A., L. K. Sowls, and D. L. Spencer. 1947. A survey of over-populated deer ranges in the United States. *The Journal of Wildlife Management* 11:162–177.

Redding, J. R. 1995. History of deer population trends and forest cutting on the Allegheny National Forest. In *Proceedings, 10th Central Hardwood Forest Conference. Morgantown, WV. General Technical Report NE-197.* eds. K. W. Gottschalk and S. L. C. Fosbroke. Radnor PA: U.S. Department of Agriculture, Forest Service, Northeastern Forest Experiment Station.

Santayana, G. 1905. *The Life of Reason: Reason in Common Sense.* New York: Scribner's and Sons.

15 Deer Density, Carrying Capacity, and Impact on Forest Resources

David S. deCalesta

CONTENTS

Rarely in the field of resource management has a term been so frequently used to the confusion of so many.

John Macnab (1985)

...beyond a certain point (MSY), the absolute number of fawns recruited goes down.

Dale McCullough (1979)

MANAGER SUMMARY

Deer can be managed for varying densities that are related to carrying capacities for different classes of forest resources and different groups of stakeholders. The highest carrying capacity, termed maximum sustained yield (MSY), occurs when production of deer forage is maximized and deer reproduction and recruitment are maximized to produce the highest deer densities, which is highly desired by a segment of deer hunters who prefer high deer density to high deer quality (body weight and antler characteristics). At such high deer densities, abundance and species richness of understory vegetation, including tree seedlings, shrubs, herbaceous vegetation, and wildlife habitat and dependent wildlife species are reduced, some species are missing, and understory vegetation is characterized by ferns, grasses, and woody species less palatable to deer and of low commercial value. At deer densities below MSY, species richness and abundance of other forest resources, such as shrubs, herbs, and tree seedlings, wildlife habitat and dependent wildlife species, increase as deer density decreases. Deer densities associated with the ability to regenerate commercially desirable tree species may be classified as regeneration carrying capacity and include increases in abundance and species richness of other forest resources. Deer density sufficiently low to favor high abundance and species richness of these other forest resources may be classified as diversity carrying capacity, which may approximate the species richness and abundance characteristic of forest ecosystems prior to the vast expansion of deer densities resulting from human impacts on forest resources. Increasing amounts of available deer forage sustainably, by harvesting/thinning overstory trees, may allow a rise in deer densities associated with carrying capacities for producing timber and the desires of hunters for higher deer densities, but likely not for abundance and species richness of other forest resources. Different silvicultural practices affect the amount of deer forage created and may affect carrying capacities. Even-aged management produces more landscape forage than uneven-aged management, resulting in higher deer densities associated with the various classes of carrying capacity than under uneven-aged management. It does not appear to be possible to increase deer density associated with diversity carrying capacity by increasing quantity and quality of deer forage regardless of silvicultural practice. Attempts to increase deer density while reducing deer impacts by increasing amounts of forage may succeed for regenerating seedlings of commercial timber species with even-aged management, but will likely not benefit herbaceous vegetation or wildlife habitat nor be viable with uneven-aged management.

15.1 INTRODUCTION

Biological carrying capacity for wildlife species was defined by Dasmann (1964) as "the maximum number of individuals a habitat can sustain in a healthy and vigorous condition." McCullough (1979) defined (deer) carrying capacity as the maximum number of deer that can be supported at equilibrium in a steady environment, calling it *K carrying capacity*. At K, there is no recruitment of fawns into the adult population—reproduction is essentially zero. McCullough described maximum sustained yield (MSY) carrying capacity as the greatest number of deer recruitable into the population as fawns, and pegged it at 50%–60% of K (Figure 15.1) as did others (Caughley 1979, Macnab 1985).

McCullough (1979) speculated that trophy deer could be produced at a deer density slightly lower than that at MSY, and Jenks et al. (2002), evaluating deer densities within a fenced forested area in Tennessee, proposed that trophy deer are produced at MSY carrying capacity. However, Jenks et al. (2002) defined trophy deer as exhibiting the maximum number of antler points and heavy body weight, and their data set included only deer abundance from MSY to K and went no lower on the deer abundance

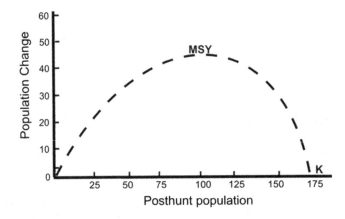

FIGURE 15.1   K and MSY deer abundance with associated recruitment levels (population change) from McCullough (1979).

curve. Data from the Kinzua Quality Deer Cooperative (KQDC) tracked deer density downward from K to MSY to levels approaching 1/4 of MSY (deCalesta 2013). As deer density dropped below MSY, antler characteristics (number of points, spread, beam diameter) and body weights continued to increase until they were maximized and maintained at a deer density $\leq$2/3 of MSY.

15.2 MULTIPLE CARRYING CAPACITIES

The definition of carrying capacity was expanded by wildlife scientists to include *cultural carrying capacity*, the desired population density of a wildlife species that individual stakeholder groups consider ideal (Ellingwood and Spignesi 1986, Decker and Purdy 1988). In addition to K and MSY, other carrying capacities have been identified as being associated with maintaining sustainable levels of forest resources other than deer as defined by the values of different stakeholder groups. For example, D'Angelo and Grund (2015) studied hunters and farmland owners in Minnesota, reporting that hunters wanted higher deer densities and farmers wanted lower deer densities. The relationship among deer density, deer impact, and stakeholder-related cultural carrying capacities (Figure 15.2) was described by deCalesta and Stout (1997) based on a 10-year study of deer impact on forest resources at varying deer density levels. The study, conducted in a northern hardwood forest in northwestern Pennsylvania, produced an amount of forage provided at sustainable levels of clearcutting and thinning typical of a northern hardwood forest.

15.2.1 K CARRYING CAPACITY

At this carrying capacity, maximum deer density as permitted by forage created in the study reported on by deCalesta and Stout (1997) was 25 deer/km²: Higher densities could not be maintained due to winter starvation losses. No stakeholder group supports this deer density.

15.2.2 DIVERSITY CARRYING CAPACITY

Historic deer density for the northeastern United States, as maintained by natural processes, including predation, and prior to European colonization of North America, was about 4–6 deer/km² (Dahlberg and Guettinger 1956, McCabe and McCabe 1984, Alverson et al. 1988). At this density, the ecosystem was intact and plant and animal species were mutually tolerant—all coexisted. Ecosystems with this deer density exhibit the highest diversity, including species richness. Deer quality, including trophy animals, and deer forage would be as high as provided by the quality and quantity of habitat components, but abundance of deer, including trophy animals, would be low. This carrying capacity is preferred by the stakeholder group composed of ecologists and landowners interested in restoring

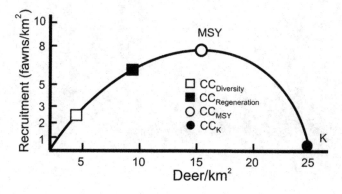

FIGURE 15.2 Cultural carrying capacities for diversity, regeneration of tree species, maximum sustained yield (MSY), and maximum population density (K).

forests as closely as possible to precolonial North America to provide the highest species richness and abundance of native plants and animals.

15.2.3 Regeneration Carrying Capacity

Progressing up and to the right on the deer abundance curve to about 8 deer/km^2, we arrive at the deer density associated with maintenance of trees of commercial value for timber harvest, and only slightly diminished deer quality, but with about twice the deer density as at diversity carrying capacity. This carrying capacity is more preferable to most deer hunters than diversity carrying capacity because of higher deer density. Also, this carrying capacity will produce the highest number of trophy class deer. Landowners interested in maintaining a sustainable allowable cut of timber for economic reasons are comfortable with this carrying capacity. However, the impact on nontimber resources of a burgeoning deer herd begins to be felt: diversity, abundance, and vertical structure of shrubs, forbs (seed-bearing plants without woody stems that die back to the growing point after flowering and are not grasses), and seedlings of tree species favored by deer as forage decline. Species richness and abundance of associated wildlife species (e.g., forest songbirds; deCalesta 1994) dependent on a diverse, abundant, and structured understory forest vegetation begin to decline.

15.2.4 Maximum Sustained Yield Carrying Capacity

Progressing still farther to the right and higher on the deer density curve, we arrive at the deer density associated with the highest fawn recruitment and thus highest "sustainable" deer harvest: MSY carrying capacity, at approximately 50%–60% of K (~16 deer/km^2). At MSY, the ratio of fawns per doe may be lower, but the number of does is so high that number of fawns recruited is still the highest of any carrying capacity. The proportion of trophy class deer is lower than at regeneration carrying capacity. Shrub and forb understories are mostly replaced by ferns, grasses, and seedlings of tree species resistant to deer browsing, and habitat for ground- and shrub-dependent wildlife species is simplified, truncating their abundance and species richness. There is a well-defined "browse line" where most vegetation within the reach of deer (<2 meters) is eliminated (Figure 15.3). This carrying

FIGURE 15.3 Browse line (and fern understory) typical of deer herds at MSY carrying capacity. (Photo by Dan Aitchison.)

capacity has also been called nutritional carrying capacity—the equilibrium between deer and food supply at maximum sustainable population (Bishop et al. 2009).

While deer abundance is high at MSY, body weights are smaller and antler characteristics are restricted to minimal sizes exhibited by yearling bucks, as most of the older bucks have been harvested in preceding years and remain scarce. Although defined as a carrying capacity, MSY may be unsustainable because deer density will drift lower if forage resources become less abundant and nutritious. Additionally, MSY (a.k.a. nutritional carrying capacity) is "a miserable place for a deer herd to be" as "nutritional stress and mortality are high while fecundity is low and population growth is effectively zero" (DeYoung 2011). MSY carrying capacity is the goal of the stakeholder group composed of hunters wishing to maximize number of deer seen during hunting season and maximize chances of harvesting a deer.

15.2.5 POPULATION CRASH

At deer density above MSY, plant and animal communities are in a declining state and the situation is unsustainable. Ground cover is primarily grasses and ferns of low wildlife or diversity value. Deer begin to starve and are characterized by small body weight and poor antler characteristics. Fewer does carry fawns to birth, and recruitment plummets. This situation occurs because deer have eliminated most browse species they prefer; capacity for regeneration of trees by seedlings is suppressed by reduction of species richness and abundance and because there is not enough deer browse available. Thus, the number of deer the habitat can maintain declines, as does recruitment, and the situation proceeds to the last condition—*population crash,* typically at K.

Population crash results in reduced deer density, as many deer do not survive starvation conditions. This stage usually occurs when deer run out of food during cold and snowy winters, use up their energy reserves (body fat), and begin to catabolize muscle protein as a remaining source of energy. Blood glucose levels continue to drop and fall below the level required to sustain life (deCalesta et al. 1975). No stakeholder group favors population crash.

After a population crash and a few deer are left, the understory vegetation begins a slow recovery under reduced browsing pressure. However, because some of the forb and shrub species may have been extirpated by decades of excessive deer browsing, they may be lost unless reintroduced by human restoration efforts. Also, changes in overstory tree species engendered by repetitive clearcutting of the forest in the 1800s and early 1900s produced a generally even-aged forest dominated today by moderately shade-intolerant species such as red maple (*Acer rubra*), black cherry (*Prunus serotina*), pin cherry (*Prunus pensylvanica*), and birches (*Betula* sp.), shifting the species mix away from the previously dominant shade-tolerant sugar maple (*Acer saccharum*), beech (*Fagus grandifolia*), striped maple (*Acer pensylvanicum*), and hemlock (*Tsuga canadensis*).

Under the scenarios described above, diversity carrying capacity is ~1/3 of MSY and 1/5 of K, regeneration carrying capacity is 2/3 of MSY and 2/5 of K, and MSY is ~1/2 of K.

The above values were derived for forests managed by even-aged management. Forests under uneven-aged management produce far less landscape forage. There is no research to support such an assertion, but it seems reasonable to assume that deer densities associated with different carrying capacities would be lower for each category under uneven-aged management.

15.2.6 CHANGES IN CARRYING CAPACITY RELATED TO DEER DENSITY

Unfortunately, there has been little research directed at determining whether increasing forage availability permits an increase in deer density without a concomitant increase in negative impact on understory vegetation. Crimmins et al. (2010) reported that increasing forage availability by harvesting overstory trees and opening the understory while simultaneously reducing deer density resulted in significantly reduced impact on browsing impact on seedlings.

Marquis et al. (1992) and deCalesta and Stout (1997) hypothesized that deer impact on forest resources is a joint function of deer density and forage availability, suggesting that deer density associated with specific carrying capacities can be increased by increasing abundance and quality of forage. The implicit assumption was that forest/deer managers could increase deer density but not deer impact by increasing quality and quantity of deer forage. Under this scenario, harvesting more trees and/or creating deer food plots (including agricultural fields) may increase deer density but not deer impact on forest resources. If true, the stakeholder group composed of hunters who want to maximize deer density would benefit the most if deer were managed for MSY carrying capacity and forest vegetation management included providing more forage as sustainable under timber harvesting practices. Figure 15.4 illustrates the concept as drawn from limited data sets.

There are no studies where forage availability was manipulated to determine whether increasing levels of deer forage results in increased deer density levels for cultural carrying capacities as proposed by Figure 15.4. Data for determining K at four levels of forage availability were drawn from four sources: (1) the 2012 Kinzua Quality Deer Cooperative annual report (deCalesta 2013), with two levels of timber harvest in a northern hardwood forest; (2) description of deer density fluctuations, including crashes, on the Allegheny National Forest 1900–1995 (Redding 1995); (3) a demonstration area featuring open fields (26%) and forest openings in a northern hardwood forest (McCullough 1979); and (4) a national park combining agricultural fields (50%) and northern hardwood forestland (50%) (Porter and Underwood 1997).

Data for determining MSY carrying capacity at two levels of timber harvest were derived from deCalesta (2013) and Redding (1995) and assumed at 60% K as suggested by McCullough (1979) and Porter and Underwood (1997). Data for estimating diversity carrying capacity were from deCalesta (2013). Data for estimating regeneration carrying capacity were from deCalesta (2013), Royo et al. (2010), Redding (1995), and Tilghman (1989). No data were available for estimating levels of these additional carrying capacities from McCullough or Porter and Underwood wherein forage production was higher. Data for determining whether increasing forage levels can increase deer "trophy" characteristics were drawn from deCalesta (2013). Data for determining whether increasing deer forage can elevate diversity carrying capacity were not available. However, insight may be

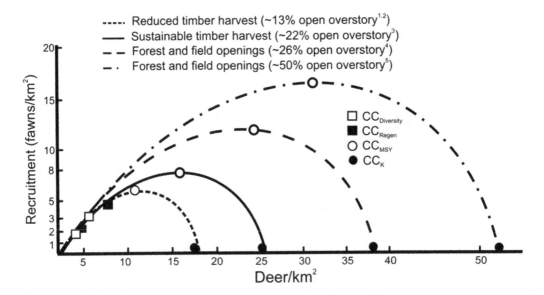

FIGURE 15.4 Cultural carrying capacities related to deer density, four levels of forage, and deer impact in managed forests. ([1]deCalesta 2013, [2]Redding 1995, [3]deCalesta 2013, [4]McCullough 1979, [5]Porter and Underwood 1997.)

drawn from comparison of species richness and abundance of herb and shrub species between a baseline study conducted in an old-growth forest in Pennsylvania prior to the increase in deer density and impact (Lutz 1930) and studies conducted 66 years later (Rooney and Dress 1997) when deer density had climbed to, and remained at, levels exceeding 10 deer/km^2. Numerous species present in 1929 were absent in 1995. Species with relative abundance values of <1% in 1929 were more likely than more abundant species to be missing in 1995. The number of families represented declined from 27 in 1929 to 10 in 1995.

Although there are no established studies to rely upon, it seems reasonable to assume that the carrying capacity curve for uneven-aged forests would most likely approximate that reflecting low production of deer forage identified for diversity carrying capacity in reduced timber harvest from Figure 15.4.

15.2.7 Can Carrying Capacity Be Changed by Altering Forage Quantity/Quality?

Assuming that MSY is 50%–60% K, it is apparent from Figure 15.4 that landscapes with higher levels of deer forage (optimizing timber harvest and/or surrounded by landscapes featuring open noncrop and/or crop fields) result in higher MSY carrying capacity: there will be more deer that can be harvested sustainably when a higher proportion of the landscape provides deer forage. It remains to be seen whether maintaining increased timber harvest and deer density levels at MSY is truly sustainable or if MSY creeps downward to the left if deer browsing reduces production and/ or quality of forage.

15.2.8 Increasing Regeneration Carrying Capacity

As deer density declines below MSY under differing forage availability scenarios and approaches 0, I hypothesize that the impact curve lines converge (Figure 15.4). Deer densities below MSY carrying capacity are much closer under the reduced timber harvest/forage production curve than for the sustainable timber harvest curve.

This hypothesis can be evaluated with data from the KQDC Demonstration Project (deCalesta 2013). Two landscapes within the demonstration area had differing and stable levels of deer density and forage production prior to and following reduction in deer density. One, a 6000-ha block composed of a private timber company and a public watershed, had an average density of 13.4 deer/km^2 prior to herd reduction, followed by an average density of 7.8 deer/km^2 after density had been stable for 5 years following herd reduction. The other, a 15,140-ha area on the abutting Allegheny National Forest, had an average density of 9.2 deer/km^2 prior to herd reduction, followed by an average density of 3.9 deer/km^2 after deer density had been stable for 5 years following herd reduction.

Both areas had stable levels of forage production prior to and following herd reduction (22.1% of plots for the high deer density area were under opened canopy vs. 12.2% of plots for the low deer density area). Two levels of deer browse impact (Pierson and deCalesta 2015) were evaluated on seedlings of five indicator species: (1) zero–moderate impact (less than half of twigs are browsed), representing a level wherein seedlings will survive to grow into saplings and larger sawtimber-class trees; and (2) heavy–severe deer impact (more than half of twigs are browsed, often so heavily that twigs are hedged back to stubs), representing a level wherein seedlings will not survive to grow into saplings and larger sawtimber-class trees—regeneration will fail. Indicator species were beech, red maple, black cherry, striped maple, and hemlock.

Reducing deer density within high- and low-forage production landscapes resulted in an increase in prevalence of plots with seedlings of zero–moderate impact and heavy-severe impact (Table 15.1). This result was consistent and significant ($P < 0.05$) for all indicator species except striped maple at the heavy–severe impact level on both landscapes and for beech at the zero–moderate level for the high deer density area and at the heavy–severe level for both landscapes. Achieving higher levels of plots with seedlings at zero–moderate impact *and* heavy–severe impact levels by reducing deer

TABLE 15.1

Impact Levels on Five Indicator Species for Two Landscapes: One with Low Deer Density and Low Forage Production, and One with Higher Deer Density and Higher Forage Production Level

	Low Deer Density, Low Forage		High Deer Density, High Forage	
Deer/km²	9.2	3.4	13.4	7.8
Impact level by species	Pre-2004	Post-2005	Pre-2004	Post-2005
Red maple: zero–moderate[a]	0.5	8.8	2.1	9.6
Red maple: heavy–severe[a]	0.7	1.6	2.6	5.6
Hemlock: zero–moderate[a]	0.2	1.5	0.3	3.7
Hemlock; heavy–severe[b]	0.2	1.0	0.3	0.7
Striped maple: zero–moderate[a]	5.8	8.9	3.6	6.2
Striped maple: heavy–severe[b]	1.3	0.7	1.4	1.2
Black cherry: zero–moderate[a]	1.6	4.7	3.3	13.0
Black cherry: heavy–severe[b]	0.0	0.3	1.1	1.0
Beech: zero–moderate[a]	13.2	20.9	18.3	16.3
Beech: heavy–severe[c]	3.8	1.4	2.9	2.2

Note: Impact data are percent plots per impact category per species by period.

[a] Percent plots zero–moderate and heavy–severe increase significantly (P < 0.05) when deer density was reduced.

[b] Change in percent plots zero–moderate and heavy–severe not significant (P > 0.1) when deer density was reduced.

[c] Change in percent plots zero–moderate and heavy–severe decreased significantly.

density while forage production was stable is an expected result, especially with preferred browse species (red maple and hemlock).

The shift in preponderance of plots from heavy–severe impact to plots with zero–moderate impact is another indicator of the reduction of deer impact on regeneration following reduction of deer density. Reduction in deer impact as deer density dropped was observed for all but striped maple heavy–severe impact within both landscapes and beech zero–moderate impact on the high deer density–high forage production landscape and on zero–moderate and heavy–severe impact levels within both landscapes.

Comparison of impact levels between the high forage/high density landscape and the low forage/low density landscape prior and subsequent to density reduction lends support to the hypothesized increase in carrying capacity (regeneration) related to higher forage production. Percent plots with indicator seedlings at zero–moderate and heavy–severe impact levels was consistently and significantly (P < 0.05) higher on the high deer density landscape for preferred forage species (red maple, hemlock, and black cherry) than on the low deer density landscape prior to and following reduction in deer density. Results were mixed for less-preferred species (beech and striped maple), where percent plots with these indicator species were sometimes lower and sometimes higher on the low-density landscape prior to and following population reduction. These latter differences may have been related to the higher prevalence of preferred species in the high density/high forage area wherein with a higher proportion of preferred browse species may have resulted in reduced browsing by deer on less-preferred species.

Thus, reducing deer density generally may result in an increase in abundance of seedling species within landscapes of differing levels of forage production—regeneration carrying capacity definitely increases with reductions in deer density below MSY carrying capacity. Landscapes with higher forage production may be able to support higher deer densities and continue to provide advance regeneration of desirable tree species with little, if any, loss in amount of forage available (in the form of tree seedlings within browsing reach of deer).

However, it should be noted that if landowners/stewards are practicing uneven-aged management using small (group selection) or single-tree openings (single tree selection), they will not be able to create enough early succession habitat to provide additional forage and permit an increase in deer density related to impact on tree regeneration.

15.2.9 Increasing "Trophy" Deer Characteristics by Increasing Forage

Changes in characteristics of deer harvested and brought to checking stations from the landscapes with (assumed) high and low forage prior to and following reduction in deer density support the contention that increasing forage is associated with increases in deer body weights and antler characteristics. Field-dressed body weights and antler characteristics were drawn from male and female deer >2½ years old—numbers of fawns and yearling does harvested were too low for comparisons and changes in harvest regulations during the demonstration period. Data were analyzed for two periods: prior to deer density reduction (2001–2003); and for 5 years (2008–2012) after deer density had been stable for 4 years.

Weight and antler characteristics were significantly ($P < 0.05$) higher after density reduction on both landscapes, indicating that reducing deer density while maintaining stable forage production results in increases in deer weight and trophy (antler) characteristics (Table 15.2).

Weight and antler characteristics for buck and doe deer were not significantly different ($P > 0.1$) between high-density/high-forage and low-density/low-forage landscapes prior and subsequent to density reduction. This result suggests that weight and antler characteristics are affected by forage availability and that the effect of higher deer densities on deer weight and antler characteristics can be offset by increasing forage availability. Deer weight and antler characteristics increased as deer densities dropped below MSY on both landscapes and were higher at regeneration carrying capacity than at MSY. Therefore, the notion that one need not drive deer density below that at MSY to obtain trophy deer is not supported by these data—deer weighed more and had better antler characteristics at densities lower than those at MSY.

15.2.10 Increasing Diversity Carrying Capacity

As indicated in Chapter 1, deer and forest communities evolved together when deer densities hovered around 4–5 deer/km^2 and were controlled by predation and weather. Studies in the Heart's Content

---

TABLE 15.2

Deer Body Weights and Antler Characteristics from Landscapes with High Deer Density/High Forage Production and with Low Deer Density/Low Forage Production Prior to and Subsequent to Reduction in Deer Density

Comparisons[a]	Deer/km^2	♀ weight[b]	♂ weight[b]	Beam[c]	Spread[d]	∑ points
High Pre-2004	13.4	43.0	55.5	25.6	32.2	6.7
High Post-2007	7.8	50.6	63.1	29.6	34.8	7.6
Low Pre-2004	9.2	42.1	52.0	25.4	32.4	6.8
Low Post-2007	3.4	47.1	61.3	28.9	34.6	7.7

[a] High pre-2004 = area with high deer density, high forage production, prior to population reduction; high post-2007 = area with high deer density, high forage production, after population reduction; low pre-2004 = area with low deer density, low forage production prior to population reduction; low post-2007 = area with low deer density, low forage production after population reduction.

[b] Kg.

[c] mm diameter at antler base.

[d] cm.

old-growth area within the Allegheny National Forest (Lutz 1930, Rooney and Dress 1997) indicated that shrub and herb species declined 59%–80% within different areas of the old-growth stand in the 1929–1996 span. During this time, deer density increased from approximately 0–6 deer/km² at the turn of the century to the rapid increase beginning around 1920 to a somewhat stable 10 deer/km² in the 1935–2000 span in the Allegheny National Forest (Redding 1995).

Following introduction of liberal antlerless deer hunting seasons and concurrent buck-doe seasons, deer density was reduced from about 10 deer/km² to about 5 deer/km² in the KQDC demonstration area of the Allegheny National Forest and maintained at that level for the following 7 years (Royo et al. 2010). Reduction in deer density to presumed pre-European colonization levels was not followed by an increase in species richness or abundance of forbs. However, reproductive status and plant vigor increased for three forb species trillium (*Trillium sp.*), Canada mayflower (*Maianthemum canadense*), and Indian cucumber-root (*Medeola virginiana*) not extirpated by decades of overabundant deer browsing improved significantly, indicating that forbs can flourish when deer density is returned to levels with which the species coevolved with deer.

Clearly, decades-long deer density exceeding 5 deer/km² can eliminate many forb species that are limited in the height they can reach and thus are constantly exposed to deer browsing. Once locally extirpated, forbs have no potential for recovery even after deer density has been reduced to levels associated historically with diverse and flourishing forbs species sensitive to deer browsing, as demonstrated by Royo et al. (2010)—sources of propagules are absent. Increasing availability of deer forage in the hopes of reducing deer impact on forbs while increasing deer density above the level historically associated with coexistence of forbs will not work in the absence of parent plants to produce seed for future generations. Increasing forage availability to permit higher deer densities without increasing deer impact may not work, as deer may preferentially browse on exposed and nutritious herbaceous vegetation even in the presence of increased alternative sources of forage. There is no scientific evidence that increasing forage levels can increase deer density associated with diversity carrying capacity and result in restored species richness and abundance of the forb layer.

Indeed, Royo et al. (2010) and Pendergast et al. (2016) voiced caveats against efforts to restore forb species extirpated by browsing by overabundant deer herds. Species eliminated by deer, by nature of their limited seed dissemination distances and likely extirpation from the surrounding landscape, might require herculean efforts to reintroduce and protect them for re-establishment even in the presence of reduced deer density.

15.2.11 Can Food Plots or Winter Deer Feeding Increase Carrying Capacity?

Although there are no studies evaluating whether increasing deer forage production with food plots or supplemental winter feeding can increase diversity and other carrying capacities, observations of both situations as well as a third wherein foresters attempted to overwhelm the local deer herd by creating a 200-ha clearcut are instructive. In the first instance, deer had been fed approximately 10 tons of shelled corn annually on a small (<1 ha) area for 10+ years in a northern hardwood forest landscape on the Frost Valley YMCA Camp in the Catskill Mountain area of New York State. Over 100 deer annually fed on the corn. The ground was compacted by the deer and completely covered with deer pellets. There were virtually no understory herbaceous or woody plants more than 350 m in any direction from the feeding grounds (personal observation, David S. deCalesta), save beech seedlings, which are resistant to deer browsing.

The second situation was a 2-ha food plot planted with clover (*Trifolium* sp.), brassica (*Brassica oleracea*), and chicory (*Cichorium intybus*) within a 200-ha forested quality deer–forest management area in southwestern New York. Deer density was approximately 12 deer/km². Advance regeneration (seedlings of sufficient height and vigor likely to advance to maturing trees) for chestnut oak (*Quercus prinus*) within a 50-m radius of the food plot was heavily browsed by deer (personal observation, David S. deCalesta). Winter feeding areas such as winter feeding grounds and food plots concentrate and hold deer for extensive periods of time, and the locally high deer density can result in excessive deer browsing on native vegetation in the surrounding landscape (Cooper and Ginnett 2006).

The third situation occurred within a 200-ha clearcut in a northern hardwood forest in north central Pennsylvania (personal observation, Jeffrey Kochel, forest land manager). Managers attempted to overwhelm the local deer herd with the large amount of forage created by the large clearcut—figuring that the deer would be unable to suppress regeneration by seedlings existing on the site prior to the clearcut. The deer *were* overwhelmed, but unfortunately there was a large seed bank of pin cherry existing on site prior to the clearcut. Pin cherry is a shade-intolerant tree species with seeds that can exist for 50+ years on a site and delay germination until a disturbance, such as clearcutting, occurs that opens up the overstory and stimulates germination. In this situation, local deer were unable to forage on sufficient numbers of germinated pin cherry to prevent that species from overtaking and dominating the understory and crowding out more desirable seedlings of black cherry and red maple. This example makes the point that deer play an important ecological role in regulating species composition of the understory and that there can be too few as well as too many deer for balance between deer and other forest resources.

15.3   INTERPRETATION AND APPLICATION

Forestland managers must determine which stakeholder group they are responsible to for managing forest resources, including deer, and attempt to manage deer density and forage availability accordingly. If more than one stakeholder group has a controlling interest in how deer and other forest resources are managed, managers have the difficult task of striking a balance among desires and values of competing stakeholders and levels of deer density and forage availability for which they are to manage.

If the predominant stakeholder values include regeneration of commercially valuable timber or optimizing number of trophy deer or deer density, managers should attempt to increase regeneration and MSY carrying capacities by increasing forage production, but only with judicious placement of 5–10 ha-sized clearcut/thinned areas distributed evenly across forested landscapes instead of concentrating them in large, isolated patches. Furthermore, such increased production of foraging areas must be maintained on a recurring basis and cannot result in exceeding a sustainable level of timber harvest (clearcutting and thinning). Efforts to reestablish extirpated forbs, even in the presence of heightened forage availability and reduced deer density, should wait until research provides guidance.

Finally, the quantitative values presented for deer densities associated with varying carrying capacities were drawn from research in northern hardwood forests under even-aged silviculture. Development of management strategies for deer and other forest resources under uneven-aged management and/or in other forest types, as identified in Chapter 3, should be based on research conducted in those forest types, under even-aged and uneven-aged management, if possible, as carrying capacity values are dependent on weather, soils, moisture, temperature, silviculture practice, and other factors as they affect and determine vegetation characteristic of the different forest types.

REFERENCES

Alverson, W. S., D. M. Waller, and S. L. Solheim. 1988. Forests too deer: Edge effects on northern Wisconsin. *Conservation Biology* 2:348–358.

Bishop, C. J., G. C. White, D. J. Freddy et al. 2009. Effect of enhanced nutrition on mule deer population rate of change. *Wildlife Monographs* 172:1–28.

Caughley, G. 1979. What is this thing called carrying capacity? In *North American elk: Behavior and management*. ed. M. S. Boyce and L. D. Hayden. Laramie WY: University of Wyoming.

Cooper, S. M. and T. F. Ginnett. 2006. Effect of supplemental feeding on spatial distribution and browse utilization by white-tailed deer in semi-arid rangeland. *Journal of Arid Environments* 66:716–726.

Crimmins, S. M., J. W. Edwards, W. M. Ford et al. 2010. Browsing patterns of white-tailed deer following increased timber harvest and a decline in population density. *International Journal of Forestry Research* 2010:ID592034.

Dahlberg, B. L. and R. C. Guettinger. 1956. *The white-tailed deer in Wisconsin. Technical Bulletin 14*. Madison WI: Wisconsin Conservation Department.

Dasmann, R. F. 1964. *Wildlife biology*. New York: MacMillan Co.

D'Angelo, G. J. and M. D. Grund. 2015. Evaluating competing preferences of hunters and landowners for management of deer populations. *Human-Wildlife Interactions* 9:236–247.

deCalesta, D. S. 1994. Impact of white-tailed deer on songbirds within managed forests in Pennsylvania. *Journal of Wildlife Management* 58:711–718.

deCalesta, D. S. 2013. Unpublished data from 2012 annual KQDC Report to KQDC Leadership team.

deCalesta, D. S., Nagy, J. G., and J. A. Bailey. 1975. Starving and refeeding mule deer. *Journal of Wildlife Management* 39:663–669.

deCalesta, D. S. and S. L. Stout 1997. Relative deer density and sustainability: A conceptual framework for integrating deer management with ecosystem management. *Wildlife Society Bulletin* 25:252–258.

Decker D. J. and K. G. Purdy. 1988. Towards a concept of wildlife acceptance capacity in wildlife management. *Wildlife Society Bulletin* 16:53–57.

DeYoung, C. A. 2011. Population dynamics. In *Biology and management of white-tailed deer*. ed. D. G. Hewitt, 147–180. Boca Raton FL: CRC Press.

Ellingwood M. R. and J. V. Spignesi. 1986. Management of an urban deer herd and the concept of cultural carrying capacity. *Transactions of the Northeast Deer Technical Committee* 22:42–45.

Jenks, J. A., W. P. Smith, and C. D. DePereno. 2002. Maximum sustained yield harvest versus trophy management. *Journal of Wildlife Management* 66:528–535.

Lutz, H. J. 1930. The vegetation of Heart's Content, a virgin forest in northwestern Pennsylvania. *Ecology* 11:1–29.

McCabe, R. E. and T. R. McCabe. 1984. Of slings and arrows: An historical perspective. In *White-tailed deer ecology and management*. ed. L. K. Halls, 19–72. Harrisburg PA: Stackpole Books.

McCullough, D. R. 1979. *The George Reserve deer herd*. Ann Arbor MI: The University of Michigan Press.

Macnab, J. 1985. Carrying capacity and related slippery shibboleths. *Wildlife Society Bulletin* 13:403–410.

Marquis, D. A., R. L. Ernst, and S. L. Stout. 1992. *Prescribing silvicultural treatments in hardwood stands of the Alleghenies (revised)*. U.S. Department of Agriculture Forest Service Northeast Forest Experiment Station General Technical Report NE-96.

Pendergast, T. H., S. H. Hanlon, Z. M. Long et al. 2016. The legacy of deer overabundance: Long-term delays in herbaceous understory recovery. *Canadian Journal of Forest Research* 46:362–369.

Pierson, T. G. and D. S. deCalesta. 2015. Methodology for estimating deer browsing impact. *Human-Wildlife Interactions* 9:67–77.

Porter, W. F. and H. B. Underwood. 1997. Case history: Deer populations are [sic] Saratoga National Historical Park. In *The science of overabundance: Deer ecology and population management*. eds. W. J. McShea, H. B. Underwood, and J. H. Rappole, 185–200. Washington: Smithsonian Institution Press.

Redding, J. R. 1995. History of deer population trends and forest cutting on the Allegheny National Forest. In *Proceedings, 10th Central Hardwood Forest Conference. Morgantown, WV. General Technical Report NE-197*. eds. K. W. Gottschalk and S. L. C. Fosbroke. Radnor PA: U.S. Department of Agriculture, Forest Service, Northeastern Forest Experiment Station.

Rooney, T. P. and W. J. Dress. 1997. Species loss over sixty-six years in the ground-layer vegetation of Heart's Content, an old-growth forest in Pennsylvania, USA. *Natural Areas Journal* 17:297–305.

Royo, A. A., S. L. Stout, D. S. deCalesta et al. 2010. Restoring forest herb communities through landscape-level deer herd reductions: Is recovery limited by legacy effects? *Biological Conservation* 143:2425–2434.

Tilghman, N. G. 1989. Impacts of white-tailed deer on forest regeneration in northwestern Pennsylvania. *Journal of Wildlife Management* 53:524–532.

Section II

Planning and Assessment

The most serious mistakes are not being made as a result of wrong answers. The truly dangerous thing is asking the wrong question.

Peter Drucker (2010)

Section I chapters provide managers with background information for setting goals and developing plans for managing deer and other forest resources. Integration of biological factors with the people management side of deer management (human dimension factors) is an essential dimension of comprehensive deer management. Before management plans can be developed and enacted, the status of resources and relationships defined by goals must be assessed as benchmarks to determine whether active management is required, and, if so, for which components. Monitoring is required to establish baseline status of identified components and to determine whether progress toward achieving those goals is occurring.

Goal statements should be established for managed resources, and associated management actions required to make progress toward achieving the goals must be identified. Chapters in Section II outline the process for defining goals, establishing objectives for reaching the goals, assessing status of objectives and goals, and identifying monitoring parameters required to ascertain progress toward achieving goals. Wherever possible, quantitative standards rather than subjective valuations should be identified as objectives by which progress towards goals can be assessed. Rather than setting as an objective "Increase understory biodiversity," a quantitative objective could be, "Increase species richness and abundance of songbird community significantly." Instead of setting as an objective "Reduce deer density," a quantitative objective could be, "Reduce deer density to below 6 deer/km^2." Setting as an objective, "Increase percent plots with zero to moderate deer impact on seedlings of desirable species to greater than 50%" is more quantitative and specific than, "Reduce deer impact on seedling regeneration."

As with many natural resource situations, goals for individual components within deer management plans must be mutually compatible.

Monitoring includes: deer density, health, and sex and age composition; deer impact on forest resources; forest vegetation changes; financial and human resource costs; and stakeholder (hunters and others) concerns. Methodologies for monitoring status of management components are presented in Chapter 17. Topics include parameters sampled; sampling regimes; and recording, analyzing, and interpreting monitoring data.

Like techniques chosen to mitigate negative deer impacts, monitoring techniques must be economical and practical, be of proven capability to determine quantitatively whether objectives have been achieved, and should not require extensive training or expertise (e.g., can be performed by landowners and managers) or specialized and expensive equipment.

Manager's Summaries are not included for chapters in Section II, as managers need more than a passing knowledge of these components of deer management—they are absolutely essential and must be understood and practiced in depth.

REFERENCES

Drucker, P. F. 2015. *Men, Ideas, and Politics* (Drucker Library). Cambridge MA. Harvard Business Review Press.

16 Goals, Objectives, and Management Plans

David S. deCalesta

CONTENTS

> People with goals succeed because they know where they are going.
>
> **Earl Nightingale (2007)**
>
> If you don't know where you're going, how will you know when you've got there?
>
> **Leigh Ashton (2013)**

16.1 INTRODUCTION

Managers identify the forest resources they want to manage and what they want these resources to look like. Answering fuzzier questions such as, "How many?" or, "What is the purpose—income, recreation, aesthetics or other?" or, "What financial and manpower resources are required to manage them?" requires setting goals and identifying objectives and the associated management steps to achieve them. Hovering over all these questions is the primal one—how will deer affect the ability to achieve goals for other forest resources, and what must be done to address deer impacts?

As we learned in Chapter 8, deer can affect the status of species within plant and animal forest communities, and vice versa. Status of deer, other wildlife, and plant communities can influence achieving goals for forest resources and must be reflected by goals for managing deer.

Managers' goals must integrate desired deer status with the status of other goals, like biodiversity or sustainable timber harvest. Some natural resource goals may conflict with others (e.g., enhancing and maintaining diversity of understory forest vegetation while maximizing deer density for hunting) and be mutually exclusive. Some deer densities are in harmony with achieving goal(s) for other forest resources, but more likely, they make achieving other goals impossible without first reducing deer density. Chapter 15 identifies deer densities compatible with goals for natural resources, and in almost all cases, that density ($\leq$4–5 deer/km^2) corresponds to deer density under natural conditions before the country was developed.

Managers of larger forestlands may establish multiple goals that require managing for different deer densities. For example, within a large forest managed primarily for producing timber, there could be special habitats harboring unique and possibly threatened species, such as wetlands with rare aquatic plants sensitive to deer browsing. The timber-specific landscape could be managed for a higher deer density than the wetland, requiring aggressive deer harvest in the landscape enclosing the wetland. Or, the wetland could be protected from deer browsing by enclosing the entire area within a deer-proof fence. Managers for two of the case histories (Chapters 33 and 34) managed for different deer densities on different parts of the landscape within their boundaries as related to differing management goals or different habitat conditions.

16.2 GOALS, OBJECTIVES, AND DEER DENSITY

As identified in Chapter 11, managers must address and integrate values and goals of varying stakeholder groups that affect deer management. Achieving goals and associated objectives requires managing deer density specific to goals of stakeholder groups. Once the goal for deer density is established, other goals related to managing vegetation composition and structure may be developed and conducted with some assurance they will be successful.

Because different stakeholder groups have differing goals for deer management, identification of such goals and associated objectives is required so that managers can manage for the appropriate deer density. Goals, objectives, requisite deer density, and supportive management activities for individual stakeholder groups are identified in Table 16.1.

16.2.1 Simple Management Areas

The majority of managers' properties will be managed for one set of goals with associated objectives. An example would be a forest landowner whose primary goal is growing timber for sale with an associated objective of managing for deer density in the ≤ 8 deer/km^2 range to establish and ensure progression of advance regeneration to mature trees. Using public hunting by alpha and beta hunters (or locavore hunters) to achieve desired deer density could be met with the same objective: obtaining and maintaining deer density in the range of ≤ 8 deer/km^2.

16.2.2 Complex Management Areas

Large management areas such as national forests or forestland cooperatives composed of multiple abutting landowners are examples of diverse landscapes wherein there may be multiple natural resource goals and objectives. Such landscapes may be managed for more than one deer density with similarly different management actions. These individual submanagement areas (call them "special management areas") must be clearly demarked to facilitate hunter identification and use of these areas to ensure increased deer harvest occurs where desired. Because of the likelihood that deer from forestland abutting special management areas will move onto these areas, managers need to keep close tabs on deer density and impact within the areas and direct hunters to these border areas to minimize deer infiltration by identifying these areas as "hot spots" for hunters to intensify hunting effort. If possible, obtaining regulations permitting additional, postseason deer damage hunts in these areas is an effective management tool (deCalesta 1985).

Aggregating abutting forestlands into cooperative management areas (Chapters 31 and 34), as recommended by the Quality Deer Management Association (QDMA) (Chapter 27), for managing deer within larger landscapes provides more flexibility and sharing of financial and human resources.

16.3 DEVELOPING A MANAGEMENT PLAN

The following Chapters 17 through 26 describe biological and human dimension factors that must be compatible and integrated as essential components in deer–forest management plans used to achieve deer management goals and objectives. The case histories in Section V describe how managers from a wide variety of deer–forest management areas (DFMAs) (see Chapter 9) identified and integrated these factors in developing and executing successful deer management operations. With

TABLE 16.1

Representative Stakeholder Goals, Objectives, Requisite Deer Densities, and Supportive Actions for Deer Management

Stakeholder	Goal	Objective(s)	Deer Density	Supportive Actions[a]
Alpha, beta and QDM hunters	Quality hunt[b]	Optimize deer quality, control hunter access	≤8 deer/km^{2,c}	Antlerless permits, antler point restriction
Omega hunters	Good old days hunting[c]	Optimize deer density	>15 deer/km^2	No antlerless harvest, all antlered deer legal
Locavore hunters	Obtain source of wild protein	Healthy, abundant deer herd	≤8 deer/km^2	Antlerless permits, special seasons for locavore hunters
Timber growers	Sustainable harvest of desirable tree species	Successful advance regeneration	≤8 deer/km^2	Antlerless permits, special hunts
Lease hunters and owners of leased lands	Quality hunting and sustainable timber harvest	Optimize deer quality and advance regeneration	≤8 deer/km^2	Antlerless permits, antler point restriction
Small woodlot owners	Sustainable timber harvest, enhance biodiversity	Successful advance regeneration	≤6 deer/km^2	Antlerless permits, deer damage hunts, deer-proof fencing
Park managers	Optimize forest benefits for all stakeholders	Produce diverse, abundant plant and animal communities	≤6 deer/km^2	Antlerless permits, special hunts[d], sharpshooters and/or deer-proof fencing
Residents in forested communities	Safe, tick-free environment, low impact on landscaping	Minimal deer for minimal impact	As few deer as possible	Special hunts[d] and/or sharpshooters and/or deer-proof fencing
Farmers adjacent to forests	Produce variety of crops at minimal cost	Minimal deer impact	<6 deer/km^2	Antlerless permits, special damage hunts
Environmentalists	Balanced, sustainable ecosystems	Abundant, diverse plant and animal communities	<6 deer/km^2	Antlerless permits, special hunts, deer-proof fencing
Natural resource administrators	Professional, noncontroversial management of natural resources	Satisfy all stakeholders	Specific to deer–forest management area	Enact regulations to allow landowners to control deer density

Source: Adapted from deCalesta, D. S., R. Latham, and K. Adams. 2016. In *Managing oak forests in the Eastern United States.* ed. P. D. Keyser, T. Fearer, and C. A. Harper. Boca Raton FL: CRC Press.

[a] Hunting regulations related to sex and antler characteristics of deer legal for harvest.

[b] Opportunity to hunt in habitat supporting trophy/quality deer with low hunter density.

[c] Deer hunting typical of 1930–1970 era—high deer density, shared camaraderie in hunting camps with friends and family, excitement for young hunters by seeing many deer, primarily hunting for antlered deer.

[d] Special hunts designated for identified properties, at specified times, and using public hunting and/or deer removal by sharpshooters.

one exception. Managers for the unsuccessful case history (Chapter 40) wished to manage for MSY and sustainable timber production. They ignored the advice of consulting deer and forest managers regarding deer biology and silvicultural practices related to producing deer habitat and regeneration of tree seedlings and were unable to produce high deer density, quality deer, or sustainable tree regeneration.

REFERENCE

Ashton, L. 2013. http://leighashton.wordpress.com/2013/11/20/if-you-dont-know-where-youre-going-how-will-youknow-when-youve-got-there%E2%80%8F/

deCalesta, D. S. 1985. Influence of regulation on deer harvest. In *Symposium on game harvest management.* ed. S. L. Beasom and S. F. Roberson, 131–138. Kingsville TX: Texas A&I University Press.

deCalesta, D. S., R. Latham, and K. Adams. 2016. Managing deer impacts on oak forests. In *Managing oak forests in the Eastern United States.* ed. P. D. Keyser, T. Fearer, and C. A. Harper. Boca Raton FL: CRC Press.

Nightingale, E. 2007. *The essence of success.* Chicago. Nightingale-Covenant corporation.

17 Monitoring

David S. deCalesta

CONTENTS

> You can't manage what you can't measure.
>
> **Attributed to Peter Drucker (2010)**
>
> In God we trust, all others must bring data.
>
> **Anonymous**

17.1 INTRODUCTION

There are characteristics of local deer herds and forest resources impacted by deer that are of interest to stakeholders and of critical knowledge for landowners and managers of forestlands. Measuring these characteristics and components is essential for determining whether proactive management steps are required to rectify undesirable deer impacts, whether designed management actions for resolving identified problems are effective, and for keeping stakeholder groups properly informed. Prior to enacting management steps, baseline conditions of forest resources, including deer density and condition, and status of other forest resources of management interest must be assessed and described to guide deer management.

Unfortunately, as discovered by Shissler and Grund (2009) in their survey of deer management programs by state agencies in the continental United States: (1) most states did not have quantitative goals for deer management programs; (2) the states that did have quantitative goals used population density goals not centered on monitoring and managing impacts caused by deer; (3) only two states indicated that forest vegetation or ecologically based goals were established for statewide deer management decision-making; managers for three others collected forest vegetation data but did not have a quantifiable goal to measure their data against; (4) a limited number of states considered the effects deer have on habitat at the landscape level or that the effects might conflict with agency mission statements; (5) no states had specific, quantitative goals related to impacts deer have on forest vegetation or forest ecosystems; (6) it was common for biologists to indicate that impacts deer had on forest vegetation were not an issue (without having data to support their position); (7) most state wildlife agencies focused on monitoring the production of deer to a greater extent than monitoring what deer depend on and the impacts deer have on ecosystems; (8) few (five) states had formal definitions of overabundant deer; (9) most states did not have an established policy to determine what overabundant deer should mean to the public and how deer would be managed if they were deemed overabundant; (10) most states did not invest much money or staff time into collecting data that would be used to make science-based decisions; (11) many deer biologists relied on anecdotal observations through hunter discussions and "gut feelings" while making deer management decisions; and (12) many states used observations collected opportunistically without an experimental approach. Shissler and Grund (2009) were dismayed to learn that some biologists indicated that crop damages were stable or decreasing even though they did not collect any data (including monitoring damage permits) that could support their position. Shissler and Grund (2009) concluded that most deer programs operated with little or no data and that much of deer management tended to operate from values and not science.

However, not to be lost in this sea of disappointing findings is the fact that Shissler and Grund (2009) surveyed state agency biologists instead of deer managers for deer–forest management areas (DFMAs),- (see Chapter 9)—the ones tied to specific forested landscapes they were actually managing by manipulating vegetation and deer numbers. As clarified in Chapter 9, with the exception of landscapes specifically administered and managed by state natural resource agencies, stage agency biologists/managers do not manage deer habitat or deer abundance as tied to specific landscapes and so have neither incentive nor resources to monitor deer density or impact on deer administratie units (DAUs—the large administrative units for which deer harvest regulations are identified, see Chapter 9). Actual management and respective monitoring can only occur with any assurance of relevance and specificity on DFMAs on private and public forestlands. Without DFMA-specific goals to evaluate by monitoring, monitoring of deer density, health, or impact within DAUs is meaningless and immensely difficult to administer, fund, characterize, evaluate, and provide specific management activities for, especially regarding habitat.

17.2   REQUISITE DEER PARAMETERS FOR MONITORING

Common hunter requests for information are: deer density, deer condition, and size of deer–forest management areas (to guarantee consistent condition of forest vegetation, deer density and health, and hunting regulations). Managers seek information on the level of deer impact on forest resources, some with emphasis on status of tree seedling regeneration, others with emphasis on status of herbaceous and shrub vegetation, especially vulnerable plant species threatened by deer browsing.

Additionally, managers involved with deer hunting and deer hunters need feedback from hunters regarding their concerns and satisfaction about hunting on forestlands.

Some state natural resource agencies, such as the Pennsylvania Game Commission (PGC; Rosenberry et al. 2009) maintain that, "Deer management objectives are no longer defined by deer densities. Instead, deer management objectives are defined by measures of deer health, forest vegetation health, and deer-human conflicts." Moreover, Rosenberry et al. (2011) cited assertions by population biologists (Caughley 1977, Hayne 1984, McCullough 1984, Morellet et al. 2007) in stating that, "The position of PGC wildlife managers that knowing the number of deer in a wildlife management area (WMU) is not needed to have sound deer management is consistent with findings from decades of wildlife research and management experience from around the world."

There is a major disconnect of philosophies and management strategies between those of forest managers of individual public and private forest properties (DFMAs) and those of agency biologists and population modelers such as those cited above. The disconnect is directly related to landscape and time aspects of deer management (deCalesta 2017). Agency personnel and modelers relate to deer management over large landscapes (e.g., statewide or regional DAUs) that encompass multiple small and large forestlands of multiple ownerships with multiple deer and forest management goals and objectives wherein no "one size fits all." Their views and philosophies reflect the interests of agency managers as related to landscape size (large). Similarly, the time frame of agency personnel and modelers is usually annual, whereas managers of individual forestlands must deal with long-term and accumulative realities of deer density and impacts.

As described in Chapter 9, actual deer management is practiced on landscapes under individual or joint ownership as defined forestlands managed as single units (DFMAs). Such forestlands include private lands owned by managers, public lands such as parks and state and national forests, and multiple forestlands owned by abutting landowners who manage deer under a unified management plan.

State natural resource agencies define deer management units and regulate deer hunting specific to such units. However, deer management within properties located in these units is practiced by individual landowners or public agencies on smaller landscapes embedded within deer management units. Deer management within these areas is guided by goals for forest resources established by the managers: these goals may not be coordinated with or similar to those developed by state natural resource agencies, such as on state forestlands, within the same management units.

State natural resource agencies rarely evaluate deer density or impact (or deer and habitat "health," for that matter) on DAUs, as the cost is prohibitive and results are highly variable due to differing levels of management on embedded management areas. It is understandable that state natural resource agencies would decline to measure deer density, impact, or health on management units because of costs and variability, but it is disingenuous to state, as some do, that it is not necessary to estimate deer density or impact for management purposes. Granted, it would be a waste of time and money on DAUs, as most state natural resource agencies do not manage forest vegetation or deer on management areas within management units (excepting on state forest and state park agencies). However, it is important for managers to monitor deer density and impact (Curtis et al. 2009, Waller et al. 2017) often on an annual basis (Morellet et al. 2007): what other sources of information are available to help them determine whether their management efforts are succeeding or in need of adjustment (Lancia et al. 1994, Rutberg and Naugle 2008)? Additionally, when managers apply to state natural resource agencies for permits to harvest antlerless deer and/or for extended hunting seasons on their respective properties, they often must document quantitatively the need for additional deer reductions. Without quantitative estimates of deer density and impact, how can they justify the need for additional removals of deer from their properties?

Removing deer density as a deer management objective eliminates a major controversy of deer management faced by state natural resource agencies. However, eliminating the association of deer density with level of impact on forest and other resources introduces ambiguity and nullifies cause-and-effect analysis. Knight et al. (2009) stated that conserving understory plant populations requires quantifying a sustainable level of deer herbivory, noting in addition that most population projection models consider only deer presence and absence in evaluating their impact on herbaceous and other vegetation.

Without estimates of deer density associated with status of other resources, it may be argued that deer are not the causal agent for negative impacts on these resources, thus clouding the process for correcting undesired conditions of these other resources. For example, it was asserted by some professionals (Mulhollem 2002) in the northeastern United States that failure of seedling regeneration was caused by acid rain, rather than by deer overbrowsing. No estimates of deer density were available in the areas with affected vegetation. However, vegetation outside fenced deer exclosures exhibited reductions in species richness and abundance of understory vegetation not matched by conditions inside the exclosures. It was concluded that acid deposition did not somehow exclude vegetation within isolated, rectangular patches and that deer likely were the agent of impact.

17.3   WHEN AND WHERE TO MONITOR

17.3.1   WHERE TO MONITOR

Where should monitoring be conducted? For monitoring to represent deer impact across landscapes, a system for collecting data from plots along transects located representatively across landscapes is essential. Arranging parallel transect lines in groups (grids) located randomly across landscapes ensures that random and representative samples are obtained. Placing transect lines >300 m apart ensures independence of data: Pierson and deCalesta (2015) determined that deer density and impact estimates between adjacent parallel transect lines spaced 300 m apart provided independent estimates of density and impact. Additionally, if composition and vertical structure of vegetation vary considerably across landowner landscapes and are known (e.g., different successional stages resulting from timber management; different habitats such as forest, grassland, wetland) grids of transect lines placed within these different habitats provide more representative estimates of deer density across the landscape as well as providing the basis for determining whether density and impact are significantly different among grids/habitats. Instructions for locating and placing transect grids are presented in Appendix 1.

17.3.2   WHEN TO MONITOR

When and how often should monitoring be conducted? To gauge success of management steps taken to reduce deer density and impact, these parameters should be monitored at least every other year if not yearly after initial baseline monitoring is conducted. The best time is in spring, when indicators of deer density and impact on seedlings and herbaceous vegetation are not obscured by other vegetation such as ferns. Deer density and impact data can be collected from the same plots. The best springtime period is after snow-melt (usually sometime in April) and before fern green-up and while herbaceous vegetation may still be in flower (usually late April to early June). To repeat, monitoring should be conducted in the years after application of corrective management steps are taken to reduce deer density and impact to determine whether changes need to be made in those steps. Monitoring of deer density and impact should be made a permanent and recurring component of deer management.

17.4   WHAT AND HOW TO MONITOR

Resources typically monitored on DFMAs include: deer density; deer impact on plant and animal species/communities; deer herd sex and age structure, recruitment, and health; stakeholder (including hunter) satisfaction with deer management; and costs of management. Data for establishing baseline conditions of these parameters must be collected over entire DFMA ownerships in a framework that representatively samples landscape position, forest vegetation, and forest and successional stages over seasonally relevant time frames (e.g., forb and seedling impacts in spring, herd sex and age and health characteristics in fall prior to winter severity, hunter satisfaction during hunting season). To track changes and establish cause-and-effect relationships, deer density and impacts must be monitored over years during which changes in density and impact may occur.

Measures of deer density and impact must provide information that produces unequivocal assessment of progress toward, and attainment of, established natural resource goals. Most managers are not experienced foresters or wildlifers and most are limited by financial and human resources regarding selection and use of methods for monitoring deer density and impact. Ideally, parameters selected require little skill in identification of parameters and their condition, do not require extensive equipment, may be applied over entire ownerships to provide representative status of parameters, and require a minimum of personnel and time.

17.4.1 Deer Density

Wildlife biologists require density estimates for deer to facilitate management (Anderson et al. 2012, Urbanek et al. 2012): techniques chosen should produce unbiased estimates with acceptable precision (Beaver et al. 2014). As discussed in Chapter 15, impact level of deer on forest resources is directly related to deer density. Impact level on various categories of forest resources (e.g., tree seedling regeneration, abundance and species richness of wildlife and understory plant species, and deer herd health) can be predicted by level of deer density. There are several established deer density estimation techniques, each with associated costs and levels of confidence. Estimators of deer density include pellet group counts; photographs obtained with trail cameras, aerial counts including infrared imaging; deer drives, distance measures (visual counts along roads or transects), and deer herd sex and age composition obtained from harvest records.

For greatest utility, density estimates should be representative of DFMAs over time and space, quantitative, amenable to analysis, and proven as actual estimators of density rather than as trend estimators (deer density is going up, down, or remaining constant). Weaknesses common to most of the techniques include lack of measures of accuracy (densities obtained from techniques not tested against known deer densities); inherent bias; and collection over short time frames (<several days), thus producing only "snapshots in time" of deer density at the time of data collection subject to daily/annual differences in habitat occupancy by deer. If management areas are not sampled representatively over time and space, the resulting estimates may have little correlation with actual deer density.

17.4.1.1 Deer Drives

Deer drives were the first organized attempts to estimate deer abundance within defined areas. The idea was to estimate deer numbers within small, confined areas such as individual private forestlands or a portion of a public forestland (e.g., national or state forest). Drives are conducted within areas with boundaries that either prevent deer egress, such as a cliffs or fence lines, or that are crewed by a line of observers positioned sufficiently close to each other that deer exiting the area are observed and counted. Drives are usually conducted when viewing is best, after tree and shrub leaves have fallen in the fall and on clear days without precipitation. Because drive counts are conducted within a single day, they represent only deer present on the area censused: such estimates may differ drastically given weather condition or habitat use of deer on that particular day. However, if the area is totally enclosed by a deer-proof fence, or is an island or other geographic feature that deer cannot enter or leave, deer drives can provide accurate counts of resident deer. Deer drives require little in the way of equipment, but do require a large number of personnel to drive and count the deer.

An example is a deer drive conducted within the fenced 445-ha Glendorn Estate in north-central Pennsylvania (deCalesta and Witmer 1990). Approximately 50 observers lined up 15 m apart along the southern edge of the area enclosed by a 3-m-tall deer fence erected and well-maintained for decades (Figure 17.1). Additional observers were arrayed 100 m apart along parallel sides of the enclosed area to count deer exiting the fence line. Another line of ~50 observers was arrayed along the northern border of the end fence line to count any deer pressed by the moving line of observers to pass through or over the fence. Movement of the 50 observers traversing the area was coordinated by section leaders with walkie-talkies to keep the line straight and maintain correct spacing between observers. The drive took approximately 3 hours to conduct. No deer were counted exiting the fence along the parallel sides, one leapt through the fence at the northern end, and one broke its neck and died leaping into the fence at the northern end of the property. The number of deer counted was within one deer of the number estimated by a pellet group count (see "Pellet Group Counts," described below) conducted the week prior. A subsequent drive conducted 10 years later resulted in a similarly accurate count of deer when compared with a density estimate derived from pellet group count data (deCalesta 2013).

17.4.1.2 Pellet Group Counts

The pellet group technique was developed in the 1950s (Eberhardt and Van Etten 1956) and despite its detractors (e.g., Fuller 1991), it provides estimates of deer density accurately, quickly, representatively, and economically (deCalesta 2013). Pellet groups are counted within fixed-sized plots distributed along parallel transect lines arrayed in grids located to sample the area of interest representatively.

Estimates of deer density calculated from adjacent transect lines yield independent estimates of deer density when placed 300 m apart (deCalesta 2013). The number and length of transect lines are determined by the dimensions of the area of interest. In smaller areas (e.g., <300 m long), transect lines should be arranged such that they extend from one end of the area to the other. For larger areas (100+ ha), deCalesta (2013) determined that grids of five transect lines, spaced 300 m apart and located randomly across the areas, produced estimates amendable to statistical analysis for determining the significance of differences in density estimates among distinct locations within areas and among years. Instructions for locating transect lines/grids and collecting and analyzing deer pellet group data are detailed in Appendix 1. Advantages of pellet group counts, in addition to being accurate estimators of deer density, are: they require minimal training and associated equipment and personnel costs; they produce estimates representative of deer densities over periods of time (winter) when deer impact is high on vegetation: are more than "snapshots in time;" and they may be collected on the same plots at the same time as data for deer impact on understory vegetation (see Section A1.4.2, Appendix 1), providing considerable savings in time and costs of data collection.

FIGURE 17.1   Drivers lined up at beginning of deer drive.

An oft-mentioned concern with the technique is that pellet groups may deteriorate due to consumption by small mammals and insects or weathering (Wallmo et al. 1962), especially in southern states where freezing winters are rare and some pellet groups might disintegrate before data collection. Harestad and Bunnell (1987), evaluating pellet group deterioration in moist conditions of coastal Washington, noted that 16%–48% of pellet groups persisted, but the period they used for evaluation was 1 year, rather than the 3–4 months typically elapsing between deposition and counting (when, it must be added, weather is colder and pellet groups are less apt to deteriorate). In a pilot study testing this concern, deCalesta (2016) reported that of 50 pellet groups randomly located after leaf-off (mid-November) on a forested hillside in Tennessee, no pellet groups were removed and none had deteriorated when relocated and observed in March (pellet groups normally are counted in early spring before green-up and while ambient temperatures are still cool). One pellet group, located in a seasonal stream bed, was washed away, but pellet group plots typically are not located in streambeds or rivulets.

Another concern about the technique is the potential problem of confusing pellet groups deposited in one year with those from another year. This problem is avoided in eastern deciduous forests, as leaves falling in autumn cover pellet groups deposited in that year: pellet groups counted for estimating over-winter deer density are deposited on top of leaves covering groups deposited before leaf-fall. In forests dominated by coniferous trees without leaf-fall, the potential exists for confusing pellet groups from previous years (not covered by leaves), but these pellet groups deposited in year(s) prior display obvious signs of deterioration. Also, pellet groups from previous years generally are detectably lighter in color than pellets deposited in the current year (Hibert et al. 2011).

Calculations derived from pellet group counts provide estimates of deer/unit area such as deer/km^2 or deer/$mile^2$.

Pellet group counts are best suited for landscapes >100 ha. On smaller landscapes, pellet group counts may not be obtainable from abutting forestlands, which may harbor deer using the smaller landscape. Such counts restricted to smaller properties may underestimate the number of deer potentially impacting forest resources.

Pellet groups require little equipment or training and may be conducted quickly (two observers can count pellet groups on grids of five transects in about 4 hours). On extremely small landscapes(<50 ha), estimates of deer density are less useful as they do not account for deer in the landscape surrounding small forestlands and may not represent density/abundance of deer causing impact on forest resources. On these small landscapes, use of trail cameras may be the best technique to obtain information on deer abundance.

17.4.1.3   Trail Cameras

Cameras that can be affixed to trees and take pictures of animals have been used for years by persons capturing images of wildlife attracted to bait placed in front of the cameras. Many of the pictures are taken at night—the cameras are paired with infrared monitors to detect deer moving and flash attachments for taking night-time pictures. Wildlife scientists have described how to use trail camera pictures of deer to calculate deer density (Curtis et al. 2009, Hamrick et al. 2013, Jacobson et al. 1997), which appear best suited for smaller properties (<50 ha).

Moore et al. (2014), comparing numbers of marked buck and doe deer within enclosures with numbers obtained from trail cameras, reported that although the cameras "captured" >90% of marked buck deer and >84% of marked doe deer, calculations developed for the technique resulted in underestimations of number of deer known to be present in six 81-ha enclosures by an average of 32.2%, primarily attributable to higher sightability of bucks vs. does. When the technique was tested on an "open range" condition (2299 ha rangeland) comparing number of deer counted from 4399 photographs obtained from 29 trail cameras and number of deer observed in a helicopter (aerial) survey, trail cameras captured 2× as many antlered bucks obtained from the helicopter survey, and the associated estimated doe:buck ratio was 280% higher on the helicopter survey, likely due mostly to overcapturing of bucks (by trail cameras) and undercapturing of does. For the open range study, the scientists placed, on average, one trail camera for every 80 ha. Simply locating and visiting trail cameras to extract the photographs requires considerable expenditure of time. In order to get representative "snapshots" of deer, trail cameras need to be placed in representative portions of forestlands and maintained for lengths of time sufficient to capture a large sample. In large forestlands (e.g., over 500 ha) placing, maintaining, and collecting photographs taken by the large number of cameras to get representative coverage poses a challenge in addition to significant investment of financial and human resources. Theft of cameras is an additional problem, as is damage by wildlife such as bear, and the cameras are not cheap.

Trail camera photographs of deer have been used as selling points to demonstrate quality of (primarily) antlered deer to attract hunters to hunt on forestlands producing the images (Curtis et al. 2009, Koerth et al. 1997). If pictures are collected over time frames representing improvements in quality and quantity of forage provided for deer and/or reductions in deer density, photographs depicting increases in deer quality may be used as proof of the value of forage improvements/deer reductions (although collecting and reporting such data from deer measured at checking stations produces more quantitative and compelling information).

17.4.1.4 Aerial Surveys

Decades ago, aerial observation of deer from fixed-wing and helicopter aircraft was developed to estimate deer population density. Aircraft flew along established transect lines with one or more observers counting deer, generally within a fixed distance of transect midlines. The best conditions for observing deer included snow cover and little tree overstory to optimize observability of deer. Improvements in the technique included detecting deer with infrared photography capable of detecting deer under overstory cover by their body heat.

Potvin and Breton (2005) evaluated the accuracy of visual counts of two observers counting deer within 60-m-wide transects and counts obtained with infrared imagery. They concluded that although surveys with two observers could provide valid estimations of densities for management purposes, the observers missed an average of 25% of deer. They stated that because of closed forest canopy, thermal infrared sensing of deer along systematic survey lines was not a reliable technique. A general rule of thumb regarding aerial counts of deer on sites with continuous forest cover is that infrared imagery accounts for about 80% of true population density (Beringer et al. 1998, Williams et al. 2014).

Urbanek et al. (2012) compared deer density estimates obtained from aerial survey counts and deer pellet group counts and concluded that if researchers knew rates of pellet group decay, pellet group deposition rates, and collected a large number of pellet groups, then the pellet group technique had less bias, was not dependent on snow cover, had lower survey costs, and did not require special and costly equipment or trained wildlife biologists to conduct and analyze survey data.

Aerial counts are best used on large landscapes (>100 ha) and are expensive, generally well beyond the means of owners of smaller forestlands.

17.4.1.5 Distance Sampling

Distance sampling provides estimates of deer density by counting the number of deer observed from motor vehicles moving slowly along roads within areas of interest. Roadside counts are converted to density estimates using conversion factors incorporating distance to observed deer, length of routes, and observability of deer. Anderson et al. (2012) compared deer density estimates derived from distance sampling (spotlight counts at nights along roads) with pellet group counts and found that distance sampling consistently produced estimates higher than derived from pellet group counts, especially in forested environments (vs. agriculture/forestry habitats). They attributed higher overestimates from distance sampling in forested environments to location of roads used to obtain samples being along riparian corridors (routes were not representative of range of habitats including nonriparian habitats) where deer were more likely to congregate, resulting in biased estimates.

Beaver et al. (2014) compared density estimates obtained from vertical-looking infrared imagery with distance sampling using ground-based detection of deer with thermal imaging obtained along roads. Counts obtained from aerial infrared imagery were higher that counts obtained along roads: detection was higher on aerial counts. They recommended using aerial counts rather than roadside counts in spite of higher costs and did not recommend use of nonrandom road-based surveys for estimating deer density.

Collier et al. (2012) evaluated distance sampling to test the assumption that repeated surveys along roads provide reliable data for use in calculating deer density estimates. They concluded that, although recommended by many state, federal, and nongovernmental agencies, the benefit of distance data for monitoring deer populations is limited and likely represents a waste of resources with no appreciable management information gained.

Because estimates of deer density obtained from distance counts have not been tested against actual density (e.g., counts obtained from fenced deer enclosures where deer density was known), there is no measure of the accuracy or precision of such estimates. If the counts are collected from the same routes over time, and if observability of deer has not changed over the same time period(s), distance measures may be useful in terms of detecting trends in deer abundance (e.g., whether deer density/abundance has gone up, down, or remained the same). For managers with limited funding and human resources, distance counts do provide a qualitative measure of response of deer herds to efforts to reduce deer density and may be compared to differences (if any) in measures of deer impact on forest vegetation collected over the same time frame.

Distance sampling has minimal requirements for equipment and training of personnel.

17.4.1.6 Roadside Counts

A variation of distance sampling is simple roadside counts, usually conducted prior to hunting season. Deer are counted by a driver and one or more passengers along roads located representatively within forestland. When compared with density estimates among years, roadside counts seem highly correlated with actual estimates of deer density (Figure 17.2) and may serve as an index for tracking annual variation in deer density.

Like distance sampling, roadside counts have minimal requirements for equipment and training of personnel.

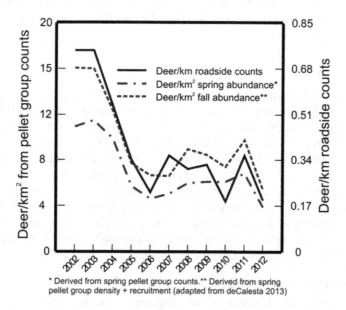

* Derived from spring pellet group counts.** Derived from spring pellet group density + recruitment (adapted from deCalesta 2013)

FIGURE 17.2 Comparison of deer density counts derived from pellet group counts and roadside counts.

17.4.1.7 Sex–Age–Kill Estimates

Eberhardt (1960) developed a model for estimating deer density using harvest data. Some of the data were obtained from hunters at deer checking stations and the majority from report cards sent in by successful hunters denoting sex and age (primarily fawns, yearlings, 2½ + year-old deer) of harvested deer. Termed the sex–age–kill (SAK) method and described by Creed et al. (1984) and Skalski and Millspaugh (2002), SAK uses sex and age data from harvested deer in a set of formulas that make certain assumptions about representativeness of the harvest (e.g., sex and age classes of harvested deer are reported proportionate to actual harvest rates). Many state natural resource agencies use the method to track changes in deer populations from year to year because the data represent deer populations from large areas (statewide and deer management units), can be collected easily at low cost (hunters are required to report harvest data), and require little field work in data collection (the hunters harvesting deer collect and send in the information). However, SAK estimates of deer density must be collected from large areas (DAUs and statewide) due to the requirement for a large sample of harvested deer: such numbers cannot be obtained from typical deer–forest management areas (small woodlots, private forests, even state and national forests). The Wisconsin Department of Natural Resources (2013) recommended, "Limiting the use of SAK style models to monitoring deer population size and trends at the state and regional levels" and that "estimates at the state level likely reflect actual conditions and this is the scale at which most other states report estimates of population size." According to Rosenberry et al. (2011), "Smaller WMUs (DAU in Pennsylvania Game Commission terminology) will not improve SAK estimates. Sample sizes needed to estimate populations will only increase with more and smaller WMUs. Without an increase in data, variation of population estimates will increase. Collecting sufficient data for a large number of small management units is often not possible for wildlife agencies." Also, only state natural resource agencies have the authority to require hunters to submit the harvest records required to make density estimates. For these reasons, SAK estimation of deer density is not a viable method for estimating deer density on forestlands of large and small private managers, commercial forests, forested residential developments, state and national forests, and state and national parks, all of which form the intended audiences of this book. It should also be noted that SAK estimates of deer density have never been compared with actual deer density such as obtained from aerial counts or pellet group counts.

Finally, McDonald et al. (2011) stated that legal harvest may not be the dominant form of deer mortality in developed landscapes and that estimates of populations or trends relying on harvest data will likely be underestimates.

17.4.2 Deer Impact

One of the most difficult tasks in deer management is characterizing and monitoring deer impact on other forest resources, including composition, abundance, and diversity of tree seedlings; composition, diversity, and structure of wildlife communities, shrub communities, and herbaceous communities; and understory vegetation dynamics. Determining the status of forest resources for which managers have goals requires monitoring technology that establishes baseline conditions and assesses progress toward and achievement of goals for individual resources. Part of the challenge lies in identifying components of each of the affected communities that can be monitored and determining how to measure them. Technology for evaluating status of many representatively is lacking (Morellet et al. 2007), including methodology for collecting quantitative data that can be analyzed and interpreted in ways that relate to deer density. Collecting impact data on multiple resources requires additional/considerable expenditure of financial and human resources.

17.4.2.1 Tree Seedling Regeneration

One of the more difficult tasks in selecting and using measures of deer impact is whether measures relate to goals. Foresters need measures that predict successful regeneration of desired tree species to replace harvested trees. Initially, foresters measured the potential for regeneration of desirable tree species by counting density of seedlings on plots and predicting success based on number of seedlings in height classes presumed to be above browsing suppression by deer. Marquis et al. (1992) identified five deer density categories (1–5; 1 = lowest impact; 5 = highest impact) and associated seedling density required for successful regeneration in each density category, introducing the concept of integration of deer density and impact. However, deer density categories were subjective and based on observer experience rather than on analysis of quantitative data.

Shafer (1963) provided one of the earliest measures of browse use by deer wherein counts of twigs were used to provide estimates of weight of woody deer browse, but these were not meant as indicators of deer impact, nor were they associated with deer density.

Waller et al. (2017) developed a technique for estimating deer browse impact by characterizing mean age of twigs browsed by deer, as determined by counting number of bud scars per twig (every year of growth, twigs add an additional bud scar, each scar representing an additional year for each twig). Waller et al. (2017) also measured height of twigs for which they counted bud scars. Twigs inside deer exclosures (protected from browsing) were compared with twigs outside exclosures and exposed to deer browsing. They found age and height of twigs inside fences were significantly greater than twigs inside, indicating that deer browsing eliminates years of growth by removing portions of twigs as well as reducing twig height. There was less variability among age of twigs than among twig heights, likely a function of influence of light level on twig height. Because deer density was not measured as an independent variable, a quantitative gradient of deer impact related to deer density was not measureable. Also, age of twigs was not related to survival or length of time required to escape deer browsing, reducing the usefulness of the technique for quantifying deer impact as a function of deer density. Data collection for the technique would not include comparison of twig age inside and outside deer exclusion fences, so interpretation of twig age as a determinant of level of deer impact would be difficult.

Sullivan et al. (2017) developed a qualitative protocol for characterizing four levels of deer impact on tree seedlings, recommending selection of ≥2 indicator seedling species including those of high and low preference by deer. The categories were: *low/ideal* = presence of variety of tree seedling species (including those preferred by deer as forage) over 1.5 m tall; *moderate to high* = preferred seedlings <0.3 m tall, none of any species >1.5 m tall; *high* = only seedlings not preferred by deer present and heavily browsed; and *very high* = forest floor totally devoid of tree seedlings of any species. Deer density was not measured as an independent variable influencing level of impact. The protocol recommended collecting data under more open overstories to increase height and number of seedlings measured and is most applicable to small properties, as sampling recommendations seemed designed for properties ∼4 ha. Observer ability to know and differentiate between seedlings of preferred and nonpreferred species is required. Small sample size and nonrandom selection of data collection sites violates assumptions for providing statistically valid estimates of deer impact; the technique would work best for properties <4 ha (small woodlot owners) and used to gauge changes in deer impact over time. Sullivan et al. (2017) did not recommend management actions to consider for estimates of deer impact at any level of intensity.

Pierson and deCalesta (2015) developed a protocol (Appendix 1) for quantifying coarse and fine measures of deer impact on tree seedlings. Both measures were obtained from 1.2-m radius plots spaced 60 m apart on transects 1610 m long spaced 300 m apart. Coarse measure was percent plots with no impact on any woody seedling or shrub. Fine measures of impact were obtained by recording levels of impact on five indicator seedling species ranging from high deer preference to no deer preference. Fine measures were collected for three impact levels: *zero–light impact* = <50% of twigs browsed, moderate impact = >50% of twigs browsed on indicator seedlings without hedging, and *heavy-severe* = >50% twigs browsed and stunted by hedging. Zero–light level impact was associated with successful progression from seedling to sapling trees. On larger properties (>200 ha), grids of five transect lines were randomly located within surveyed properties: depending on the size of the property, a number of grids could be placed to obtain representative samples of deer impact.

Deer impact data were collected from the same plots at the same time as deer density data (see Appendix 1) to facilitate sampling and reduce time spent in the field. The Pierson and deCalesta (2015) technique is one of a few reported in the literature wherein quantitative estimates of deer density were collected concurrently with impact levels to permit correlating deer impact with deer density, which was significant for coarse and fine measures (Pierson and deCalesta 2015; Figure 17.3). Data were collected from woodlots typical of forest management, with a preponderance of plots falling under closed canopy overstories—reduced light levels are associated with lower rates of seedling germination and development than in sites with open overstories representing final harvest or thinned sites. Collection of deer density and impact data concurrently from the same plots represents a large savings in expense and time.

As with pellet group counts, the impact methodology of Pierson and deCalesta (2015) requires little equipment, is representative of deer impact over entire ownerships, and can be performed inexpensively with minimal personnel requirements. However, persons collecting impact data must be able to identify seedlings of species selected as indicators.

17.4.2.2 Herbaceous Vegetation

Landowners interested in enhancing and preserving biodiversity on their forestlands include abundance and condition of herbaceous vegetation (generally referred to as wildflowers) as a natural

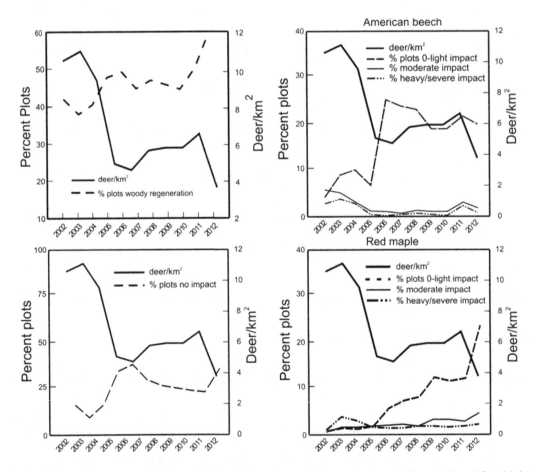

FIGURE 17.3 Comparison of deer density with measures of impact: coarse (left graphs) and fine (right graphs—for red maple, a preferred species; and beech, a nonpreferred species). (From Pierson, T. G. and D. S. deCalesta. 2015. *Human-Wildlife Interactions* 9:67–77.)

FIGURE 17.4 Height and leaf size between *Trillium erectum* exposed to deer browsing (right panel) and protected from deer browsing (left panel).

resource of interest. Scientists have used characteristics of herbaceous growth (whether flowering or not, height, and leaf size) as indicators of deer browsing impact (Figure 17.4). Browsing by deer can reduce size and reproductive success of wildflowers such as *Trillium* (Knight 2003). Anderson (1994) reported that stem height of *Trillium grandiflorum* is useful as an indicator deer browsing, and Koh et al. (2010) documented a significant negative relationship between deer density and maximum *Trillium* height over a 15-year period.

Pierson and deCalesta (2015), based on data from Royo et al. (2010), illustrated changes in height, flowering, and leaf size of three herbaceous species with changes in deer density over time (Figure 17.5).

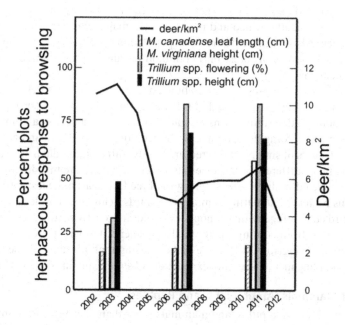

FIGURE 17.5 Comparison of changes in deer density with changes in leaf length, plant height, and percent flowering in three herbaceous species. (Adapted from Pierson, T. G. and D. S. deCalesta. 2015. *Human-Wildlife Interactions* 9:67–77; Royo, A. A. et al. 2010. *Biological Conservation* 143:2425–2434.)

The problem with assessing deer impact on wildflowers is the precarious state of wildflower presence, abundance, and distribution—decades of overbrowsing may have eliminated wildflower species from forest understories (Rooney and Dress 1997). Augustine and Jordan (1998) noted that while herbaceous species such as *Trillium*, *Clintonia*, and *Chelone* have been used to document deer impacts, they may disappear with persistent herbivory. In addition, abundance or condition of wildflowers may vary among plant communities (Frerker et al. 2013), making it difficult to correlate impact on herbaceous plant species with deer density. Royo et al. (2010) noted that the legacy effect of long-term deer browsing reduces the abundance and dispersal of potential herbaceous indicator species, making it difficult to find enough indicator herbaceous plants for quantitative analysis of deer impact. Finally, Kirschbaum and Anacker (2005) were unable to demonstrate spatial variation in browsing-related characteristics of two potential herbs (*Trillium* and *Maianthemum*) across a gradient of deer density within a large, typical forested landscape in northwestern Pennsylvania, hypothesizing that the deer/herb relationship may have been confounded by environmental variables not measured.

Sullivan et al. (2017) proposed a qualitative protocol for characterizing four levels of deer impact on wildflowers, including *Trillium*, jack-in-the-pulpit (*Arisaema triphyllum*), Indian cucumberroot (*Medeola virginiana*), and Canada mayflower (*Maianthemum canadense*). The levels and characteristics were: *low/ideal* = wildflowers grow and flower regularly; *moderate-high* = wildflowers present but do not grow tall or flower; *high* = wildflowers absent; and *very high* = no wildflowers and invasive understory species, such as garlic mustard (*Alliaria petiolata*) or ferns take over ground cover. Sullivan et al. (2017) recommended their method for properties ~3 ha in size, but did not note that the patchy distribution and low abundance of wildflowers may affect ability to locate many, if any, for characterization of deer impact. Adding to the difficulty of assessing deer impact on wildflowers, there is no clear, consistent, and quantitative (or qualitative) method for relating deer density to impact on identified/indicator herbaceous plants established in the literature.

17.4.2.3 Forest Birds

Studies evaluating the impact of overabundant deer herds on forest birds have focused on smaller species, such as woodpeckers and woodland songbirds, and have ignored birds of prey, including hawks, owls, vultures, and eagles. The larger birds have large home range sizes, often including forestlands of the size typically owned and managed by a majority of landowners as smaller parts of larger home ranges. Also, because of their low population density, the larger birds often are not encountered in typical surveys of forest bird communities and the low numbers obtained are not amenable to statistical analysis.

Methods for evaluating the presence and abundance of the smaller forest birds fall into two categories: observers counting singing male birds during breeding season (spring-early summer), usually at stations along predetermined census routes, and use of mist nets (nearly invisible mesh nets used to trap flying birds in foraging habitats). Both techniques have disadvantages and limitations (Bibby et al. 2000). Counts of singing males require highly trained individuals who can identify, and distinguish among, many different bird species, and even then there are misidentifications. Usually, such counts are conducted multiple times during the breeding season, necessitating multiple trips to sites with census routes. Mist netting samples only birds below the middle-upper forest canopy, thereby missing birds that forage in that canopy level, requires costly netting, and users must obtain permits for mist netting from administering wildlife agencies. On larger properties encompassing hundreds of hectares, the logistics of locating and censusing multiple and representative landscape sites for both bird sampling methods require expenditure of significant financial and human resources.

17.4.2.4 Forest Mammals, Amphibians, and Reptiles

Censusing forest amphibians, reptiles, and mammals other than deer faces the same problems and shortcomings of censusing forest birds: it's time consuming, may not sample all species equally if at all, and requires expenditure of significant financial and human resources. These species are furtive and not easily observed, requiring trapping them to collect requisite data. Some trapping methods require placing arrangements of objects on top of the ground (groups of boards placed on top of the litter level to capture amphibians, metal panels covered with soot to record mammal tracks), and some capture individuals of some species live (box traps for small mammals). Others may result in injuries (leghold traps and snares for bears, coyotes, foxes, bobcats) or death (pitfall traps wherein captured animals are drowned to prevent their escaping the traps). Because these species are protected under state/federal law, permits must be obtained to trap them and trapping records must be kept and submitted to the agencies. Larger mammals, such as coyotes, bobcats, raccoons, opossums, and porcupines may include landscapes surrounding individual forestland properties in their home ranges and may not even be trapped on such properties. In summary, there are no inexpensive methods for monitoring reptile, amphibian, or mammal population responses to deer density and impact (on habitat).

17.4.3 MONITORING DEER IMPACT ON FOREST RESOURCES WITH PROXIES

If landowners wish to census major forest natural resource groups, including tree seedlings, deer density, herbaceous plants, and wildlife groupings in addition to deer (birds, mammals, amphibians, and reptiles), they must invest considerable financial and human resources to determine whether deer are affecting forest resources of interest on their properties. And, such investments are not a one-time effort but rather must continue on an ongoing basis to determine whether management is successful in reducing deer impacts to acceptable levels and, if so, is maintaining impact at those levels.

Collecting data characterizing deer impact on herbaceous vegetation and other communities (birds, mammals, salamanders, and reptiles) consumes significant financial and human resources. An emerging solution to the need for collecting deer impact data on multiple plant and animal communities is to use a limited number of indicator plants to represent deer impact on a broader spectrum of forest resources. The idea is to identify a short list of woody species of varying degrees of palatability to deer and categorize levels of deer impact on them according to frequency, abundance, and destructiveness of impact and relate those levels to intensity of impact on other forest resources.

Sullivan et al. (2017) suggested using a limited number of parameters that can serve as proxies for evaluating deer impact on multiple forest resources, specifically herbaceous, shrub, and animal species of interest other than deer. Deer impact on seedlings can serve as a proxy for characterizing impact on other forest resources.

Managers might consider the various carrying capacities identified for deer in Chapter 15 as a guide for using proxies to represent deer impact on multiple forest resources. Regeneration carrying capacity (Figure 15.2, Chapter 15) identifies a specific deer density (8 deer/km^2) where seedlings of tree species of commercial value can become established and progress to producing marketable trees. At this density, diversity, abundance, and vertical structure of shrubs, forbs, and seedlings of tree species favored by deer are in decline, as are wildlife species (e.g., forest songbirds) dependent on diverse, abundant, and multistructured understory and midstory forest vegetation. Forested landscapes at 8 deer/km^2 will produce valuable timber, but communities of other forest resources will be negatively impacted.

When deer density is ~5/km^2 (diversity carrying capacity) and seedlings of all overstory tree species are sufficiently abundant and of a condition (moderate or less browsing impact) to guarantee progression into the overstory, all plant and animal communities/species have the potential for continued and sustainable presence (and recolonization if extirpated). The word "potential" is important—some plant and animal species may have been extirpated due to long-term deer overbrowsing and unable to repopulate the forested property, in which case heroic management steps, such as reintroduction (with protection) of extirpated herbs and shrubs may be required. Reintroduction of extirpated wildlife species is far more difficult and costly and rarely done, one exception being the case of gray wolves intentionally reintroduced by state/federal agencies to counter impacts of overabundant deer (*Odocoileus* sp.) and Rocky Mountain elk (*Cervus canadensis*) on large (>1000 ha) areas in western/midwestern states.

Managers might also consider collecting data on a restricted number of wildflower species such as *Trillium* sp. and *Maianthemum* sp. to serve as proxies for other herbaceous species. Williams et al. (2000) recommended using an assemblage of herbaceous browse indicator species rather than reliance on a single species. Data on wildflower proxies should be characterized in the four categories detailed above by Sullivan et al. (2017).

Deer density and level of deer impact on forest seedlings and wildflowers may be used as proxies for characterizing levels of deer impact on other forest resources. However, managers cannot estimate deer density using impact level as a proxy if actual density is a requirement by state natural resource agencies for issuing permits to increase harvest of (antlerless) deer. Hunters on these properties might also require estimates of deer density as a condition for their continued support of lowering deer densities by harvesting requisite numbers of deer.

Abundance and quality of deer forage can result from heightened level of timber harvest, with creation of early succession forage, or by creating food plots of high quality deer forage. Increasing quality and quantity of deer forage can raise deer density associated with successful seedling regeneration required for sustained timber production and harvest (see Chapter 15). However, the same does not seem to apply to other forest resources, including herbaceous vegetation and wildlife species, wherein optimal health and abundance are associated with density in the range of 5–6 deer/km^2.

17.4.4   SAMPLING FRAMEWORK FOR PROXIES

It is likely that forest managers are more interested in whether forest resources in addition to deer and tree seedlings are present in sufficient abundance and distribution to represent sustainable populations than in gradients of these other resources in response to deer density. In essence, this is an either/or situation: either populations of natural resources of interest are present, or have the potential to be present, in numbers to assure their continuance within the landscape or they do not. In this case, deer density <6 deer/km^2 and/or impact level on indicator seedlings of zero-moderate represent the desired proxy level. Reduction of deer density from that at MSY (approximately 15 deer/km^2) and the associated heavy/severe impact level on indictor seedlings through hunting/culling represent progress toward the desired status of natural resources, as demonstrated by reduction in density and impact on proxies. Collection of proxy data for smaller properties can be conducted in the same way as for larger properties: from plots arrayed along transect lines (see Appendix 1 for details).

17.4.5   SELECTION OF PARAMETERS BASED ON PROPERTY SIZE

For properties >200 ha, I recommend using the deer pellet group technique for estimating deer density and proxies for other forest resources (seedling regeneration and wildflowers) as detailed above.

For properties smaller than the typical deer home range (~200 ha), and especially those under 100 ha, deer impacting forest resources will include these properties as part of their home ranges incorporating the surrounding landscape, especially if it is forested. Deer pellet group information collected from these properties can be used to estimate the number of deer impacting the property (see Appendix 1), but density estimates (deer/km^2) cannot be obtained, as they do not include a sufficiently large area (>200 ha) unless permission is granted from abutting property owners to include their lands in transects sampled for deer density. Owners of these smaller properties might prefer instead to use a small number of trail cameras placed representatively across their properties to capture visual images of deer that will allow identification by sex (male or female for adults) and age (fawns and older animals) if collected while bucks have antlers and fawns have spots (June–September). Comparing data from trail cameras over years, if data are collected the same way and at the same intensity, will allow property owners to determine if the number of deer using their property changes from year to year, including years when efforts to reduce deer density by hunting/culling are performed.

17.4.6 DEER HERD CONDITION

The condition of deer within their forestlands may be of low interest to landowners more interested in other natural resources, including harvestable timber, unless they have a personal interest in deer health or if hunters hunting their lands (as a way of reducing deer density) indicate that their willingness to hunt the owner's land is based at least partly on production of quality deer for hunting.

17.4.6.1 Deer Health/Quality

Deer health/quality is usually evaluated by deer weight by sex and age classes and by antler characteristics. These data are best obtained from checking stations where hunters bring deer harvested during hunting season. There are four management-related age classes: fawns (the age class most vulnerable to overwinter starvation losses if they have been unable to store enough fat to survive winter), yearlings (1½-year-old deer that have survived their first winter and may exhibit significant differences in weight and antler characteristics directly related to nutrition), young adults (2½- and 3½-year-old deer that have yet to attain mature body weight and antler characteristics), and mature adults (deer $\geq$ 4½ years of age that display maximum conditions of weight and antler characteristics). Data for the four age groups are collected and analyzed separately by sex. The four age groups can be differentiated reliably based on the appearance of deer teeth (incisors, premolars, and molars—see Appendix 2).

An additional bonus of checking stations is that they provide a way for managers to interact with hunters and exchange information (e.g., handing out maps of hunting area including access road networks, obtaining harvest loci from hunters to characterize harvest locations). Because the benefits of checking stations extend beyond merely collecting data on deer health/quality, a separate chapter (18) is devoted to them.

17.4.6.2 Deer Herd Sex and Age Ratios

In addition to the need for estimates of deer density and impact to determine whether deer population control efforts are working, managers need information on deer age and sex ratios (Curtis et al. 2009). Ratios of adult male to adult female deer and fall fawn recruitment, expressed as number of fawns (surviving weather and predation) per adult doe and/or per adult deer (bucks and does) may be obtained from roadside counts of deer conducted while fawns still have spots and before hunting season (late summer/early fall). These counts may be used to derive the ratio of adult bucks to adult does (optimal is 1:2) for expression of breeding potential (ratios of >3 does per buck indicate herd composition that will result in less than optimal breeding of does). Additionally, estimation of fall fawn recruitment can provide managers with an evaluation of herd increase prior to hunting season and provide a basis for estimating desired harvest to drive deer density in the desired direction through hunting.

These estimates are generally derived by driving along forest roads at ~0.5 km/hour and counting fawn and adult deer (antlered bucks and does). Driving routes should traverse representative habitats (different successional stages, including clearcuts and meadows, sapling stands, and maturing forest stands) across properties and be conducted several times during late summer/early fall. Counts obtained from such censuses do not represent numbers of deer within the properties but rather provide trending estimates of buck:doe ratios and fawn recruitment for evaluation of how deer management is affecting potential for reproduction and recruitment. Only one of the case histories (Chapter 34) used such roadside drive counts.

Evaluation of numbers of adult and fawn deer brought to checking stations should be guarded, as they do not provide reliable estimates of buck:doe and fawn:adult ratios—fewer hunters harvesting antlerless deer (and especially fawns) bring these deer to checking stations than do hunters harvesting antlered deer.

17.4.7 DEER HABITAT CHANGES

Quality and quantity of deer forage and cover are not static but rather change over time due to natural or human-induced changes in species composition and structure of overstory and understory vegetation. In landscapes managed for timber production, regular, sustainable removals of forest overstory canopy (thinning and final harvests) and manipulation of understory vegetation composition (controlled burns; removal of interfering plants such as ferns, undesirable woody species, and exotic plants through herbicides; and/or mechanical removal via chainsaw) result in increases in deer forage quality and quantity. Natural disturbances (ice storms, windstorms, fire) also create openings in the overstory and changes in understory vegetation that can result in higher quality and quantity of deer forage production in the understory. These increases in deer forage can result in increased regeneration carrying capacity on properties managed for commercial tree management and possibly higher sustained yields of deer production for hunting purposes (see Chapter 15).

Contrarily, reductions in opening of the overstory canopy through natural disturbance or management designed to favor old-growth forest characteristics (emphasis on closed overstory canopy and conversion to uneven-aged status both featuring irregular occurrence of small-sized overstory openings and dominance by shade-tolerant tree species such as sugar maple, American beech, and eastern hemlock) may result in reduced quality and quantity of deer forage and reduced carrying capacity for diversity. Conversion of forestland to agricultural land or residential developments will greatly curtail production of deer forage and cover and result in much-reduced carrying capacity for deer.

If forest managers record quantifiable changes in deer forage and cover conditions resulting from timber management (e.g., proportion of property featuring overstory removals from thinnings or final harvest, herbicide applications on sites with interfering plants), they can relate these potential changes in deer forage to level of deer impact. Such records do not include openings in the overstory and midstory forest canopy resulting from natural disturbance. Managers for the Kinzua Quality Deer Cooperative (Chapter 34) tracked potential for change in deer forage by recording whether plots used to collect deer density and impact data were under closed forest canopies (maturing timber, sapling-pole stands) or under openings created by timber management (thinnings, final removals) or by natural disturbance (wind-shear, tornado, ice storm, mortality from insect/pathogen infestation). Although recording of open/closed overstory canopy provides only a rough approximation of conditions related to deer forage, it does allow evaluation of the relationship between deer density and (potential) forage availability. Properties with a higher proportion of overstory forest openings exhibited impact levels similar to those of other properties with lower deer density and lower proportion of overstory canopy opening (Chapter 15, "Deer Density and Carrying Capacity"). Appendix 1 provides an example of how to collect data on potential deer forage and cover at the same time, and on the same plots, for collecting data on deer density and impact.

Similarly, recording whether plots used to collect deer density and impact data fall under closed overstory conifer (e.g., white pine [*Pinus strobus*] or hemlock [*Tsuga canadensis*]) cover can serve as a proxy for indicating whether properties provide more or less thermal cover for deer.

Finally, recording potential for abutting properties to provide deer with forage/cover (e.g., crops produced on agricultural lands, food plots created for deer within forestlands) may be helpful in refining and understanding the relationship between deer density and impact on forestland resources. Such knowledge may help in refining deer management (e.g., directing hunting/culling efforts at property boundaries common with abutting properties providing deer forage and/or cover).

17.4.8 MONITORING STAKEHOLDERS

The stakeholder group of most importance to managers relying on public hunting to reduce deer density and impact is the hunters who hunt their forestlands. Alpha and beta hunters are the most successful and probably harvest the majority of deer. Managers should be monitoring the satisfaction and concerns of these hunters so that they can respond to hunter concerns and address them to encourage the hunters to continue hunting and reducing deer density and impact. However, other stakeholder groups' concerns about deer management should also be considered and monitored. D'Angelo and Grund (2015) compared preferences of hunters and farmland owners for managing deer and found that the landowner group wanted lower deer density while the hunters wanted higher deer density. To better manage deer, D'Angelo and Grund (2015) recommended conducting frequent surveys of primary stakeholders to allow managers (state natural resource agency managers of DAUs, not DFMA managers) to monitor trends of stakeholder satisfaction related to developing objectives for deer management. It is even more important for DFMA managers to monitor concerns of stakeholders within DFMAs, as these groups affect deer management within DFMAs and their perceptions can have major impacts on the success of deer management (e.g., if hunters are not satisfied, they may cease or greatly curtail their hunting of specific DFMAs).

The key is to survey and retain successful alpha and beta hunters—the ones most likely to harvest deer and bring them to checking stations. Asking these hunters to answer survey questions that relate to their satisfaction level lets managers know what it might take to retain these hunters for future deer harvests.

The surveys need not be complex nor require professionals to design and interpret them—the idea is to get a quick, unequivocal assessment of how satisfied successful hunters are and what they might need as inducement to continue hunting the property. The advantage of having successful hunters take the survey at checking stations is that it can be collected there, avoiding the problems of nonresponse and expenses of mail-in surveys. Surveying successful hunters at checking stations is an additional reason for having checking stations to collect data and enhance hunter/manager communications.

17.4.9 MONITORING BASED ON PROPERTY SIZE

Smaller properties, mostly owed by small woodlot owners who may not have timber management/harvest as primary goals, are generally too small to provide all resources for deer. Deer will make use of their properties, but their home ranges will incorporate lands outside those of the smaller properties. For these smaller properties, especially those under 100 ha, pellet group counts will not produce representative estimates of deer density. However, use of the plot/transect/grid format (Appendix 1) for collecting information on deer impact on forest resources and forage related to overhead canopy closure will provide a stable, repeatable methodology useful for estimating deer impact and may be used to estimate the number of deer causing impacts. Because persons hunting smaller properties will more likely be family, friends, and others well known to property owners, it may not be necessary to monitor deer health, herd sex, and age ratios or hunter satisfaction.

For larger properties, it is recommended that managers use pellet group counts to estimate deer density, percent overstory cover, percent conifer cover, impact on seedlings and herbs to serve as proxies for deer impact on forest resources, trail cameras to provide evidence their forestlands support high-quality deer for alpha and omega hunters to hunt their properties, checking stations to gauge deer quality, roadside counts to estimate adult doe:buck ratios and fawn recruitment, and surveys of successful hunters at checking stations to monitor hunter satisfaction.

REFERENCES

Anderson, C. W., C. K. Nielsen, C. M. Hester et al. 2012. Comparison of indirect and direct methods of distance sampling for estimating density of white-tailed deer. *Wildlife Society Bulletin* 37:146–154.

Anderson, R. C. 1994. Height of white-flowered Trillium (*Trillium grandiflorum*) as an index of deer browsing intensity. *Ecological Applications* 4:104–109.

Augustine, D. J. and P. A. Jordan. 1998. Predictors of white-tailed deer grazing intensity in fragmented deciduous forests. *Journal of Wildlife Management* 62:1076–1085.

Beaver, J. T., C. A. Harper, R. F. Kissell et al. 2014. Aerial vertical-looking infrared imagery to evaluate bias of distance sampling for white-tailed deer. *Wildlife Society Bulletin* 38:419–427.

Beringer, J., L. P. Hansen, and O. Sexton. 1998. Detection rates of white-tailed deer with a helicopter over snow. *Wildlife Society Bulletin* 26:24–28.

Bibby, C. J., N. D. Burgess, D. J. Hill et al. 2000. *Bird census techniques.* Cambridge MA: Academic Press.

Caughley, G. 1977. *Analysis of vertebrate populations.* London: Wiley-Interscience Publications.

Collier, B. A., S. S. Ditchkoff, C. R. Ruth Jr. et al. 2012. Spotlight surveys for white-tailed deer: Monitoring panacea or exercise in futility? *Journal of Wildlife Management* 77:165–171.

Creed, W. A., F. Haberland, B. E. Kohn et al. 1984. Harvest management: The Wisconsin experience. In *White-tailed deer: Ecology and management.* ed. L. K. Halls, Harrisburg PA: Stackpole Books.

Curtis, P. D., B. Boldgiv, P. M. Mattison et al. 2009. Estimating deer abundance in suburban areas with infrared-triggered cameras. *Human-Wildlife Interactions* 3:116–128.

D'Angelo, G. J. and M. D. Grund. 2015. Evaluating competing preferences of hunters and landowners for management of deer populations. *Human-Wildlife Interactions* 9:230–247.

deCalesta, D. S. 2013. Reliability and precision of pellet-group counts for estimating landscape-level deer density. *Human-Wildlife Interactions* 7:60–68.

deCalesta, D. S. 2016. Can deer pellet groups be used to estimate deer density in Tennessee: A pilot study. Poster. *39th Annual Conference of the Tennessee Wildlife Society.* Montgomery Bell State Park, Burns TN.

deCalesta, D. S. 2017. Bridging the disconnect between agencies and managers to manage deer impact. *Human-Wildlife Interactions* 11:112–115.

deCalesta, D. S. and G. W. Witmer. 1990. *Drive line census for deer within fenced enclosures.* Forest Service Research Paper. NE-643. Radnor PA: United States Department of Agriculture Forest Service.

Drucker, P. 2010. Quoted by Larry Prusak. What can't be measured. *Harvard Business Review*, https://hbr.org/2010/10/what-cant-be-measured

Eberhardt, L. L. 1960. *Estimation of vital characteristics of Michigan deer herds.* Report 2282. East Lansing MI: Michigan Department of Conservation, Game Division.

Eberhardt, L L. and R. C. Van Etten. 1956. Evaluation of the pellet group count as a deer census method. *Journal of Wildlife Management* 20:70–74.

Frerker, K., G. Sonnier, and D. G. Waller. 2013. Browsing rates and ratios provide reliable indices of ungulate impacts on forest plant communities. *Forest Ecology and Management* 291:55–64.

Fuller, T. K. 1991. Do pellet counts index white-tailed deer numbers and population change? *Journal of Wildlife Management* 55:393–396.

Hamrick, B., B. Strickland, S. Demarais et al. 2013. *Conducting camera surveys to estimate population characteristics of white-tailed deer.* Publication 2788. Starkville MS: Mississippi State University Extension Service.

Harestad A. S. and F. L. Bunnell 1987. Persistence of black-tailed deer fecal pellets in coastal habitats. *Journal of Wildlife Management* 51:33–37.

Hayne, D. W. 1984. Population dynamics and analysis. In *White-tailed deer: Ecology and management.* ed. L. K. Halls. Harrisburg PA: Stackpole Books.

Hibert, E., D. H. F. Maillard, M. Garel et al. 2011. Ageing of ungulate pellets in semi-arid landscapes: How the shade of colour can refine pellet-group counts. *European Journal of Wildlife Research* 57:495–503.

Jacobson, J. A., J. C. Kroll, R. W. Browning et al. 1997. Infrared-triggered cameras for censusing white-tailed deer. *Wildlife Society Bulletin* 25:547–556.

Kirschbaum, C. D. and B. L. Anacker. 2005. The utility of *Trillium* and *Maianthemum* as phyto-indicators of deer impact in northwestern Pennsylvania. *Forest Ecology and Management* 217:54–66.

Knight, T. M. 2003. Effects of herbivory and its timing across populations of *Trillium grandiflorum* (Liliaceae). *American Journal of Botany* 90:1207–1214.

Knight, T. M., H. Caswell, and S. Kalisz. 2009. Population growth rate of a common understory herb decreases non-linearly across a gradient of deer herbivory. *Forest Ecology and Management* 257:1095–1103.

Koerth, B. H., C. D. McKown, and J. C. Kroll. 1997. Infrared-triggered cameras versus helicopter counts of white-tailed deer. *Wildlife Society Bulletin* 25:557–562.

Koh, S., D. R. Bazely, A. J. Tanentzap et al. 2010. *Trillium grandiflorum* height is an indicator of white-tailed deer density at local and regional scales. *Forest Ecology and Management* 29:1472–1479.

Lancia, R. A., J. D. Nichols, and K. H. Pollock. 1994. Estimating the number of animals in wildlife populations. In *Research and management techniques for wildlife and habitats*, Fifth edition. ed. T. A. Bookhout. Washington: The Wildlife Society.

Marquis, D. A., R. L. Ernst, and S. L. Stout. 1992. *Prescribing silvicultural treatments in hardwood stands of the Alleghenies. (Revised).* General Technical Report NE-96. Broomall PA: U. S. Department of Agriculture, Forest Service, Northeastern Forest Experimental Station.

McCullough, D. R. 1984. Lessons from the George Reserve, Michigan. In *White-tailed deer: Ecology and management*. ed. L. K. Halls, Harrisburg PA: Stackpole Books.

McDonald, J. E., S. DeStefano, C. Gaughan et al. 2011. Survival and harvest-related mortality of white-tailed deer in Massachusetts. *Wildlife Society Bulletin* 35:209–2019.

Moore, M. T., A. M. Foley, C. A. DeYoung et al. 2014. Evaluation of population estimates of white-tailed deer from camera survey. *Journal of the Southeastern Association of Fish and Wildlife Agencies* 1:127–132.

Morellet, N., J. M. Gaillard, A. J. M. Hewison et al. 2007. Indicators of environmental change: New tools for managing populations of large herbivores. *Journal of Applied Ecology* 44:634–643.

Mulhollem, J. 2002. *Penn State Expert Blames Forest Problem on Acid Rain, Not Deer*. State College, PA: Penn State News. http://news.psu.edu/story/185931/2002/05/17/penn-state-expert-blames-forest-problem-acid-rain-not-deer

Pierson, T. G. and D. S. deCalesta. 2015. Methodology for estimating deer impact on forest resources. *Human-Wildlife Interactions* 9:67–77.

Potvin, F. and L. Breton. 2005. Testing 2 aerial survey techniques on deer in fenced inclosures-visual double counts and thermal infrared sensing. *Wildlife Society Bulletin* 33:317–325.

Rooney, T. P. and W. J. Dress. 1997. Species loss over sixty-six years in the ground layer vegetation of Heart's Content, an old-growth forest in Pennsylvania USA. *Natural Areas Journal* 17:297–305.

Rosenberry, C. S., J. T. Fleegle, and B. D. Wallingford. 2009. *Management and biology of white-tailed deer in Pennsylvania*. Harrisburg PA: The Pennsylvania Game Commission.

Rosenberry, C. S., J. T. Fleegle, and B. D. Wallingford. 2011. *Monitoring deer populations in Pennsylvania*. Harrisburg PA: The Pennsylvania Game Commission.

Royo, A. A., S. L. Stout, D. S. deCalesta et al. 2010. Restoring forest herb communities through landscape-level deer herd reductions: Is recovery limited by legacy effects? *Biological Conservation* 143:2425–2434.

Rutberg, A. T. and R. E. Naugle. 2008. Deer-vehicle collision trends at a suburban immunocontraception site. *Human-Wildlife Interactions* 2:60–67.

Shafer, E. J. 1963. The twig count method for estimating hardwood deer browse. *Journal of Wildlife Management* 27:428–437.

Shissler, B. and M. Grund. 2009. Managing deer in the commonwealth: A study of Pennsylvania & other states. Discussion Paper, Washington: Pinchot Institute for Conservation.

Skalski, J. R. and J. J. Millspaugh. 2002. Generic variance expressions, precision, and sampling optimization for the sex-age-kill model of population reconstruction. *Journal of Wildlife Management* 66:1308–1316.

Sullivan, K. L., P. J. Smallidge, and P. D. Curtis. 2017. *AVID Assessing Vegetation Impacts from Deer: A rapid assessment method for evaluating deer impacts to forest vegetation. (Draft)*. Ithaca NY: Cornell Cooperative Extension

Urbanek, R. E., C. K. Nielsen, T. S. Preuss et al. 2012. Comparison of aerial surveys and pellet-group distance sampling methods for estimating deer density. *Wildlife Society Bulletin* 16:100–106.

Waller, D. M., S. E. Johnson, and J. C. Witt. 2017. A new rapid and efficient method to estimate browse impacts from twig age. *Forest Ecology and Management* 404:361–369.

Wallmo, O. C., A. W. Jackson, T. L. Hailey et al. 1962. Influence of rain on the count of deer pellet groups. *Journal of Wildlife Management* 26:50–55.

Williams, C. E., E. V. Mosbacher, and W. J. Moriarty. 2000. Use of turtlehead (*Chelone glabra*) and other herbaceous plants to assess intensity of white-tailed deer browsing on Allegheny Plateau riparian forests, USA. *Biological Conservation* 92:207–215.

Williams, S. C., L. E. Hayes, M. R. Short et al. 2014. *Results and Discussion of the Redding Aerial Deer Survey 24 January 2014 0945-1430 hours*. Storrs CT: Connecticut Agricultural Station.

Wisconsin Department of Natural Resources. 2013. *Final Report: Public Proposals to Implement Recommendations from the Deer Trustee Report*. http://dnr.wi.gov/topic/wildlifehabitat/documents/DTR/FinalReport/DTR%202013%20Action%20Team%20Final%20Report_Final.pdf

18 Deer Checking Stations

David S. deCalesta

CONTENTS

> The check station was part of the deer hunting culture. It was a place for hunters to record their harvest, show off their deer, tell stories, enjoy camaraderie, and gain information about the deer activity. It was a part of the deer hunting experience.
>
> **Dan Stefanich (2015)**

18.1 INTRODUCTION

Although the topic of deer checking stations belongs in the chapter on monitoring, managers accomplish much more than simply monitoring deer health with checking stations. Checking stations can be an effective way to gather information on hunter attitudes. They are an excellent way to communicate with hunters, seek their feedback, and disseminate information (e.g., maps detailing access points and areas of high deer density, summary information on results of deer management). By their presence at checking stations, managers demonstrate their commitment/dedication to hunting and hunters. Checking stations are a unique opportunity for managers to provide incentives/rewards to hunters for harvesting deer. Because of these multiple benefits offered by checking stations and their importance to the success of deer management programs based on public hunting, especially on large forestlands, they are accorded this separate chapter.

Before natural resource agencies began developing contact lists of hunters for the purpose of soliciting hunting information from them, deer checking stations were about the only way to gather information on characteristics of the deer harvest, including numbers and health of deer harvested. These initial checking stations were designed to gather information at the deer management unit (DAU) level—few if any individual managers had the means or expertise to collect harvest data at the deer–forest management area (DFMA) level. Consequently, harvest data collected and compiled by natural resource agencies were used to characterize health and numbers of deer harvested from

DAUs and to describe deer population trends over time. Data derived from DAUs were of little use to managers managing deer at the DFMA level.

Using checking stations to gather information on deer health/quality on DFMAs (individual private forestlands and state and federal public forestlands) is a fairly new notion—managers have only recently become aware of the benefits of checking stations to their deer–forest management. When state natural resource agencies began issuing permits for harvesting additional deer from DFMAs as special damage hunts, they usually required managers to collect harvest information, enhancing the notion of collecting checking station data from DFMAs.

State natural resource agencies have been substituting mail-in harvest reports from successful hunters for deer checking stations of old. These agencies are able to use mail-in reports because the reports are usually mandatory and accompany hunting licenses. DFMA managers are not able to require successful hunters to mail in harvest reports, but because checking stations provide so much more than simple reports on characteristics of harvested deer, they are the preferred method for obtaining harvest information and additional opportunities for useful communications with hunters. Additionally, deer health characteristics collected at checking stations are more standardized, comprehensive, and reliable than information mailed in by hunters.

18.2   WHO SHOULD CONDUCT DEER CHECKING STATIONS?

Managers of DFMAs may want to consider including deer checking stations as part of their overall deer management program. The larger the forestland, the more important it is to conduct checking stations, as benefits increase with increase in size of forestland.

Small woodlot owners (<100-ha woodlots) will receive minimal benefits because their hunters more likely are family and friends who do not need additional incentives to hunt these properties, because value of deer health information gleaned from very small sample sizes is extremely variable and often is lacking completely for some sex-age groups, and because the landowner/hunter interface during deer hunting season is established, continuous, and secure: checking harvested deer at an owner checking station provides little extra in the way of communications/information exchanges between landowners and hunters.

Managers of medium-sized forestlands (>100 < 500-ha woodlots) likely need a hunting base composed of more than family, neighbors, and friends. For these managers, checking stations can enhance communications with participating hunters, increase hunter efficiency by providing maps of access points and areas of high deer density, provide incentives for harvesting deer (especially antlerless), and provide the manager/steward an interface with hunters that boosts mutual trust and communication.

Checking stations are most beneficial for managers of large forestlands (>500 ha) who depend on a hunter population that extends far beyond friends, family, and local hunters. There will be enough deer checked at these stations to characterize deer health/quality for most if not all age and sex classes—trends in weight and quality related to changes in deer density can be displayed with confidence. Pictures taken of quality deer at these checking stations can attract needed hunters from outside the local area, and resources should be available to facilitate soliciting hunters via social media including websites and blogs. Checking stations for large forestlands can be valuable vehicles for providing hunters with information facilitating their success on the forestlands, providing information on trends in deer herd condition, and collection of hunter satisfaction data used to direct management of deer hunting.

18.3   PERMANENT, MOBILE, AND ROVING CHECKING STATIONS

18.3.1   PERMANENT CHECKING STATIONS

Permanently located checking stations are better than mobile and roving checking stations (more on that below). Because hunters become accustomed to taking harvested deer to permanent locations, these stations can be used to collect data from representative portions of forestlands, and set-up and maintenance are easier.

FIGURE 18.1 Deer checking station shed. (Photo by Brad Nelson.)

Because hunting season is usually cold and snowy in the Northeast, checking stations should feature an enclosed, heated structure (Figure 18.1) where data collection can be standardized, checking station personnel can be protected from weather extremes, and information (verbal and written) is most easily provided to hunters.

Checking stations should be heated (small propane heaters work well) and lighted (propane lanterns) and enclosable to maintain heated facilities for checking station workers. Checking stations that can be moved (e.g., small sheds, trailers) and stored in secure locations are best, as they are not exposed to off-season vandalism. Desks and chairs provide working surfaces, and interior walls may feature pictures of harvested deer and charts displaying number, quality, and harvest locations of deer brought to checking stations.

Checking stations with a hanging arm, block and tackle, weight scale, and sling (Figures 18.1 and 18.2) facilitate collecting data on sex, weight, age (by wear and eruption patterns on teeth—Appendix 2), and measurement of antler characteristics (number of points, beam diameter, spread at widest point of main beams). Steel hanging arms for permanent checking stations (sheds) and mobile checking stations (pick-up trucks) can be fabricated by local welding shops or ordered from hunting/forestry supply catalogues for ~$150. The sling can be adjusted to hoist and hold deer at waist level for easy recording of data.

18.3.2 Mobile Checking Stations

Lacking enclosed checking station buildings, managers can operate mobile checking stations from pick-up trucks parked at road intersections/restaurants/sporting goods stores. Steel hanging arms can be inserted into the trailer hitch for use with a block and tackle and sling (Figure 18.3). Hanging arms may be fabricated by local welding shops or can be purchased from sporting goods catalogues

FIGURE 18.2   Sling apparatus for collecting deer data. (Photo by David S. deCalesta.)

FIGURE 18.3   Mobile checking station with hanging arm. (Photo by David S. deCalesta.)

for ~$150. Equipping the pick-up truck with a canopy protects data-collecting material, including data sheets, from weather. Checking station operators can stay warm inside the cab of the pickup truck while waiting for hunters.

Permanent and mobile checking stations should be placed at road intersections representing major access/travel routes to and from the forestland. If possible, they should be located near/in parking lots of restaurants/sporting goods stores for ready access to restrooms and food and drink (for checking station personnel). Additionally, restaurants/sporting goods stores receive much hunter traffic and may enhance data collection from additional harvested deer/opportunities and interactions with hunters. Store owners benefit from the increased customer base provided by hunters visiting checking stations.

18.3.3 Roving Checking Stations

Roving checking stations, where check station personnel travel to hunting camps to collect information on deer health (Figure 18.4), may be employed to collect data from harvested deer, but they represent manager efforts to reach hunters rather than vice versa and usually do not collect as much or as representative data on deer harvest as permanent checking stations. Limited deer herd information may be taken, such as sex and age and antler characteristics, but roving checking stations usually cannot provide deer reliable weights (deer may be skinned and dismembered) or survey hunters.

Persons operating roving checking stations may use chest girth (measured by encircling the deer's body just behind the front legs) to approximate deer weight and refer to conversion tables

FIGURE 18.4 Roving checking station. (Photo by Michael C. Eckley.)

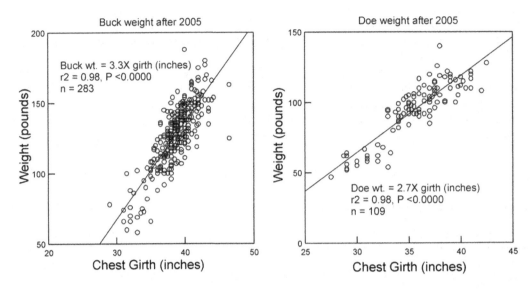

FIGURE 18.5   Relationship between chest girth and body weight for field-dressed buck and doe deer.

to convert chest girth to field-dressed weight (some states provide girth tapes that are marked in pounds rather than inches and are determined by comparing chest to field-dressed weight from large samples of harvested deer). These tapes generally do not differentiate between male and female deer. Comparison of hundreds of male and female deer brought to checking stations for the Kinzua Quality Deer Cooperative (Chapter 34) reveals that the ratios are different between sexes: if one estimates deer field-dressed weight from girth measurements, the conversion factors are different for male and female deer (Figure 18.5).

18.4   EQUIPPING CHECKING STATIONS

Besides propane heaters and lanterns to heat and light permanent checking stations, and the hanging arm, weight scale, block-and-tackle, and sling to position harvested deer for collecting data, there is little additional equipment required. Data sheets, clipboards, and pencils (not ballpoint pens, which don't always work when wet or cold) are used to record data, a sharp knife is used to slit open cheek skin to expose teeth for aging, a flashlight is used to illuminate back teeth for estimating age, a caliper is used to measure antler beam diameter, and a flexible tape for measuring antler spread and chest girth. If the time interval between harvesting and checking deer is long enough for the jaw muscles to stiffen, making it difficult to open the jaws sufficiently wide to view teeth for aging, a jawbone spreader (Figure 18.6) is used to open the jaw. The rounded end of the tool is inserted between the lower and upper jaws just behind the incisor teeth and twisted 90°, forcing the jaws apart for easy inspection of teeth. A flashlight is usually required to illuminate the teeth (many deer are brought to checking stations after dark and the interior of deer mouths are dim under the best of lighting conditions). Warm clothing and insulated, waterproof boots are essential comfort items for checking station operators.

18.5   WHEN TO OPERATE CHECKING STATIONS

Days and hours of checking station operation represent a balance between number of harvested deer checked and time spent checking deer (and waiting for hunters to bring harvested deer to checking stations). The greatest number of deer is brought to checking stations on opening day of

FIGURE 18.6 Jawbone spreader. (Photo from Forestry Suppliers catalogue.)

deer season, unless weather is too extreme (heavy rain or snowfall, dense fog, warm temperatures). The next highest number usually occurs on the second day of hunting season, and the third highest number occurs on the first Saturday after the season opener (Figure 18.7). The number of deer brought to checking stations is many fewer on other days; the best days are the two days of opening and the following weekend days (Figure 18.7): 85% of deer brought to checking stations were from these days.

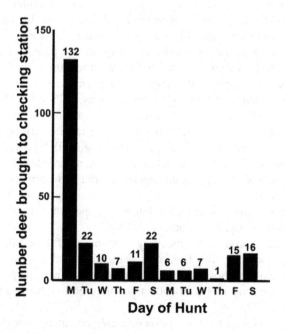

FIGURE 18.7 Number of deer brought to checking station by day of hunt. (Data from Chapter 34.)

Hunters rarely bring deer to checking stations before noon, and sometimes wait until after dark to do so. The most efficient open time runs from 10 a.m. until 1 hour after it is completely dark. The rule of thumb is that ~10% of harvested deer are brought to checking stations, so the number of deer brought to checking stations is not a good predictor of deer density.

18.6   DATA COLLECTION AND PROCEDURES

Data are recorded on waterproof (write-in-the-rain) paper with a pencil. Data collected may include sex, age, and weight of deer; antler characteristics (spread, number of antler points, beam diameter); day and time of harvest; weather conditions (precipitation, wind, temperature, cloud cover); and name(s) of data collector(s). Data sheets may also be used to record hunter contact information for subsequent correspondence including information on herd health. Hunter information could include name, license number, antlerless permit category—if any, town/county/state of residence, and contact information including email address and telephone number.

The back sides of data sheets can include maps for marking harvest locations—tracking harvest locations over the years can pinpoint areas where deer harvest needs to be facilitated, such as by increased access or identification on maps as areas of high deer density. Harvest location data may be used to identify areas of the forestland where harvest is low and may indicate a need for better hunter access and/or better communication between manager and hunter regarding where to hunt. If prehunting season deer density has been obtained through pellet counts or other methods for individual portions of the forestland, and if it has been compared with harvest loci over successive years, managers can pinpoint areas where deer harvest needs to be facilitated, such as by increased access or identification on maps as areas of high deer density. Managers can use this information to encourage hunters to hunt in areas with consistently high deer density and low harvest rates. The back sides of data sheets may also include protocols for recording harvest/hunter information.

18.6.1   HANDLING OF DEER/RECORDING DATA

Checking station operators should remove deer from the hunter's vehicle (hence the minimum required number of two checking station operators) and place it on the sling for weighing. One operator makes the measurements, and the other records them on data sheets. Deer are weighed, then one of the operators opens the jaws and estimates deer age based on the wear-and-eruption protocol (Appendix 2). The best viewing of teeth for age estimation is obtained by slicing through the skin over the jawbone the length of the jawbone, but permission from the hunter must *always* be obtained—if the deer head is to be mounted as a trophy, slicing the skin over the jaw ruins the appearance. If rigor mortis has set in, the jaws must be forced apart as described above. Deer age should be recorded by management-related age classes: fawn, yearling, young adult (2½- and 3½-year-olds), and mature adults (deer ≥4½ years old)—Figure 18.8 provides an example of how to display the data. Number of antler points (left and right) is recorded, as is antler spread (greatest distance between outer beams) and beam diameter (measured with calipers ~3 cm above there the antler is attached to the head).

The recording operator obtains contact information from the hunter and records them on the data sheet along with date, time of day, and weather conditions.

Digital pictures of deer brought to checking stations should be taken (with the hunter's permission) and, together with pictures from past years, displayed on posters inside the checking station. The pictures, including those of "trophy-quality" deer, may be provided to hunters and other stakeholders for each season of data collection and used to advertise the DFMA as a place to hunt for quality deer.

If hunters are being surveyed for their observations/recommendations/satisfaction, checking stations are the time and place to administer and collect written questionnaires on clipboards with

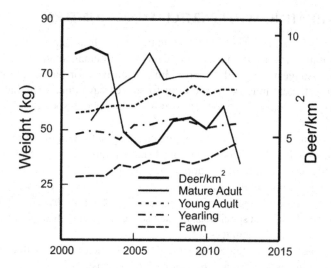

FIGURE 18.8   Example display of deer harvest characteristics by four age classes for buck deer (field-dressed weight changes with reduction in deer density from Chapter 34—KQDC case history).

pencils. Ancillary data provided by hunter anecdotes at check stations may also be recorded to enhance hunter feedback.

Once all data are collected, checking station operators place the deer back on/in the hunter's vehicle and make it a point to thank the hunter for bringing the deer to the checking station.

18.6.2   STAFFING CHECKING STATIONS

There should be a minimum of two staff per checking station (fixed, mobile, and roving) and they either should be prepared to work the whole day or arrangements made for additional staff on shifts. On small to medium-sized forestlands, there generally are enough friends and family members to fully staff checking stations. On large forestlands, the number of checking station operators required for multiple checking station locations may exceed the available pool and volunteers/paid staffers may need to be solicited from other sources, such as state agency field biologists.

Checking station staff should receive training in protocols for collection of data and estimation of deer age. They should be familiar with management operations on the forestland, respective deer management objectives, and the need for control of deer density through harvest removals, including use of permits to harvest additional antlerless deer.

18.6.3   SAFETY CONCERNS AT CHECKING STATIONS

Guns and checking stations do not mix. All it takes is one or two incidents (hunter-carried rifle inside checking station discharges round through roof, discharging rifle in parking lot shatters vehicle window) to sensitize checking station personnel to gun safety—hunters must leave guns in vehicles at checking stations. Belligerent hunters, disgruntled at not seeing as many deer as they expect, may threaten checking station staff—there must be a minimum of two checking station personnel. Checking station staff should always carry cell phones with numbers of enforcement personnel to call for assistance. Checking station personnel should call contacts after leaving stations to certify their safety, and checking station sheds should be padlocked to discourage theft or vandalism. It should be emphasized that checking station staff do not have the authority to arrest hunters for illegal actions and that their duties do not include law enforcement.

18.7 COMMUNICATING WITH PARTICIPATING HUNTERS

Checking stations present a rare opportunity for managers to have face-to-face interactions with hunters, for disseminating information related to deer–forest management, for conducting surveys of hunter satisfaction/suggestions, and for providing incentives for harvesting deer, particularly antlerless deer. Raffle tickets may be handed out to successful hunters (perhaps two tickets for antlerless deer, one for antlered deer) for after-season drawings of sporting goods equipment or other prizes.

Tables and walls of checking stations may feature charts, graphs, and literature featuring past and ongoing hunts, pictures of harvested deer in previous years, annual deer–forest management reports for the forestland, and associated literature supportive of reducing deer density to reduce deer impact. When data collection is complete, harvest characteristics can be averaged by sex and age category and distributed to participating hunters (if hunter contact information has been collected). Hunters are usually interested in the largest specimens in each sex and age class brought to checking stations; these data can also be provided to hunters.

Copies of current deer hunting regulations should be available together with map handouts of the forestland emphasizing access roads and areas of high deer density.

18.8 ATTRACTING HUNTERS

In addition to informing repeat hunters about locations and times of checking stations, solicitations for hunters to bring deer to checking stations may attract new hunters to the forestland, increasing the number of persons hunting the DFMA. Information about hunting on the DFMA should be part of information posted on social media (e.g., websites, blogs, and Facebook), as well as advertised in local newspapers. This information may also be sent to past hunters for whom the manager has contact information.

If the state natural resource agency has identified the forestland as a place where hunters may request permits (e.g., DMAP) to harvest additional antlerless deer, this information should be included in that distributed to past and future hunters. If state natural resource agencies are made aware of the opportunity for public hunting on DFMAs, they may pass the information on to hunters looking for places to hunt.

Providing other sources of information to hunters (e.g., outdoor writers for local newspapers, local radio hosts) of hunting seasons and checking stations on DFMAs may also increase the number of hunters hunting on them.

18.9 USING CHECKING STATION PARTNERS

The more people who are aware of hunting opportunities on DFMAs, the greater the likelihood of reducing deer density and impact. Local field representatives of state wildlife agencies, often identified as Wildlife Conservation Officers (WCOs), can provide expertise, direct hunters to DFMAs, and patrol DFMAs to discourage illegal hunting activities. WCOs should be solicited for membership on any organizing board overseeing and recommending deer management operations on DFMAs. Wildlife/natural resource departments in local colleges may be a source of instructors and their students seeking wildlife field experiences. The same may be said for field staff at local state and federal parks.

REFERENCE

Stefanich, D. 2015. Return of the deer check station in Illinois—for CWD, http://www.chicagonow.com/dan-stef-outdoors/2015/11/return-of-the-deer-check-station-in-illinois-for-cwd/

19 Financial and Human Resources

David S. deCalesta

CONTENTS

> We must always talk in the market place of what happens to us in the forest.
>
> **Nathaniel Hawthorne (1850)**

> Beware of little expenses. A small leak will sink a great ship.
>
> **Benjamin Franklin (1758)**

19.1 INTRODUCTION

Deer management, like that of other forest resources, incurs recurring annual maintenance and associated expenditures of financial and human capital that may be limited, especially for owners of small woodlots. Some of the costs, such as monitoring status of deer density and impact and maintenance of access roads, may be performed in conjunction with other forest resource maintenance requirements (e.g., monitoring of deer impact and status of advance regeneration, application of silvicultural practices such as thinning or final removals for forage creation and timber harvest). Some are specific to deer management (e.g., conducting checking stations, communicating with stakeholders).

Costs of timber management are offset by income generated by harvesting timber (final removals and thinnings), but in most cases (excepting lease hunting, see Chapter 29), the costs of deer management are borne by the managers/landowners and are not offset by income generated by deer hunting. With few exceptions, deer managers are left to their own devices for paying for deer management. Those exceptions include state natural resource agencies absorbing the costs of overseeing/regulating deer hunting at deer administrative unit (DAUs—see Chapter 9) levels, control and distribution of permits required for harvesting additional deer from deer–forest management areas (DFMAs—see Chapter 9), cooperative arrangements for improving deer forage (e.g., mowing fields, pruning apple trees) on lands open to public hunting, and production and distribution of information relative to deer and forest management (which are too superficial to provide DFMA-specific management guidelines and/or specifics). Some agencies support research on deer management topics and the resulting publications add to the body of knowledge on deer biology, but few if any provide guidelines for managing deer at DFMA levels.

The objectives of this chapter are: (1) to make managers aware of costs, including personnel time, required to carry out tasks they will face in managing deer on forestlands; (2) to make hunters allowed to hunt on those forestlands aware of the magnitude of costs owners of forestlands incur in managing deer for their benefit; and (3) to help state natural resource administrators to understand the costs forestland managers incur in providing sites for deer hunters so that they, the administrators, will be more amenable to legislating/regulating deer hunting in ways to help managers defray and/or reduce the costs of deer management.

19.2 START-UP EXPENDITURES OF DEER MANAGEMENT

There are many one-time, start-up costs managers/owners of public and private forestlands incur in managing deer density and impact (Table 19.1). Not all managers will use all of them, but all will use some of them, incurring funding for materials, equipment, and personnel for applying deer management activities. When the decision is made to manage deer and deer impact, some of the start-up items may already have been performed as part of timber or other forest management operations. Regardless, costs associated with their establishment, which will benefit hunting, will have been expended.

TABLE 19.1
One-Time Requirements for Establishing Deer Management Programs

Requirement	Material	Equipment[a]	Personnel[b]
Post boundaries	Signage, wire[c]	All-terrain vehicle for woods travel	Persons to post boundary signs
Identify stands	Forestry equipment	Foot travel	Professional timber cruisers
Build access roads	Crushed rock	Bulldozer, trucks to disperse rock	Heavy equipment operators
Gate roads	Gates and posts	Post-hole diggers and/or power augers	Professional gate builders or experienced staff/volunteers
Build log decks/ parking areas[d]	Crushed rock	Bulldozer and/or other clearing and road-building equipment	Heavy equipment operators
Bridges and culverts	Bridging material[e]	Heavy equipment	Heavy equipment operators
Deer-proof fencing[f]	Fencing materials	Fence-building equipment	Professional fence builders
Check station	Weighing, measuring harvested deer	Building, tables/chairs, heating devices	Staff/volunteers

[a] Equipment required to construct/place required structure on-site.
[b] Personnel required to perform tasks to establish structure(s).
[c] Some managers elect to place several strands of smooth wire along property boundary lines.
[d] Log decks are areas where harvested logs are stored prior to pick-up—these sites form parking areas for hunters using DFMAs.
[e] Access roads invariably cross streams—bridging structures of some kind amenable to hunter vehicle travel are required.
[f] Deer-proof fencing for managers who cannot or will not use hunting to reduce deer density/impact.

Financial and personnel commitments for each management requirement are too variable for estimating representative financial and personnel cost ranges. Rather, managers will have to estimate costs per requirement by consulting with material providers and with equipment operators/consultants. Some, but not all, personnel requirements for individual components may be met by employees/volunteers. For public agencies with established personnel positions and associated budgets, individual line items dedicated for personnel to perform deer management tasks must be budgeted and retained annually, generally ensuring annual potential for eliminating or severely reducing budgeting for deer management activities requiring personnel, materials, and equipment.

19.3 RECURRING COSTS OF DEER MANAGEMENT

As with start-up costs of deer management, there are many recurring costs of deer management, and not all managers will use all of them, but all will use some of them. As is typical of long-term management, these requirements recur at annual or longer intervals, requiring commitment to long-term staffing, funding, and budgeting. And, as with one-time requirements, financial and personnel commitments for each are too variable for estimating representative financial and personnel costs, representing another task for managers. However, managers should be aware of these costs and communicate them to hunters allowed to hunt deer on their properties. They should also communicate their costs to natural resource administrators so they will understand the costs of deer management borne by managers and be sympathetic to assisting them in their efforts to reduce deer density and impact.

19.4 FINANCIAL/HUMAN RESOURCE ASSISTANCE

There are sources of financial/human resource assistance available to managers. Some are one-time start-ups, some are fixed-term grants from charitable institutions, some are long-term continuing sources from state/federal agencies or endowments from charitable organizations, some are in the form of reduction of tax burden granted to DFMAs meeting certain land use characteristics, some are cooperative arrangements with state natural resource agencies for sharing resources to improve wildlife habitat, and some result from pooling of resources/funding provided by DFMA managers who have joined their properties in deer–forest management cooperatives.

19.5 COST–BENEFIT ANALYSIS OF DEER MANAGEMENT COSTS INCURRED BY MANAGERS

The bottom line for DFMA managers regarding implementing deer management practices is, "Do they pay for themselves?" Addressing assessment of deer management (in this instance building of deer-proof fencing) cost effectiveness, Jacobson (2001) stated that (deer) management costs should return benefits, including better wildlife habitat, future timber revenues, and other forest values important to the landowner. He noted that deer management (regarding fencing) is a long-term investment and that financial benefits might occur for many years after application of management activities.

If the decision to manage deer is made purely for future timber revenues, managers may use cost–benefit analysis to estimate returns from the investment. Jacobson (2001) provided an example of carrying costs of building a deer fence until offset by revenues from timber harvest. Income from harvested timber had to exceed a high dollar threshold to justify building a fence and carrying the costs forward for decades until the timber was harvested. The problem with cost-effective analysis is that it is difficult to apply it to recurring expenses, as described by Table 19.2. None of the managers of the nine case histories in Section V of this book used cost–benefit analysis to justify expenses of deer management, and it is doubtful whether any managers would do so, given the uncertainties regarding costs of deer management and returns, especially if harvested timber is not the source of off-setting income. Also, costs of individual requirements occur over differing time periods, and have different off-setting sources. The bottom line is, it is impossible to use cost–benefit analysis to address costs of deer management.

TABLE 19.2

Recurring Requirements for Maintaining Established Deer Management Programs

Requirement	Material	Equipment	Personnel
Access[a]	Road/parking surfaces	Front-loaders	Heavy equipment operators
Monitoring[b]	n/a	Data recording, trail cams	Staff/volunteers
Deer impact abatement[c]	Herbicides, chemical protective devices	Forestry equipment	Forest technicians
Timber harvest[d]	Fuel for harvesting equipment	Specialized harvesting equipment	Forestry technicians
Special forage production[e]	Seed, fertilizer	Mowing, pruning, fertilizing equipment	Staff/volunteers
Winter road plowing, salting/sanding	Plow fuel, salt/sand	Truck(s) with plow	Staff
Protective device inspections[f]	Repair materials, deer-proof fencing	All-terrain vehicle for inspecting fencelines	Staff
Communications[g]	Printer ink, paper	Computers, tablets	Staff, consultants
Permits[h]	Application forms	Computers	Staff

[a] Maintenance of roads, parking areas, gates, property boundary lines.
[b] Monitoring (deer density and impact, checking stations, hunter satisfaction).
[c] Silvicultural practices (herbicide use, mechanical removal to mitigate development/dominance of undesirable herbaceous and woody vegetation caused by high deer density).
[d] Timber harvest to create deer forage, early succession habitat.
[e] Food plots.
[f] Fence maintenance.
[g] Maintain websites, blogs, Facebook/Twitter accounts.
[h] Application for, distribution of antlerless/depredation deer permits.

Instead, managers reduce costs by: (1) seeking outside sources of funding for material, equipment, and personnel costs; (2) stretching their financial and personnel resources; and, (3) pooling material and equipment with other DFMA managers and using/sharing volunteers.

19.6   FINANCING DEER MANAGEMENT WITH OUTSIDE FUNDING SOURCES

19.6.1   OUTSIDE FUNDING SOURCES: GRANTS FROM CHARITABLE ORGANIZATIONS

Charitable organizations interested in supporting deer management by DFMAs designed to enhance/preserve/maintain diversity and abundance of forest resources may provide seed money to get DFMA operations off the ground. This source of funding generally is available to larger DFMAs with the potential for demonstrating the effect of comprehensive deer management on deer density and forest resources. The case history in Chapter 34 involving a large (30,000 ha) cooperative was funded for 10 years by the Sand County Foundation, a nongovernmental organization. Deer density was reduced to the goal level, managers no longer had to use deer fencing to protect forest regeneration, and deer impact was significantly reduced. Such charitable organizations exist in many states in the eastern United States and may provide support similar to that of Sand County Foundation.

19.6.2   FINANCIAL ASSISTANCE FROM FEDERAL PROGRAMS

The United States Department of Agriculture Forest Service offers two programs (FLEP and EQUIP) to provide funding in support of forestry operations for qualifying DFMA manages/

landowners. Funding generally is for vegetation manipulation, including habitat improvements and herbicide use to control unwanted vegetation. Funding does not cover costs of managing human dimensions components of management plans such as checking stations, communications with hunters/stakeholders, and hunter recruitment and recognition.

19.6.2.1 Forest Land Enhancement Program

The Forest Land Enhancement Program (FLEP) supports long-term sustainability of nonindustrial private forestlands by providing financial, technical, and educational assistance by state forestry agencies to assist private landowners in managing their land. FLEP cost-share payments for management practices may amount to 75% of cost. The acreage limit of 1000 acres may be increased to 5000 acres if it is determined that the treatment of additional acres will result in significant public benefit. The aggregate payment to any one landowner may not exceed $100,000 (as of 2007)—changes may have been made since then.

To be eligible for cost-share assistance, owners of nonindustrial private forests (NIPFs) must develop and implement a management plan (addressing certain criteria) that provides for the treatment of their forestlands. The management plan must cover a period of at least 10 years and must be approved by State Foresters in participating states. All NIPF lands are eligible for technical and educational assistance. The State and Private branch of the Forest Service has offices in participating counties; staff members can assist DFMA managers in developing a management plan that meets requirements. The mission of the State and Private Forestry program is to provide technical and financial assistance to private landowners, state agencies, tribes, and community resource managers to help sustain the United States' urban and rural forests and to protect communities and the environment from wildland fires, insects, disease, and invasive plants. The program is located at 17 sites throughout the USA. The delivery of the State and Private Forestry program is carried out by eight National Forest System regions and the Northeastern Area. Information about FLEP may be obtained from the website: https://timbertax.org/getstarted/costshare/programs/flep/

19.6.2.2 Environmental Quality Incentives Program

An additional Forest Service program, under the auspices of the Natural Resources Conservation Service (NRCS), is the Environmental Quality Incentives Program—EQUIP. This program helps agricultural (and forest products) producers resolve management challenges while conserving natural resources like soil, water, and air. The program is designed to conserve natural resources for the future while also improving agricultural (and forestry) operations. Through EQIP, NRCS provides agricultural (and timber) producers with financial resources and help in planning and implementing improvements (conservation practices designed to provide cleaner water and air, healthier soil, and better wildlife habitat).

Conservation practices for which EQUIP assistance is sought must conform to standards developed state by state. Funded practices include: fence construction, access road development and maintenance, forest stand improvement, forest trails and landing construction and maintenance, herbaceous weed treatment, integrated pest management, prescribed burning, restoration of rare or declining natural communities, development of riparian forest buffers and riparian herbaceous cover, construction of stream crossings, building structures for wildlife, tree and shrub establishment, and upland wildlife habitat management. Managers can determine whether EQUIP can provide funding for their DFMA by contacting local NRCS offices for guidance on applying for assistance. NRCS works with managers to develop a conservation plan that meets management goals. Financial assistance covers part of the costs from implementing developed conservation practices. NRCS offers successful applicants an EQUIP contract to receive financial assistance for the cost of implementing practices. Payment rates for conservation practices are reviewed and set each year. Information on this process is available on the NRCS website: https://www.nrcs.usda.gov/wps/portal/nrcs/main/national/programs/financial/eqip/. State websites for NRCS are: https://www.nrcs.usda.gov/wps/portal/nrcs/sitenav/national/states/.

NRCS provides cost-share payments to landowners for programs that are usually 5–10 years in duration, depending on the practices to be installed. NRCS provides greater cost-share assistance to landowners who enter into agreements of 15 years or more for practices to enhance/improve essential plant and animal habitat. There are shorter-term agreements available to fund installation practices for wildlife emergencies that may emerge. Unlike the FLEP program, NRCS does not place limits on the number of acres that can be enrolled in the program or the amount or payment to be made for EQUIP funding. However some states may set limits on assistance, depending on the management problems addressed.

One of the small woodlot case histories (Chapter 38) provides an excellent example of how FLEP and EQUIP programs significantly reduced costs of forest management practices used to combat deer impact and improve wildlife habitat.

19.6.2.3 Forest Stewardship Programs

The Forest Stewardship Program (FSP) of the U.S. Forest Service works in partnership with state forestry agencies, cooperative extension, and conservation districts to provide private landowners with the information and tools they need to manage their forests and woodlands. The program helps managers develop realistic, achievable forest resource goals for their forest resources and identify and develop management activities needed to meet them. Managers can obtain Stewardship program information from their state coordinator, identified on the website: https://www.fs.fed.us/about-agency/contact-us/forest-stewardship-state-coordinators.

The FSP program does not provide funds for carrying out management activities.

19.6.3 State-Relief Programs for Private Forest Landowners

Some states (e.g., Pennsylvania's "clean and green" program) provide for lower property tax assessments of land capable of producing timber or providing open space for public use.

The Clean and Green program, established by the Pennsylvania Farmland and Forest Land Assessment Act, provides for lower property tax assessments of land capable of producing timber. The intent of the act was to protect farmland, forest, and open space by allowing for land taxation according to its use value rather than the prevailing market value. DFMAs qualify for Forest Reserve Land classification if they are composed of at least 10 contiguous acres, including any farmstead land, which is stocked by forest trees of any size and capable of producing timber or other wood products. Based on data from the Pennsylvania Department of Agriculture, the average reduction in fair market assessed value is nearly 50%. These tax savings can increase the amount of money available to managers to pay for deer management and may be available in other states.

19.7 POOLING EQUIPMENT, MATERIALS, AND PERSONNEL

DFMA managers can pool equipment, material, and personnel costs by sharing them with neighboring DFMAs. Individual DFMAs can be combined with neighboring DFMAs to form cooperatives (Chapter 31). The larger size of cooperatives makes them more attractive to hunters and more amenable to receiving permits from state natural resource agencies for harvesting antlerless deer as a way to reduce deer density. DFMAs within cooperatives can share personnel for monitoring and checking station operations, and can also share in use of, and paying for, forestry and wildlife consultants (see the Chapter 34 case history for an example).

19.8 USING/SHARING VOLUNTEERS

Volunteer field personnel for performing some management activities (e.g., monitoring, checking station operation) may be obtained from stakeholder groups (e.g., local hunters, state natural resource agencies) and local colleges offering classes in forest resource management—class instructors may be able to provide students for performing some data collection as part of required class field experiences. The cooperative case history (Chapter 34) made heavy use of such volunteers to stretch its human resources.

19.9 EDUCATION

Many state extension services provide educational classes for forestland managers on a variety of forest management topics. The Pennsylvania State University Forest Stewardship (PFS) program provides classes/field trips for managers, including a 1-day deer density and impact workshop. Other educational opportunities include holding conferences for small woodlot owners and facilitation of local DFMA owner/manager groups/associations for sharing information and resources.

REFERENCES

Franklin, B. 1758. *The way to wealth.* In Poor Richard's Almanac. http://www.cashflowbridge.co.uk/blog/beware-of-little-expenses-a-small-leak-will-sink-a-great-ship.html

Hawthorne, N. 1850. *The Scarlet Letter.* Visalis CA: Vintage Press.

Jacobson, M. 2001. *Forest finance 2. Fencing for forest regeneration: Does it pay?* University Park PA: The Pennsylvania State University Cooperative Extension Service.

Section III

Managing Ecological and
Human Factors

There is no 'true'. There are merely ways of perceiving truth.

Gustave Flaubert (cited by Francis Steegmuller 1982)

Perception is reality.

Lee Atwater (1991)

Perhaps the hardest of things to reconcile in management of natural resources, including deer, is the objectivity of ecological fact and the subjectivity of human perception, especially when perception is weighted more heavily than fact, as is sometimes the case in management decisions made by natural resource agencies. Unfortunately, when management is guided by perception rather than hard reality, costly mistakes are made that may take decades or longer to correct, with catastrophic impacts on affected resources. One need only revisit the results of the misguided perceptions that outlawing doe harvest and eliminating natural deer predators such as wolves and mountain lions would restore deer health and abundance to the heights of white-tailed deer population expansion in the 1920s, as chronicled in Chapter 14.

Scientists' claims in the early 1920s that such population expansions were unsustainable were hooted down at the time but later proven correct, as exemplified by the population crash of deer and destruction of understory vegetation in the late 1930s in Pennsylvania. The resulting forest vegetation was devoid of herbaceous vegetation, and tree species diversity was reduced by elimination of species such as American yew (*Taxus canadensus*) and emerging and persistent dominance of exotic shrubs such as Japanese barberry, garlic mustard, and tree species (e.g., beech, ironwood) not favored by deer.

Once problems caused by management by perception are identified, it may be too late to recover extirpated species. Rooney and Dress (1997) attributed the loss of 33 (80% of total) herbaceous species in one stand and 16 (59% of total) herbaceous species lost in another stand in an old-growth forest in Pennsylvania between 1929 and 1995 to browsing by an overabundant deer herd.

Chapters 20 through 27 describe management actions that integrate science and perception to develop and deliver successful deer management on deer–forest management areas (DFMAs)—properties where deer habitat and numbers are actually manipulated by managers. However, these management actions are not amenable to investigation by, and establishment of standards for, scientific management of deer. Conditions among DFMAs are extremely variable, which prevents them from being placed in discrete categories for which science can establish guidelines (with assurances of success and probabilities associated with chances of success). The upshot is that management of deer and forest vegetation on DFMAs is a unique package that must be applied, refined, tested, and changed by adaptive management specific to individual DFMAs.

As with Section II chapters, Section III chapters are not provided with Manager's Summaries because, as with Section II chapters, full understanding and utilization of information within these chapters is essential for successful deer management.

REFERENCES

Atwater, Lee. 1991. *Life* magazine, February issue.

Rooney, T. P. and W. J. Dress. 1997. Species loss over sixty-six years in the ground-layer vegetation of Heart's Content, an old-growth forest in Pennsylvania, USA. *Natural Areas Journal* 17:297–305.

Steegmuller, F. 1982. *The Letters of Gustave Flaubert,* 1857–1880 (Vol. 2). Cambridge MA. Harvard University Press.

20 Reducing Deer Impact

David S. deCalesta

CONTENTS

> If we do not voluntarily bring population growth under control ... nature will do it for us in the most brutal way, whether we like it or not.
>
> **Henry W. Kendall (2015)**

20.1 INTRODUCTION

The goal of this book is to provide managers with tools for reducing deer impact on natural resources to acceptable levels on deer–forest management areas (DFMAs). By far the most preferred and cost-effective method is lowering and maintaining deer density to a level compatible with sustainable production of forest resources (Hesselton et al. 1965, Matschke et al. 1984, McCullough 1979). Hunting is the only realistic large-scale management tool available for controlling deer densities (Ward et al. 2008). The most effective and cost-efficient way to use hunting to control deer density is by public hunting and sharpshooting (Jacobsen 2006, Williams et al. 2008). Natural resource agencies use deer hunting as the most effective, practical, and flexible method for deer population management (Predl et al. 2008).

Less effective alternatives include use of exclusionary or other devices that prevent deer from damaging vulnerable forest resources. For whatever method is employed to reduce deer impact, there is one constant—that method must be conducted annually and altered if needed as dictated by monitoring. Because of deer ability to increase in population size (and impact) almost overnight, their density and impact should be monitored every year to ascertain whether deer management actions are working or need adjusting.

20.2   INFLUENCE OF LANDSCAPE SIZE

State natural resource agencies attempt to manage deer at deer administrative areas (DAUs—see Chapter 9) levels by regulating length of hunting season and number of deer hunters can harvest. However, because of the variety of ownerships and deer–forest management goals and practices on forestlands within DAUs, nothing state natural resource agencies do at DAU levels can target deer density and impact at the deer–forest management area (DFMA—see Chapter 9) level. Directed, effective management of deer density and impact only occurs on forestlands within individual DFMAs.

As detailed in Chapter 3, deer from properties abutting DFMAs may move onto DFMAs and cause negative impacts on forest resources. Management efforts to reduce deer density and impact on DFMAs by hunting may be unsuccessful if deer density within lands surrounding smaller DFMAs cannot be reduced. For large DFMAs (e.g., the 30,000-ha Kinzua Quality Deer Cooperative (KQDC) in Chapter 34), the sheer size of the DFMA will dilute the numbers and impact of deer moving from abutting lands. Protracted, persistent hunting pressure on small DFMAs may succeed in reducing deer density and impact (addressed below in the section on localized reduction of deer density) to acceptable levels.

20.3   REDUCING DEER DENSITY: HUNTING

Prior to European colonization of North America, predation (by wolves, mountain lions, bears, and Native Americans), severe weather, and forage shortages kept deer density in check: deer do not control their own numbers intrinsically. With the removal of most historic deer predators and an increase in forage creation resulting from unrestricted logging and replacement of forests with agricultural land, natural control of deer density was greatly reduced by the 1900s. To make up for the reduction in natural predator control of deer density, modern-day hunting must either take up the slack in deer mortality or other factors must come into play if forest resources are to be protected from excessive browsing by overabundant deer populations. Public hunting on private and public forestlands is limited by hunting regulations to specified time periods (hunting seasons, including damage hunts held outside of regular hunting seasons), limited numbers of deer harvested by individual hunters (bag limits), and sex and age of deer legally harvested. Removals by contracted sharpshooters (generally on private lands or on public lands such as parks where legal hunting is not allowed) are not as limited by season, bag limits, and restrictions on sex and age of deer legal for harvest. Additionally, sharpshooting programs, by their secretive, nondisruptive nature, do not cause deer to alter movement and home range patterns and are much more effective in removing deer (Williams et al. 2008). Sharpshooting programs are best suited for private forestlands, such as forested (human) residential communities, where public hunting may not be allowed or, by its nature of a limited number of hunters, may not effect desired reductions in deer density (see Chapter 37). Public hunting is better suited for the larger public forestlands where public hunting is more politically acceptable than sharpshooting, and sheer size of public forestlands may require more sharpshooters than available or affordable.

20.3.1   PUBLIC HUNTING

Increasing deer mortality by recreational/subsistence hunting can reduce and maintain deer density at sustainable levels on DFMAs if enhanced by extended hunting seasons, increased bag limits, and/ or provision of permits to harvest additional antlerless deer on DFMAs (see Chapter 34). A number of northeastern states (e.g., New York, Pennsylvania) developed deer management assistance programs

(DMAPs; see Chapter 9) whereby forest managers of private and public lands can request and receive permits for harvesting additional antlerless deer and distribute them to hunters. Generally, managers of these forestlands petition state natural resource agencies for the permits, which are issued either in good faith to the managers or are based on evidence provided by the managers of excessive deer browsing impact on forest resources. Case history Chapters 32 through 36 provide examples of successful use of public hunting to reduce deer density and impact.

20.3.2 PRIVATE HUNTING/LEASING

Owners of DFMAs may choose to restrict hunting on their forestlands to private groups through lease hunting. These managers can require lease hunting groups to harvest enough deer to reduce deer density to acceptable levels—if the groups fail to do so, their leases can be revoked and offered to other groups more accepting of hunting to reduce deer density (Chapter 29). Public deer hunting is limited to seasons and bag limits as established by the governing authorities. In some states, landowners restricting hunting on their forestlands to hunting leases do not qualify for receiving permits for harvest of additional antlerless deer through DMAP programs.

20.3.3 CONTRACTED REMOVALS BY SHARPSHOOTERS

For safety or other reasons, managers may contract for the services of sharpshooter consultants/local enforcement agencies to remove deer at night with rifles equipped with silencers. Euthanized deer are retrieved and removed by the persons performing the service. Local natural resource agencies may permit these removals to take place during and outside of regulated deer hunting seasons. As indicated by Williams et al. (2008), sharpshooting is more efficient that public hunting in reducing deer density. Chapter 37 provides a case history of successful use of contracted sharpshooters to reduce deer density and impact.

20.3.4 LOCALIZED REDUCTION IN DEER DENSITY ON SMALL DEER–FOREST MANAGEMENT AREAS

Mathews and Porter 1992 postulated a "Rose Petal" hypothesis via local area reduction in deer density, likening removal of individual matriarchal deer groups (family members consisting of a matriarchal doe, her female offspring, and subsequent female offspring) to removing individual petals of a rose. They hypothesized that by removing most if not all matriarchal deer groups from forestlands with high deer density that deer density (and impact on forest resources) would remain at lowered levels as home ranges of adjacent matriarchal groups would not expand to include the area from which the original group(s) had been eliminated or greatly reduced. McNulty et al. (1997) tested this hypothesis with a deer herd at 2–6 deer/km^2 in the Adirondack Park (AP) in northeastern New York. Removing 14 deer from a 1.4-km^2 area resulted in low to zero deer density in the removal area 2 years later—no radio-collared female deer from adjacent areas had recolonized the site. Six years after the removal, no radio-collared deer from surrounding areas moved into the removal area, and deer density remained low. Repopulating deer were offspring of females from the removal area or immigrants from adjacent areas (Oyer and Porter 2004).

However, winters are severe in the AP and serve to depress deer density to levels similar to low levels. Miller et al. (2010) tested the Rose Petal hypothesis within a 3413-ha West Virginia research site with an overall deer density of 12–20 deer/km^2. They removed 51 deer from an area of approximately 1.1 km^2. Within 2 years, deer had repopulated the area and the initially depressed impact level of deer on seedling regeneration increased.

As described by Lutz et al. (2015), when deer densities are above low levels, female deer move from such areas into adjacent areas of lower deer density, which may have been the case in the West Virginia study. The Rose Petal phenomenon appears to work when deer densities are low but may not work when they are higher, especially within the surrounding landscape. For small DFMAs, such as small woodlots surrounded by abutting forestland, managers may have to plan for annual deer harvests to retain deer density at desired low levels. Possibly several years' worth of deer removals may reduce the density of matriarchal deer groups in abutting forestlands such that immigration pressure will be low and the requirement for annual deer removals will be relaxed. Such smaller properties may be the best places to apply the Rose Petal strategy: it would be impossible to eliminate all matriarchal groups from larger properties and ecologically unwise. Deer evolved with forest vegetation and their browsing pressure under precolonial conditions likely helped maintain some kind of balance among ground-level vegetation plant species and may have prevented dominance of some species, such as beech and striped maple, that could reduce biodiversity by their competitive advantage (deer preferentially find other species more palatable).

20.3.5 Special "Deer Damage" Hunts

Some states provide permits for special, extra season deer damage hunts for identified DFMAs (e.g., Pennsylvania's Agricultural Deer Control Red Tag program, which unfortunately is not available for forestland managers). Usually the landowner must characterize deer damage and cite the failure of regular season deer hunting to reduce deer density and impact sufficiently to reduce impact to acceptable levels. Such damage control hunts can be successful for DFMAs (e.g., Roseberry et al. 1969, Winchcombe 1992). When held after normal deer hunting seasons are over, deer damage hunts attract hunters still seeking to harvest deer (deCalesta 1985) and can be heavily used.

20.4 REDUCING DEER DENSITY: ENHANCING NATURAL PREDATION

Increasing abundance of natural predators (e.g., mountain lions, wolves) to control deer herds has been hypothesized as a way to reduce deer density and impact. However, an analysis by Predl et al. (2008) provides the definitive response to the proposition of reintroduction of predators to control deer density and impact on DFMAs: "Restoration of wolves and mountain lions is infeasible in much of the United States because it is too densely populated by humans to provide suitable habitat for these species. In addition, it is unlikely that rural residents would tolerate large predators at levels dense enough to limit deer populations because such predators also readily consume livestock. Predation of non-target species including other native wildlife, livestock and pets, as well as concerns for human safety, are but a few examples of the conflicts that would arise as a result of predator reintroductions."

20.5 REDUCING DEER DENSITY: CHEMOSTERILIZATION

Research has identified a vaccine (porcine zona pellucida–PZP) that prevents fertilization of deer ova—it is an antibody-type agent extracted from female domestic hogs. The extract must be injected (usually with dart guns of limited range) into wild deer. Because wild deer generally are too skittish to be darted successfully, because does so immunized with earlier forms of the vaccine must receive annual booster shots, and because of the impossibility of sterilizing enough does in a free-ranging deer herd, chemosterilization is applicable only in small deer herds in enclosed situations and not effective with free-ranging deer as found in typical deer impact situations. Even with good access to relatively small (<30 does) and isolated populations, the approximately 70%–90% of females in a population requiring sterilization cannot be reached, and the cost may exceed $1000 per treated doe. The cost and practicality of treating enough deer to reduce free-ranging populations limit use of the technique to small, isolated populations such as on islands with little potential for colonization by neighboring deer. Sterilization cannot replace hunting for controlling free-ranging deer populations on large areas (Kilpatric and LaBonte 2007).

20.6 REDUCING DEER IMPACT BY INCREASING FOOD SUPPLY

20.6.1 TIMBER HARVEST

Marquis et al. (1992) hypothesized that impact of deer on seedling regeneration could be reduced by increased harvest of trees (final harvest removals and/or thinnings), which would produce more forage and spread out deer impact—in essence overwhelming deer by creating more forage than they can eat. However, resident deer may not use the additional forage if it is created outside of their home ranges: forest vegetation management aimed at attracting deer away from problem areas (e.g., areas with low regeneration success) or toward browse supplies during severe winters would likely be unsuccessful (Campbell et al. 2004). Furthermore, unless density of deer whose home ranges include areas with heightened timber harvest can be maintained at the pre–timber harvest level, deer may simply increase in abundance and continue to negatively impact seedling regeneration. Moreover, the impact of deer on herbaceous vegetation will increase, even if seedling impact does not (see Chapter 15). And, as noted anecdotally in Chapter 15, clearcutting large areas in attempts to overwhelm deer with forage may result in explosive growth of seedlings highly preferred by deer, such as pin cherry (*Prunus pensylvanica*), which may crowd out seedlings of commercially valuable species such as black cherry (*Prunus serotina*) or red maple (*Acer rubrum*). Finally, harvesting timber from large areas, while potentially reducing deer impact on seedlings and herbaceous vegetation, may not be sustainable if foresters run out of timber to harvest (Chapter 40).

20.6.2 FOOD PLOTS

Another way of increasing forage as a way to reduce deer impact is to establish food plots—areas within forestlands cleared and planted with deer-specific forage crops such as brassica (*Brassica oleracea*), chicory (*Cichorium intybus*), and clovers (*Trifolium* sp.). The idea is that food plots draw deer away from portions of forestlands they are impacting and/or provide them with an alternative food source that will reduce impact on understory forest vegetation. However, establishing and maintaining food plots is expensive, they must be maintained yearly, and they may actually increase deer impact in forestlands bordering food plots. Food plots did not affect the amount of deer damage in forest environments in a study in Wisconsin (Jansen 2016). Winter feeding grounds and food plots concentrate and hold deer for extensive periods of time, and the high concentrations of deer can result in excessive deer browsing on native vegetation within 900–1700 m of feeding sites (Doenier et al. 1997, Williamson 2000). Cooper et al. (2002) found that deer reduced home range size near feeding sites and that browsing was seven times higher than on sites without feeders concentrating deer. Rather than resorting to food plots, maintaining deer densities within the carrying capacity of the habitat and managing for quality habitat should be the first priorities of any deer management program (Fulbright 1999).

20.6.3 CONSTANT TENDING

Constant tending of understory woody tree species, in conjunction with deer hunting, as a deer management strategy was experimented with, seemingly successfully, in the Brubaker small woodlot case history (Chapter 39). In this case history, undesirable tree seedlings/saplings in a mixed hardwood woodlot were trimmed with a chainsaw on an ongoing basis. Slash resulting from trimming these seedlings was distributed over/around desirable tree seedling species to protect them from deer browsing. Trimming was timed to coincide with spring green-up so that the trimmed seedlings would sprout and produce deer forage, creating an attractive "salad bar" of succulent new growth. Trimming of some shrubs—hawthorn (*Crataegus* sp.) and black haw (*Viburnum prunifolium*)—produced a cagelike branching structure that protected oak (*Quercus* sp.) and other desirable species growing underneath them. One of the assumed benefits of constant tending and deer hunting was the reappearance of pink lady slippers (*Cypripedium acaule*), Indian cucumber root

(*Medeola virginiana*), and jack-in-the-pulpit (*Arisaema triphyllum*)—wild flowers not seen before application of tending and deer hunting.

20.7 REDUCING DEER IMPACT BY EXCLUSION

Exclusionary devices (fences, individual seedling protectors) prevent deer from impacting forest vegetation so protected—they prevent deer damage by placing barriers between deer and individual plants or restricted areas, such as timber harvest sites. However, exclusion only protects individual seedlings on enclosed regeneration sites and does not reduce deer impact on understory vegetation not protected. In essence, exclusionary devices do nothing to reduce deer impact on plants not directly protected and do not protect understory plant diversity or wildlife habitat outside of excluded areas. Fences to protect herbs and shrubs remain indefinitely. Fences to protect seedlings must be removed prior to timber harvest.

20.7.1 FENCING

As noted above, fencing protects only plants within fenced-in areas and does nothing to protect diversity or quality of understory vegetation or wildlife habitat outside of fenced areas. However, if entire DFMAs can be enclosed inside deer-proof fencing, then understory vegetation, including wildlife habitat, can be protected from deer impact. The Minnesota Department of Natural Resources has produced a circular on building deer fences from livestock fencing that provides comprehensive construction details (https://files.dnr.state.mn.us/recreation/hunting/deer/bovine-tb/fencing_guide. pdf). Deer fencing is generally constructed of two sections of 1.2-m-wide livestock mesh fencing, one layer atop the other with a 1+ m–wide apron attached to the bottom of the fence parallel to and lying on the ground to prevent deer from crawling under fences (Figure 20.1).

Fencing should extend at least 2.5 m above ground—adding two strands of barbed wire above the top livestock fencing section will provide a 2.5 m fence height. Because vehicular access to the interior of fenced areas is often required, deer must be prevented from entering fenced-in areas through these access points with gates that can be manually or remotely opened and closed or special "cattle guard" grates (Figure 20.2) that allow vehicles to pass but inhibit deer from crossing (VerCauteren et al. 2009). VerCauteren et al. (2009) evaluated makeshift cattle guards that were only 3 m long; deer

FIGURE 20.1   Deer exclusion fence. (Photo by David S. deCalesta.)

FIGURE 20.2 Deer guard for excluding deer from entering fenced areas. (Photo by Martin Ranch Supply.)

easily defeated them. Professionally made deer guards must be used instead; such grates are usually 5 m long, which will prevent deer from crossing them. With professional installation, fencing can cost $3–$12 per meter of fence line, and $3000–$12,000 per ha of enclosed area. Installed gates/deer guards add the additional expense of deer fence construction.

Small woodlot owners wishing to fence their DFMAs may choose to use rigid deer gates (Figure 20.3) to protect entry points—such gates are far less expensive than deer grates and the inconvenience

FIGURE 20.3 Panel deer fence for manual opening and closing. (Photo by Benner Deer Fence Company.)

of getting in and out of vehicles to open and close them may be acceptable given the low frequency of entry by humans.

Exclusion fences of 5+ strands of electrified wire have also been used to protect areas from deer browsing, but these fences are not as effective as woven-wire fences and are subject to frequent failures resulting from breaks in the lines caused by falling limbs, requiring frequent maintenance. Deer also learn to penetrate electric fences (Jacobson 2006)—woven wire deer fences are the only effective design (VerCauteren et al. 2006) for minimizing deer penetration.

Because of the high cost, deer-proof fencing is an option only for small (<100-ha, or 250-acre) DFMAs where other options, such as chemosterilants or deer harvest by public hunting or sharpshooting, are not viable and the managers have access to funding required to build and maintain such fences. Examples are few and far between: most DFMAs are fenced to prevent deer inside fenced areas from escaping or interbreeding with outside deer, such as on deer farms where deer are raised for meat production or on game preserves where deer hunting is an economic venture wherein numbers and quality of deer are tightly controlled for the benefit of paying hunters. The Fox Chapel Borough north of Pittsburgh, Pennsylvania resorted to "deer-proof" fencing to prevent deer from eradicating trillium plants (Trillium sp.) along a famed "Trillium Trail". The fence allowed the trillium to recover to previous presence after 7 years. Additionally, a section of land containing 35 townhouses within the borough was enclosed by a deer-proof fence, with a deer gate for deer-proof entry by vehicles. Deer impact on landscaping and deer/vehicle collisions were greatly reduced by the fence (Dempsy Bruce, Fox Borough lands manager, personal communication).

20.7.2 INDIVIDUAL SEEDLING PROTECTORS

Some tree species, notably oaks (Quercus sp.), do not produce seed every year, and when their seedlings do germinate, they may be eliminated in areas with high deer density, as they are highly preferred deer forage. To counter these two negative factors, foresters may plant nursery stock oak seedlings and protect them from deer browsing by enclosing them in translucent plastic tubes supported by stakes driven in the ground (Figure 20.4).

FIGURE 20.4   Tubes protecting planted oak seedlings. (Photo by David S. deCalesta.)

The tubes protect seedlings from deer browsing until they grow out of the tops of the tubes, when they may be browsed repeatedly and fail to develop into saplings that grow out of the reach of deer. Also, because these seedlings are supported by the tubes as they grow, their stems may not be as resilient to wind and are often spindly, requiring additional support when the tubes are removed from around them.

Bud caps, made of paper or plastic mesh, are a variant of tubes used to protect conifer seedlings. These seedling protectors are placed over the leaders (uppermost growing stems) to protect the tender growing points from deer browsing. As with tubes, bud caps are expensive, time consuming to apply, must be maintained frequently, only protect the uppermost growing twigs (Figure 20.5), and do not protect wildlife habitat or understory plant biodiversity. High winds can blow bud caps off terminal twigs, and deer occasionally pull them off and browse on the exposed leaders.

20.7.3 PROTECTING UNDERSTORY VEGETATION WITH SLASH

Managers place cut tops and stems (slash) of interfering woody vegetation around tree seedlings as a physical barrier to deer browsing—their interwoven branching structure provides a barrier to deer browsing (Figure 20.6). If piled high and deeply enough, slash can protect seedlings and herbaceous vegetation from deer browsing sufficiently long (before decaying) to allow protected seedlings to escape deer browsing. Slash generally is only placed above germinated seedlings on sites harvested for timber as a way to establish advance regeneration required to reforest harvested sites. Use of slash

FIGURE 20.5   Bud caps for preventing deer browsing. (Photo by Minnesota Forestry Association.)

FIGURE 20.6   Slash piled to protect germinating tree seedlings. (Photo by Sue and Jeff Hamilton.)

shares a shortcoming with other deer exclusionary devices—only seedlings, shrubs, and herbaceous vegetation on timber harvest sites are so protected, leaving understory vegetation on other portions of DFMAs (usually over 20%) vulnerable to deer browsing. Landowners of the small woodlot case histories (Chapters 37 and 38) used slash as a way to partially protect advance regeneration and other understory vegetation in conjunction with hunting and/or fencing.

20.8   REPELLENTS

Western conifers, generally ponderosa pine (*Pinus ponderosa*) and Douglas-fir (*Pseudotsuga menziesii*), have been treated with repellents to deter deer browsing, but the repellents must be applied repeatedly, especially after precipitation events that may wash off repellents. Adding an inert spreader-sticker to repellent sprays helps reduce wash-off. And, simply applying the spreader sticker material appears to reduce deer browsing. Such repellents are applied to planted conifer seedlings, which concentrates application over limited areas. However, spraying repellents on dispersed hardwood seedlings, shrubs, and herbaceous vegetation over entire DFMAs is not a viable option for preventing deer impact in eastern deciduous forestlands from economic and human resource bases.

REFERENCES

Campbell, T. A., B. R. Laseter, W. M. Ford et al. 2004. Movements of female white-tailed deer (*Odocoileus virginianus*) in relation to timber harvests in the central Appalachians. *Forest Ecology and Management* 199:371–378.

Cooper, S. M., R. M. Cooper, M. K. Owens, et al.. 2002. Effect of supplemental feeding on use of space and browse utilization by white-tailed deer. In *Land use for water and wildlife*. ed. D. Forbes and G. Piccinni. Uvalde TX: Texas Agricultural Research and Extension Center, UREC-02-031.

deCalesta, D. S. 1985. Influence of regulation on deer harvest. In *Symposium on game harvest management*. ed. S.L. Beasom and S.F. Roberson. Kingsville TX: Texas A & I University.

Doenier, P. N., G. D. DelGuiduice, and M. R. Riggs. 1997. Effects of winter supplemental feeding on browse consumption by white-tailed deer. *Wildlife Society Bulletin* 25:235–243.

Fulbright, T. A. 1999. Food plots for white-tailed deer. *Management Bulletin Number* 3:164–171. Kingsville TX: Caesar Kleberg Wildlife Research Institute Texas A&M University.

Hesselton, W. T., C. W. Severinghaus, and J. E. Tanck. 1965. Population dynamics of deer at the Seneca Army Depot. *New York Fish and Game Journal* 12:17–30.

Jacobson, M. 2006. *Forest finance 2. Fencing for forest regeneration: Does it pay?* University Park PA: The Pennsylvania State University Cooperative Extension Service.

Jansen, W. 2016. Are food plots effective at protecting row crops? An analysis of ecological and agricultural impact by whitetail deer. *Undergraduate thesis*, Carthage College, Kenosha WI.

Kendall, H. 2015. *In* Butler, T., M. Kanyoro, and W. N. Ryerson. (eds). *Overdevelopment, overpopulation, overshoot.* New York. Goff Books.

Kilpatric, H. J. and A. M. LaBonte. 2007. *Managing urban deer in Connecticut: A guide for residents and communities*, 2nd ed. Hartford CT: Division of Natural Resources/Wildlife Division, Connecticut Department of Environmental Protection.

Lutz, C. L., D. R. Diefenbach, and C. S. Rosenberry. 2015. Population density influences dispersal of white-tailed deer. *Journal of Mammalogy* 96:494–501.

Marquis, D. A., R. L. Ernst, and S. L. Stout. 1992. *Prescribing silvicultural treatments in hardwood stands of the Alleghenies (revised).* Radnor PA: Northeast Forest Experiment Station United States Department of Agriculture Forest Service. General Technical Report NE-96.

Mathews, N. E. and W. F. Porter. 1992. The rose petal theory: Implications for localized deer management. U.S. Department of the Interior. Fish and Wildlife Service No. 59.

Matschke, G. H., D. S. deCalesta, and J. D. Harder. 1984. Crop damage and control. In *White-tailed deer: Ecology and management.* ed. L. K. Halls. Harrisburg PA: Stackpole Books.

McCullough, D. R. 1979. *The George Reserve deer herd.* Ann Arbor MI: The University of Michigan Press.

McNulty, S. A., W. F. Porter, N. E. Matthews et al. 1997. Localized management for reducing white-tailed deer populations. *Wildlife Society Bulletin* 25:265–271.

Miller, B. F., T. A. Campbell, B. R. Laseter et al. 2010. Test of localized management for reducing deer browsing in forest regeneration areas. *Journal of Wildlife Management* 74:370–378.

Minnesota Department of Natural Resources. No date. *Fencing handbook for 10' woven wire deer exclusion fence.* Brainerd MN: Wildlife Damage Management Program. https://files.dnr.state.mn.us/recreation/hunting/deer/bovine-tb/fencing_guide.pdf

Oyer, A. M. and W. F. Porter. 2004. Localized management of white-tailed deer in the central Adirondack Mountains of New York. *Wildlife Society Bulletin* 68:257–265.

Predl, S., C. Kandoth, and J. Buck. 2008. An evaluation of deer management options. Northeast Deer Technical Committee.

Roseberry, J. L., D. C. Autry, W. D. Klimstra et al. 1969. A controlled deer hunt on the Crab Orchard National Wildlife Refuge. *Journal of Wildlife Management* 33:791–795.

VerCauteren, K. C., M. J. Lavelle, and S. Hyngstrom. 2006. Fences and deer-damage management: A review of designs and efficacy. *Wildlife Society Bulletin* 34:191–200.

VerCauteren, K. C., N. W. Seward, M. J. Lavelle et al. 2009. Deer guards and bump gates for excluding white-tailed deer from fenced resources. *Human-Wildlife Interactions* 3:145–153.

Ward, K. J., R. C. Stedman, A. E. Luloff et al. 2008. Categorizing deer hunters by typologies useful to game managers: A latent-class model. *Society and Natural Resources* 21:215–229.

Williams, S. C., A. J. DiNicola, and I. M. Ortega. 2008. Behavioral responses of white-tailed deer subjected to lethal management. *Canadian Journal of Zoology* 86:1358–1366.

Williamson, S. J. 2000. *Feeding wildlife—Just say no!* Washington: Wildlife Management Institute.

Winchcombe, R. J. 1992. Minimizing deer damage to forest vegetation through aggressive deer population management. *Proceedings of the Eastern Wildlife Damage Control Conference*, vol. 5, pp. 182–186.

21 Managing Vegetation to Benefit Deer Management

David S. deCalesta

CONTENTS

> Only with continued collaborations between foresters and wildlifers ... can we hope for effective solutions to the impact of deer.
>
> **Laura Kenefic, Jean-Claude Ruel, Jean-Pierre Tremblay (2015)**

21.1 INTRODUCTION

Chapter 13 dealt with the impact of overabundant deer herds on silvicultural treatments designed to enhance timber production. This chapter deals with the other side of the deer-silviculture coin: using silvicultural treatments to improve deer and habitat condition while reducing the negative impact of deer on forest resources. Silvicultural influence on deer management is of two kinds: (1) improvement of habitat, including forage, thermal, and hiding cover, and (2) influence on deer spatial use of habitat to reduce impact on forest resources. Silvicultural practices designed to influence deer management are effective only at the deer–forest management area (DFMA—see Chapter 9) level where managers have control over the placement, size, type, and timing of the practices.

21.2 FORAGE CREATION/HABITAT IMPROVEMENT

Increasing forage production can be accomplished by harvesting maturing overstory trees (final harvest removals) and reverting stands to the early succession stage, or by thinning overstory trees (shelterwood seed cuts and/or improvement cuts that remove unthrifty overstory trees) sufficiently to enhance germination and development of seedling regeneration. The resulting forage can reduce intensity of deer impact on seedling regeneration within DFMAs by enhancing growth of deer forage species (seedlings, shrubs, herbs). For this to work, deer density must be maintained at levels compatible with maintenance of diverse and abundant understory forest vegetation. The amount of overstory removals/thinnings is constrained by the requirement to retain sufficient regeneration and overstory trees to insure harvests are sustainable and is usually referred to as the annual allowable cut (AAC; see Chapter 13).

Increasing forage production via silvicultural treatments generally involves reverting maturing timber stands to early succession stands through timber harvest, and can be maintained by restricting timber harvest to sustainable levels. For increasing deer forage quality and quantity to affect deer impact, deer density must be reduced to levels compatible with the carrying capacity chosen for impact on forest resources and maintained at that level.

Early succession stands are a critical habitat type for forest gamebirds and songbirds dependent on vegetative structure (dense understory and shrub layers) for nesting, as a foraging habitat for feeding nestlings, and as critical postnesting habitat: these species will benefit from silvicultural practices designed to increase amount and quality of deer forage (Thompson and Dessecker 1997, Chandler et al. 2012, King and Schlossberg 2014, Stoleson 2013).

Managers may develop and maintain more or less permanent deer foraging areas by the practice known as coppicing. In coppicing, tree species that regenerate from sprouts growing from cut stumps (oaks, maples, birches—species preferred as deer forage) are harvested repeatedly at short (2–10-year) intervals in stands managed for coppicing. The harvested wood is used as pulpwood or wood fuel pellets for which there are sustainable markets. Coppice wood grows rapidly and is highly digestible to deer. Coppicing creates an irregular mosaic of habitats in a patchwork of stands of different ages, attracting a wide range of flora and fauna—ideal for wildlife, particularly those species requiring an open woodland habitat. The succulent young shoots from coppiced trees attract browsing deer (The Conservation Volunteers 2018).

DFMA managers might wish to consider maintaining a portion of their forestlands in coppices, as they form permanent forage locations that may reduce deer browsing pressure on seedlings in regeneration sites designed to produce lumber, which require longer growing cycles, often as much as 80 or more years. In one of the small woodlot case histories (Chapter 39), the manager repeatedly coppiced areas he called "salad bars," which he used to reduce deer browsing on seedlings of preferred tree species.

Before DFMA managers contemplate any kind of forest vegetation manipulation, including final harvesting of mature timber or thinning of maturing stands to create favorable habitat for deer, they must consult with professional foresters to ensure that conditions are favorable for harvesting/ thinning trees and obtaining desired habitat conditions. Before overstory trees may be harvested, understory advance regeneration of desired species composition and abundance of woody, shrubby, and herbaceous vegetation must be ascertained by monitoring. If barriers to successful regeneration are present, such as dense fern ground cover or dominance of interfering seedlings of species that can outcompete desired seedling species, preharvest treatments such as eliminating interfering vegetation with herbicides must be applied.

Larger DFMAs, such as state and national forests and large commercial forest holdings, may have on staff foresters who can determine whether preharvest conditions for timber harvest/thinning are met or whether preliminary steps, such as removing unwanted, competing vegetation with the use of herbicides must first be conducted. Managers of smaller DFMAs may need to seek the services of competent forestry specialists to identify and apply needed treatments, as well as plan and conduct appropriate timber harvesting/thinning.

21.3   INCREASING LANDSCAPE LEVEL CREATION AND JUXTAPOSITION OF SUCCESSIONAL STAGES

Creating a complete suite of deer successional habitat stages (fawning and forage, hiding, thermal, and multipurpose) in close proximity and near sources of water provides deer with their habitat requirements, ensuring quality deer and year-round deer presence for hunting and viewing, if deer density is maintained at the chosen carrying capacity. Favorable juxtaposition of successional stages can be arranged by judicious and sequential placement of overstory removals and thinnings.

21.4   INFLUENCE ON DEER USE OF HABITAT

Campbell et al. (2004) suggested that lack of changes in white-tailed deer (*Odocoileus virginianus*) movements before, during, and after timber harvest indicated that forest vegetation management aimed at attracting deer away from problem areas (e.g., areas with low regeneration success) or toward browse supplies during severe winters would likely be unsuccessful. Their study did not evaluate actual browse impact of deer on harvest sites; rather, it evaluated deer movements/home range changes over a short period of time. By contrast, Brown and Cooper (2006) cited several studies involving established deer supplemental winter feeding stations; deer browsing pressure was seven times that in areas without feeding stations, and heightened impact on natural browse occurred at distances 900–1200 m from feeding sites. If deer foraging on native vegetation in forestlands surrounding food plots planted to herbaceous forage exhibits an effect similar to supplemental feeding sites (likely), timber harvest sites and food plots may increase browsing pressure on seedling regeneration, leading to failures of regeneration following timber harvest in areas surrounding food plots, if deer density is not brought to and maintained at desired carrying capacity.

If deer density can be maintained within carrying capacity for diversity and abundance of forest vegetation and wildlife habitat, the above studies suggest that deer movement patterns and habitat use related to food plots/timber harvest sites can be capitalized on by focusing hunting effort in those areas and along deer trails leading to those areas. Apprising hunters (and/or sharpshooters contracted to reduce deer density) of enhanced forage locations (through maps) will make hunters/sharpshooters more effective in reducing deer density and associated impact on forest resources.

As noted in Chapter 10, hunters rarely travel more than 600 m from access points (forest roads, timber landings/parking areas, maintained trails) to hunt for deer. Hunters will hunt deer foraging areas such as clearcuts and thinnings, but only if they are made aware of them and if they do not have to travel long distances to hunt them (and drag out harvested deer). It is incumbent on managers to ensure that roads leading to recent timber harvests, and landings where logs were deposited (and which form parking areas for hunters) are well-maintained and accessible to hunter vehicles (which include passenger cars), and that information on how to reach these access routes/points is readily available to hunters and clearly marked within DFMAs.

REFERENCES

Brown, R. D. and S. M. Cooper. 2006. The nutritional, ecological, and ethical arguments against baiting and feeding white-tailed deer. *Wildlife Society Bulletin* 34:519–524.

Campbell, T. A., B. R. Laseter, W. M. Ford et al. 2004. Movements of female white-tailed deer (*Odocoileus virginianus*) in relation to timber harvests in the central Appalachians. *Forest Ecology and Management* 19:371–378.

Chandler, C. R., D. I. King, and R. B Chandler. 2012. Do mature forest birds prefer early-successional habitat during the post-fledging period? *Forest Ecology and Management* 264:1–9.

Kenefic, L., J. C. Ruel, and J. P. Tremblay. 2015. Sustainable management of white-tailed deer and white-cedar. *The Wildlife Professional* Fall 2015: 33–36.

King, D. I. and S. Schlossberg. 2014. Synthesis of the conservation value of the early-successional stage in forests of North America. *Forest Ecology and Management* 324:186–195.

Stoleson, S. H. 2013. Condition varies with habitat choice in postbreeding forest birds. *The Auk* 130:417–428.

The Conservation Volunteers. 2018. Coppicing—Why cut trees for conservation? https://www.conservationhandbooks.com/coppicing-cut-trees-conservation/

Thompson, F. R. and D. R. Dessecker. 1997. *Management of early successional communities in central hardwood forests with special emphasis on the ecology and management of oaks, ruffed grouse, and forest songbirds.* General Technical Report NC 195. St. Paul MN: USDA Forest Service, North Central Forest Experiment Station.

22 Communicating with and Educating Stakeholders

David S. deCalesta

CONTENTS

I am convinced that most Americans of the New Generation have no idea what a decent forest looks like. The only way to tell them is to show them.

Aldo Leopold (1986)

I hear I forget, I see I remember, I do I understand.

Confucius (Chinese proverb 5th century BC)

Hunters need their state wildlife agency, and the agency needs hunters. It's not good when hunters circumvent their agency, nor is it good when agencies don't listen to their hunters.

Kip Adams (2017)

22.1 INTRODUCTION

It was once believed that communicating relevant facts would convince people to manage deer based on science. However, perception is reality, and one person's facts may be another's fiction. As discussed in Chapter 11, managers must recognize and incorporate the culture and values of the disparate groups of stakeholders that will influence their deer management. Some stakeholders (legislators and administrators) may constrain or enhance deer management via legislation and regulation; some (hunters) provide an essential service (reducing deer density); some (farmers and foresters) provide political and practical support; and some (outdoor writers) may promote (or question) facts, goals, and management actions for deer. If managers are managing someone's forestland, the owners of those landscapes form another stakeholder category. All must be cultivated, informed of the requirements regarding deer management, involved, and motivated in ways that are effective.

Accordingly, managers must choose topics and related messages that resonate with stakeholder concerns and perceptions if they wish to communicate effectively with them. After they have developed and honed the messages, they must use relevant communication technology to deliver and reinforce the messages. They must enlist messengers credible with targeted stakeholder groups to help develop and deliver the message specific to these groups. And, just as deer management is an evolving and annually recurring necessity, so too is the education of, and communication with, stakeholders.

Managers may have the science right, and they may relate it to the values of stakeholders in deer management, but if they do not communicate it effectively, they will neither convince stakeholders to adopt and support the science nor teach them how to use it in deer management. They must package the information carefully and communicate it in ways that assure understanding, acceptance, and use: what they say and how they impart it are critical in deer management.

Components for an effective communication/education program for deer management are: (1) the goal of the program; (2) the messages; (3) the messengers and means by which the messages are delivered; and (4) the process to insure that recipients understand, gain confidence in, and incorporate the information in their actions related to deer management. Preceding chapters on biology, science, values and culture, and politics describe how these factors affect deer management and identify associated stakeholder groups. This chapter describes, by stakeholder group, the messages, how to package and present them, and how to involve recipients of the information to the point where they believe in and use the information contained in the messages.

As noted in this book, deer management is performed on deer–forest management areas (DFMAs—see Chapter 9) by managers/managers and not on deer administrative units (DAUs – see Chapter 9), which are political rather than management landscapes. Communication/educational programs should be conducted for DFMAs and on DFMAs. It should be recognized that these programs are designed to educate/inform stakeholders of the requirements for actively managing deer, forest vegetation, and hunting on individual DFMAs. Furthermore, it should be emphasized that while regulations for managing deer on DAUs form a framework for harvesting deer on DFMAs, deer hunting is only one of many equally important deer management activities. Finally, it should be emphasized that if hunting regulations specific to DAUs can be enhanced by specifying them for individual DFMAs, managers of DFMAs will have more effective programs for using deer harvest as an important component of deer–forest management.

A final organizational concept is that communication and education programs should be developed and delivered for two different audiences. The first is stakeholders (managers, hunters, outdoor writers, educators, and researchers) concerned with deer management on individual forested properties (DFMAs) where deer management specific to individual properties occurs. The second is natural resource administrators and politicians whose involvement in deer management is primarily from a regulatory basis relative to large landscapes (states and DAUs within states). Administrators and politicians need to be aware of issues faced by owners of DFMA properties and how regulations for harvesting deer set by them affect deer management on DFMAs. Educational packages developed for administrators and politicians should include information on how deer management decisions/regulations designated for DAUs affect success on DFMAs.

Communication programs are limited by available financial and human resources and must be scaled accordingly. Managers of large tracts of forestland, such as national or state forests, or large commercial forests, can spend far more capital on communication programs than can owners of, say, small woodlots for whom resources are far more limited. Suggestions for organization of communication programs according to resources available to categories of managers based on size are presented in the concluding section of this chapter.

22.2 GOAL OF EDUCATION/COMMUNICATION PROGRAMS

The goal of a deer management education/communication program is to gain the support and involvement of members of stakeholder groups so that they act in ways that advance the goals of forest management (e.g., reduce deer density and impact).

22.3 MESSAGES

The messages must be stakeholder specific; must address stakeholders' values, culture, and concerns about deer as affected by deer management; must be delivered by credible messengers; and must motivate stakeholders to support, promote, and actively engage in activities to advance deer management goals of managers. For example, hunters interested in harvesting trophy deer will understand and support reducing deer density if it is presented as necessary to allow managers to produce quality forage in quantities sufficient to produce trophy deer. The same hunters will not be motivated to reduce deer density if the reason given is to promote sufficient advance regeneration to allow the landowner to produce and harvest quality timber.

22.4 MESSENGERS

Information delivered to stakeholder groups must be presented by messengers that stakeholders believe in and trust. Again, taking hunters as an example stakeholder group, stakeholders will not be swayed by data-filled charts and graphs presented by university/government scientists in coats and ties giving PowerPoint presentations. But they will listen to and accept information in local newspaper articles, or outdoor newsletters written by outdoor writers who relate the information to hunter interests and who hunters have rubbed shoulders with in outdoor expos or field trip show-and-tells.

22.5 PREPARING STAKEHOLDERS TO BELIEVE THE MESSAGE

Prior to presenting the message, the messenger(s) must prepare the stakeholders to accept the message. Termed "presuasion" or "prepersuasion" (Cialdini 1994), presenters should employ established principles of persuasion: (1) indicate that they share the values of the stakeholders and include praising values of the stakeholders, (2) incorporate peer pressure by mentioning that others in the stakeholder group have supported the message, (3) obtain public commitment from stakeholders for the message so they feel engaged and involved, (4) provide evidence of expertise in the subject matter that will resonate with stakeholders, and (5) explain to stakeholders how accepting and employing actions described by the message will help them obtain the benefits they are seeking.

22.6 MESSAGE DELIVERY SYSTEMS

The most effective messages/educational programs are those that incorporate audible and visual cues, involve hands-on involvement of participants (stakeholders) and provide opportunities for real-time feedback and two-way communications.

22.6.1 Active Message Delivery Systems

22.6.1.1 Public Presentations

Public presentations feature credible messengers using visual aids to inform/educate stakeholders about the need for, and benefits of, deer management designed to manage deer density and impact. Presentations are usually held at large venues such as auditoriums and other indoor venues where weather will not affect attendance, use visual aids (slide shows, PowerPoint presentations), and can address sizeable audiences (hundreds).

The Kinzua Quality Deer Cooperative (KQDC) case history (Chapter 34) provides an example of such a public presentation. Prior to initiation of the KQDC demonstration project (reducing deer density by public hunting can restore quality deer and quality habitat), the KQDC leadership team held a well-advertised public meeting at a local university to introduce the project and generate support from local stakeholders, primarily deer hunters. The hunters' acceptance of and participation in the program were critical—they were the ones who would reduce deer density to target levels by harvesting more deer, especially more antlerless deer.

The program began with a comparison of two sets of antlers; one harvested decades earlier when deer density was lower and more deer forage was available and the other harvested under current high deer density and limited amounts of quality deer forage (Figure 22.1). Trophy and diminutive antlers were presented by two local residents who were highly respected and trusted by local hunters—one was a wildlife conservation officer from the Pennsylvania Game Commission, the other was the regional forestry extension agent. Both had given many presentations on wildlife and wildlife habitat to local hunting groups, were avid deer hunters, were personally known to many in the audience, and had forged trusting relationships with local hunters. These two reliable sources of information identified the time frames whence the deer were harvested and asked the leading question: "Which of these two antler sets would you prefer to see and perhaps harvest while hunting? We'd rather be hunting quality deer, wouldn't you?"—thereby planting the idea that things could be done to bring back the era of quality deer.

They then stated that the secret was abundant and quality deer forage and introduced a third speaker who produced the answer to the question of how to manage for quality deer by forage improvement. This person was perhaps the most respected and trusted wildlife biologist employed by the Pennsylvania Game Commission (PGC), a nationally recognized expert on bear biology and management named Gary Alt, who had been reassigned to renovate the PGC deer management program. Alt gave a slide presentation that featured side-by-side comparison of abundant and quality deer forage (inside a fenced deer exclosure) with currently available forage not protected from excessive deer browsing. He noted that the factor responsible for vegetation differences inside and outside the exclosure was high deer density outside the exclosure (>12 deer/km^2) versus zero deer/km^2 inside the exclosure. Alt went on to say that quality habitat existing in areas with target deer density also provided greater abundance and quality of other game species such as rabbits, grouse, and turkey, citing scientific studies that supported his position. And, he proposed a simple solution—reducing local deer density to levels ($\leq$5 deer/km^2)

FIGURE 22.1 Comparison of antlers harvested under low deer density and high forage availability (left panel) with antlers from period of high deer density and low forage availability (right panel). (Photos by John Dzemyan and David S. deCalesta.)

known to be associated with quality habitat and quality deer. He stated that the easiest way to achieve the desired reduction in deer density was for hunters to harvest more antlerless deer as facilitated by liberal hunting regulations for antlerless deer.

Alt's educational program was designed basically for deer management at the DAU level and did not describe costs of, or requirements for, deer management by managers of DFMAs. His message was designed to promote more liberal hunting regulations that would lead to reductions in deer density and impact on DAUs as well as on DFMAs embedded within DAUs.

Alt had given the same presentation to thousands of hunters at many similar local presentations and won over deer hunters and the commissioners of the Pennsylvania Game Commission with it—two facts he shared with the hunters and other stakeholders in the audience. He concluded by stating that harvesting more antlerless deer, as accomplished with concurrent buck and doe hunting seasons, and restricting harvest of antlered deer to bucks with at least two antler points on at least one of two antlers, would result in lowered deer density and improvements in quality and quantity of deer forage, and would allow more yearling bucks to survive to older age classes where, coupled with improved forage, they would exhibit the characteristics of quality deer (bigger, more impressive antlers, and heavier body weight and condition).

22.6.1.2 Field Trips

Show-and-tell field trips in the woods, primarily for natural resource administrators and hunters, are usually conducted around fenced exclosures providing comparison of vegetation protected from, and exposed to, deer browsing. Field trips provide the additional reinforcement of showing participants what deer browsing looks like in the field, providing indisputable evidence of deer impact. However, without the additional contexts of deer density, hunting regulations, culture, values, or follow-up, field trips rarely accomplish more than acknowledgment by administrators that deer can impact forest resources. And, because of the logistics of transporting participants to field trip sites, and the limited audience size (usually <20 participants), field trips do little more than increase awareness of the reality of deer impact among a small percentage of stakeholders.

22.6.1.3 Workshops

The best setting for integrating science and perception in deer management with maximum reinforcement of the message is workshops held on DFMAs. The target audience should include managers, hunters, natural resource biologists and administrators, consultants, university researchers, educators and extension professionals, and politicians who oversee natural resource agencies. Workshops provide the additional educational enhancement of active involvement of participants. Messages presented at workshops are likely to stick because the participants go into the woods, collect relevant data for estimating deer density and deer impact, and then evaluate the relationship between the two with analysis and interpretation of the data they helped collect.

The need and utility of such workshops was identified by Timothy Pierson, a Pennsylvania State University extension forester who was frustrated by his inability to reconcile science and perception regarding deer management among stakeholders holding opposing views on desirable deer density and deer impact.

Pierson developed a deer density and impact workshop program based on five sources of information related to sustainable forest resource management: (1) development of a Society of American Foresters Deer Committee effort to provide training for youth deer hunters, (2) evolving development of a deer density and carrying capacity workshop for the Kinzua Quality Deer Cooperative (see Chapter 34), (3) deer management programs developed by the Quality Deer Management Association, (4) development of technology for characterizing deer impact (Pierson and deCalesta 2015), and (5) SILViculture of Allegheny Hardwoods (SILVAH) workshops developed by Marquis et al. (1992) that applied research findings to solve problems (including deer impact) with establishing seedling regeneration needed for forest management. SILVAH workshop attendees were presented with lectures on forest ecology and learned how to manage vegetation to achieve forest management goals. SILVAH workshops featured field exercises wherein attendees collected forestry-related data that were crunched in classroom portions of the workshops to identify and derive management actions.

Pierson developed and delivered 1-day deer density and impact workshops for presentation on forested properties (DFMAs) where deer density and impact were high. During a morning classroom setting, instructors presented information on basic deer biology, including food habits and deer health as related to deer density, and deer impact on forest vegetation, including seedlings, shrubs, and herbaceous vegetation, and on wildlife dependent on quality habitat to meet their life needs (Figure 22.2). Methodologies (pellet group counts and estimates of level of deer impact on indicator tree seedling species) for collecting and analyzing deer density and impact were described. The concept of carrying capacity, particularly as it differed among stakeholders holding varying goals related to deer quality, abundance, and health and status of other forest resources, were presented and discussed.

FIGURE 22.2 Classroom portion of deer density and impact workshop. (Photo by David S. deCalesta.)

The intent was for hunters and other stakeholders to understand that deer density related to meeting their goals might prevent managers from achieving their goals when their goals required managing for a different deer density. Hunters and other non-landowner stakeholders learned that forest landowner goals for deer density would lead to quality deer habitat and quality deer and that it was the forestlands of these owners that most deer hunters hunted on, not having their own forestland to hunt on. Accepting lower deer density as a compromise for allowing managers to achieve their goal for managing forest resources was suggested as an easy trade-off for hunters, especially as lower deer density would result in deer of higher quality.

During an afternoon session, participants received final instructions (Figure 22.3), and were divided into five crews, each with the task of collecting deer density and impact data from plots on a 1700-m-long transect within a grid of five such transects. Each crew was led by a forester/wildlifer who could identify indicator seedlings. Two members of each crew clipped and collected deer browse occurring on the 52 plots taken per transect line. Participants took turns counting deer pellet groups, identifying and classifying deer impact on seedlings, and clipping and bagging deer forage from the plots. Protocols for collecting deer density and impact data (Appendix 1) were followed. Crews returned to the classroom following completion of data collection and the workshop instructors collected data sheets, entered raw data into a spreadsheet, and calculated deer density and impact for each transect and for the grid as a whole.

Resulting estimates of deer density and impact were displayed on a carrying capacity curve (Chapter 15) and discussed relative to goals of different stakeholder groups. Bags of clipped forage collected along with density and impact data were examined to present graphic evidence of the paucity of forage resulting from high deer density.

Management options were discussed for reducing deer density and impact to levels specific to different stakeholder goals. Workshop attendees learned that achieving goals for different stakeholder groups involved adjusting stakeholder group goals to arrive at compromise solutions.

Responses to the workshops were positive and enlightening, as some stakeholder groups (hunters and natural resource administrators) were forced to see how achieving their deer density goals prevented other stakeholders, particularly the landowners who open their forests to hunters, from achieving theirs. When the goals of hunters and landowners were reconciled by adopting as goals production of quality deer and quality habitat, hunters and administrators could see that achieving goals for deer health and forest health could be met by a compromise deer density.

Not all hunter participants were willing to compromise on reduced deer density to achieve improved deer health, even if it meant that the landowners who opened their land to hunting experienced economic losses engendered by high deer density. But many hunters could understand the need for compromise, and natural resource administrators and politicians had their eyes opened relative to the

FIGURE 22.3 Students receiving final instructions before collecting field data for a deer density and impact workshop. (Photo by David S. deCalesta.)

impacts their setting of deer harvest regulations had on the landowners who provided hunting areas to deer hunters. And, these administrators and politicians could better understand that regulations designated for managing deer density on DFMAs, in line with the needs of managers of those DFMAs, would benefit important stakeholders (hunters) by enhancing deer habitat and quantity on the lands of others where they hunted.

22.6.2   PASSIVE MESSAGE DELIVERY SYSTEMS

22.6.2.1   Public Testimony

University and government (state and federal) scientists conducting basic and applied research for resolving deer overabundance and impact on forest resources testify at (annual) deer management hearings held by natural resource agency administrators. Unfortunately, their belief that science-based testimony in support of reduced deer density will sway agency administrators to opt for deer management designed to reduce deer density and impact generally falls on deaf ears. Affected landowners may also present testimony on the impact of deer on their resources, but these testimonials, like those of scientists, generally are countered and overwhelmed by far greater numbers of emotional pleas from hunters to save their culture, values, and traditions by increasing deer density to generate, among other things, excitement of young, first-time hunters. Unsolicited letters sent to natural resource administrators by scientists and managers of DFMAs promoting reductions in deer density based on science are likewise swamped by hundreds to thousands of letters sent by hunters requesting increases in deer density. These efforts to promote science and description of the impact of deer on personal and public property as the bases for deer management, though well-intended, logical, and conclusive, fail in the face of overwhelming input from hunters based on culture and values.

22.6.2.2   News Releases (Newspaper, Television, Radio)

One-way communications from scientists and managers in the form of written material (scientific papers, handouts, leaflets, circulars) use only one medium—the written/spoken word. They provide no opportunity for reinforcement by qualified speakers or for feedback from intended audiences. Indeed, targeted stakeholders (such as hunters) have little incentive to pay any attention to information presented in this way and generally do not. Also, production of written materials is generally beyond the capability of managers—they rely on materials provided by professionals in universities, government institutions, or management organizations. These materials are generic and not specific to individual DFMAs, reducing their relevance. A last roadblock is the cost and difficulty of distributing the information to targeted stakeholders.

22.6.2.3   Social Media

There is more managers can do to produce comprehensive and interactive systems for educating and communicating with stakeholders. The simplest and least expensive way to inform and interact with stakeholders is via social media. The internet opened up a new means of communicating with/educating stakeholders through their computers and smartphones. The material is written and enhanced with photographs and opportunities for two-way communications among managers and stakeholders.

Social media systems should integrate the established communication methods described above with more interactive and media-savvy delivery mechanisms (websites; blogs; and other applications such as Facebook, Twitter, Instagram, and Snapchat). These media enable managers to create and share information about their deer–forest management with stakeholders. Websites and blogs are the most comprehensive and established means for using social media to reach a broad audience of stakeholders who likely are familiar with them and will be comfortable with using them. But they require specialized software to build them and computer-savvy personnel to run them—all requiring an outlay of time and money.

Platforms for social media interactions are e-mail, websites, blogs, Facebook, Twitter, Snapchat, and Instagram. Reliance is still on the written word, but social media outlets are formatted to enhance communication among users, including managers and stakeholders. Disadvantages of social media are that the reliability of information behind social media communications is questionable and they do not provide hands-on, in-the-field experience necessary for reinforcing messages.

Managers can build e-mail lists of stakeholders by groups (e.g., hunters, natural resource managers, administrators, and neighboring landowners) and provide them with timely information, such as maps illustrating access points to properties, advice on how to apply for antlerless permits, or annual progress reports that can be attached to e-mails. And, e-mail facilitates two-way communication—stakeholders can respond to e-mails, providing feedback and/or requesting additional information. However, because initial contact between managers and stakeholders is initiated by managers/landowners (who must first obtain stakeholder e-mail addresses), stakeholders have no way to seek out and obtain information about landowners' forestlands/deer hunting opportunities via e-mail without first being contacted by the managers/landowners. Other, more facile communication methods, such as websites and blogs, can be found by stakeholders searching the internet, which in turn will lead them to managers' information without the need for managers to initiate the contact.

In addition to educating/communicating with stakeholders, social media provide a way to recruit hunters to hunt on specified DFMAs and to enhance their hunting success by providing directions to the DFMAs, maps detailing hunting locations, and access points for individual DFMAs. An important additional benefit of social media is that it allows stakeholders (e.g., hunters) to exchange information among themselves in ways that increase potential for harvesting more deer on DFMAs.

Landowners can create additional opportunities for engaging and educating stakeholders in the field through deer checking stations and monitoring activities (e.g., collecting information on deer density and/or impact) involving hunters and other stakeholders as volunteer or paid participants. Including multiple stakeholder groups, such as hunters, natural resource administrators, legislators, and managers in these activities provides an opportunity for often-opposing stakeholder groups to interact positively and productively. If the activities conclude with informal social events such as cookouts in neutral settings, they can foster understanding and help build bridges among stakeholder groups.

22.6.2.4   Social Media: Websites

The internet provides a vehicle—websites—for businesses (deer management is a business) that allows them to attract new customers as well as keeping current ones informed, up to date, and aware of new developments/products. In the early days of website design and development, building websites was restricted to programmers familiar with programming codes such as HTML and JavaScript. Now, improved software (e.g., Sabin-Wilson 2017—*WordPress for Dummies*) works in what is called design mode that novices can use to build and maintain websites easily and intuitively with no training in computer programming.

Websites are an excellent way to inform stakeholders about forest operation and management, problems managers may be having with deer, listing of hunting seasons and regulations, detailing how hunters can obtain access to hunt DFMAs and applying for permits for harvesting antlerless deer (see Figure 22.4 for an example). Websites can provide updated road access and hunting maps specific to DFMAs (among other information) that can be downloaded and printed. They can also inform hunters of local lodging, restaurants, and sporting goods stores. Information posted on websites usually is updated as needed, sometimes as little as two or three times a year.

22.6.2.5   Social Media: Blogs

Like websites, blogs can provide stakeholders with information about deer management. Blogs provide information as posts (see Figure 22.5 for an example), wherein managers can provide updates to deer management as short notes to keep stakeholders interested and informed. Relevant pictures may be added to blogs to stimulate interest. Additionally, blogs allow viewers to respond to

Kinzua Quality Deer Cooperative Website

HOME   HUNT THE KQDC   PHOTO GALLERY   AREA INFO   CONTACT US   ROAD ACCESS MAPS   KQDC BLOG

The Kinzua Quality Deer Cooperative (KQDC), begun in 2000, is an on-going demonstration, of how hunting can be used to meet the goals of multiple publics for managing deer. A partnership of forest landowners, forest managers, biologists, hunters, and local businesses developed the program which relies on hunters to manage deer density on a representative timber management area. The program is conducted on a 74,000 acre demonstration are in northwestern Pennsylvania.

Deer management in the northeastern United States, and Pennsylvania in particular, has been controversial and difficult since the early 1900s because of the impact of overabundant deer herds on so many different publics. For some, such as forest managers and ecologists, there have been too many deer. For others, such as hunters and businesses dependent on recreation with forest resources there are not enough deer.

Deer browsing changed understory vegetation from desired shrubs and tree seedlings to undesirable seedling species and ferns - foresters had to use fences and herbicides to give desirable seedlings and other understory plants a chance to grow in the presence of too high deer density. At the same time, the abundant and famous deer herd attracted thousands hunters to the area and business catering to hunting flourished.

Combining deer science with adaptive management, the KQDC identified and integrated goals for deer management to benefit affected publics. The program centers n collecting and analyzing monitoring data to determine if the goals are being met. As deer density and impact declined following changes in management, hunting regulations were adjusted to bring deer density and impact in line with goals. Thanks to hunters and hunting, the goals were met and maintained after 5 years of management.

FIGURE 22.4   Example website. (From the KQDC case history, Chapter 34.)

Kinzua Quality Deer Cooperative Blog

HOME   HUNT THE KQDC   PHOTO GALLERY   AREA INFO   CONTACT US   PUBLICATIONS   KQDC WEBSITE

KQDC Deer Season Kick-Off October 25, 2015
The KQDC has the information you need to enjoy hunting Allegheny Whitetails even more this season. At our deer season Kick-Off event on October 1st, you'll find booths from the KQDC and info about roads, recent timber harvests, spring deer density counts and more. There will also be booths from the Federated Sportsmen of Pennsylvania, The Pennsylvania Game Commission, and local sporting goods merchants.

KQDC Pellet Group Counts to measure white-tailed deer density
As is done every spring, we collected data on deer density and impact. Deer density was up slightly (14 deer/mile^2) from 2014 (13.4 deer/mile^2) but within goal level. Impact was nearly identical but hopefully will drop lower with more timber harvest and more deer food.

| Search |

Recent Posts

Roads open for 2015 hunting season

Number of KQDC antlerless permits set

Last of deer fencing taken down

Tips on hunting from tree stands

2014 big doe contest winner

FIGURE 22.5   Example blog. (From the KQDC case history, Chapter 34.)

information which may then be displayed as archived blog posts—blogs are more interactive than websites and are updated more frequently, sometimes as often as once a week or more. Software is similar to that used to develop websites (e.g., Blair 2016—*Blogging for Dummies*, or Sabin-Wilson 2014—*WordPress for Dummies*). Some include a blog as part of their website.

22.6.3   REQUIREMENTS FOR WEBSITES AND BLOGS

Websites and blogs are usually built and maintained on computers, either desktops or laptops, so managers will need access to a computer with adequate storage and speed and will need to be able to connect to the internet through a browser (interface), the commoner ones being Internet Explorer (IE), Google Chrome, or Mozilla Firefox. Because websites and blogs are viewed on the internet, managers will need an internet provider to host their website and post it on the internet. Commonly used internet hosting companies are BlueHost, eHost, and GoDaddy. There is an annual fee for internet access (generally ~$100/year), and managers will have to obtain their own domain (name of website that must be unique, such as www.kqdc.com). The domain name for websites or blogs is the internet address that stakeholders enter on their computers (or smartphones) to get access to the information on websites/blogs. Internet hosting companies can assist managers in obtaining a domain name (easy) and there is usually a small, one-time fee (may be free or less than $25) to purchase domain names.

Stakeholders can access websites or blogs with a computer *and* with smartphones, which provide easy, instant, and portable access. They can find websites/blogs when they search the internet using key words. Key words are descriptions of websites/blogs that relate to the content. Appropriate keywords (which can be phrases in addition to single words) might be: property/operation name (e.g., *KQDC* for the Kinzua Quality Deer Cooperative, Chapter 34), *hunt deer in* (state name), or *free deer hunting at* (DFMA). All websites/blogs have a suffix like .com, .net, or .org which is set up when domain names are obtained and interfaces websites/blogs with the internet. Keywords are embedded in websites/blogs in descriptors called meta-tags. Managers can increase the odds of people searching the internet finding their websites/blogs by obtaining the services of an online facilitator service (like Google adwords). These facilitators charge a small fee to post keywords on the internet as a way for internet searchers to find websites/blogs. These services provide managers with feedback regarding how many persons view their website and the number of times they visited different pages on websites/blogs, as well as suggesting additional keywords for use.

The biggest challenge to newcomers is designing and uploading their website/blog to the internet and maintaining it (providing new information as needed). Managers can hire an internet consultant (webmaster) who can build and maintain their website/blog and interface it with a web hosting company, but their fees for developing a website (usually in excess of $500) and keeping it current (usually a small monthly fee) are an additional management expense. And, every time managers want to post new information, they have to contact their webmaster so they can effect the changes. Or, managers may have an arrangement with the webmaster that provides them with access to the website/blog such that they can provide updates. It is usually worth the investment in time for managers to teach themselves how to build and maintain a website/blog—or hire an information technology (IT) person whose responsibilities include building and maintaining the website/blog. Websites and blogs do face the very real potential for discontinuity of service/messaging if the person in charge of the program leaves without a successor being appointed, resulting in disappointment and dissatisfaction among the stakeholders you worked so hard to connect with.

22.6.4   SOCIAL MEDIA: FACEBOOK, TWITTER, SNAPCHAT, AND INSTAGRAM

These social media programs offer additional ways to interact with and inform stakeholders. Like e-mail, managers must reach out to recipients—stakeholders have no way to access these media unless managers contact them and make them aware of them. These programs require access to the internet but do not require specific software or connection through the services of webhosting companies, or even a computer—managers can set up social media accounts and provide information with a smartphone (although it may be easier to do on a computer). And, they are free. Managers simply post (or read) short information messages (that can include photographs) through e-mail that can be read instantly.

The downside of these media is that, unlike websites or blogs, managers can only provide limited information about their deer management program, and the material is not archived—readers cannot go back in time to previous posts to retrieve information such as maps of DFMAs showing access points or hunting season information, including how to obtain permits for harvesting antlerless deer. Also, managers should post information on a relatively frequent schedule (perhaps weekly or monthly) so readers will not lose interest and cease accessing the information. Mostly, these programs are a way to provide stakeholders with recent developments to keep them informed and interested and to direct them to websites/blogs. An additional feature of these media is that they allow stakeholders to interact with each other. Having hunters sharing information about managers' programs with other hunters (including photographs of deer they have harvested or recorded with trail cameras) can lead to additional hunters hunting deer on managers' property, and build interest and enthusiasm for hunting on their forestlands.

These media are an effective way to provide special groups of stakeholders, such as natural resource administrators or legislators, with short "word bites" of information (results of harvest on specific DFMAs, successful reduction of deer density and impact, satisfaction and support of hunters) on a timely and recurring basis to keep them informed about concerns and management. Given that time is at a premium with these stakeholders, providing concise, timely, and digestible information increases the likelihood they will read it and act upon it.

22.6.4.1 Social Media: Facebook

Facebook is likened to an interconnected and expanding web of contacts with the person writing the information at the center. The information posted is shared by persons on the network, who pass the information on to other people they know in an ever-widening circle of contacts. Facebook allows managers to reach an infinitely large group of people of like interests, but managers can limit the number of persons getting the information they post.

As with websites and blogs, there is a helpful how-to book for beginners (*Facebook for Dummies*—Abram 2016) that guides one through the process of setting up a Facebook account (free) and includes tips on using Facebook to increase outreach to potential users (the deer management is a business thing). One registers with Facebook by going to the website (www.Facebook.com), creating a user profile (description of the deer management program), and adding a list of "friends" (stakeholders to interact with and who can spread messages about deer management programs). Then, managers can post updates (new information), including photographs, and exchange messages with "friends" organized into groups (e.g., the administrators and legislators managers want to reach, other managers to interact with, persons hunting on specified DFMAs). Most managers have children or grandchildren who can show them how to set up a Facebook account and begin connecting with stakeholders.

22.6.4.2 Social Media: Twitter

Twitter is an online social networking service that enables users to send and read short 140-character (a single letter or space is a character) messages called "tweets" to lists of persons (called friends or followers) they communicate with. Managers must register (free) for the service to use it (www.Twitter.com). Registered users can read and post tweets and forward them to their friends as "re-tweets," but those who are not registered can only read them. Users access Twitter through the website interface, Short Message Service (SMS), as a text messaging service component of a smartphone, website, or blog. Photos, videos, and the sending person's e-mail address (or website/ blog internet address) can be attached to tweets. Twitter has apps (application programs that install Twitter on smartphones) managers can provide when registering.

The term "hashtag" is applied to tweets—a hashtag is a type of label or piece of computer language inserted into tweets that makes it easier for users to find messages with a specific theme or content. Users create hashtags by placing the hash character (#), also called a pound sign, in front of a word or phrase, either in the main text of a message or at the end. Searching for that hashtag will then present each message that has been tagged with it.

Regarding deer management, Twitter probably is best used to communicate thoughts on singular issues related to deer management, such as season and bag limits, and controversial regulations concerning deer harvest, rather than informing stakeholders about deer management program. Like all the other media formats, people wishing to learn how to use Twitter can obtain a "for Dummies" book that makes the technology easy to use (Poston 2009).

22.6.4.3 Social Media: Snapchat and Instagram

These social media formats are internet-based applications for sharing with friends (and/or stakeholders) pictures, videos, and blocks of text taken as pictures with smartphones. One takes pictures (with a smartphone), adds a caption describing the picture (which can be a block of text), and posts it to their list(s) of friends. The images are viewed for a short period of time and then disappear. Information so sent is of limited educational value but may serve to stimulate interest among stakeholders, and is a way for managers to share information with them (which they can also share with their friends, expanding managers' reach). Again, if managers infrequently use this form of communication, their stakeholders will probably disregard posts or stop looking for them.

Managers need to register online with Snapchat and Instagram to use and get guidance on how to use the programs (https://www.snapchat.com and https://www.instagram.com), but it is simple and easy. And, if managers feel the need for additional help on using these programs, it is accessible online (for Snapchat, https://support.snapchat.com/en-US/ca/howto; for Instagram, https://www.instagram.com/about/faq/).

22.7 STAKEHOLDER-SPECIFIC MESSAGES, MESSENGERS, AND DELIVERY SYSTEMS

22.7.1 HUNTERS: ALPHA AND BETA

These are the hunters managers want to attract to their DFMA to help reduce deer density and continue their impact by having them return annually to harvest deer. They are more interested in quality deer and quality of the hunting experience than seeing large numbers of deer when hunting or scouting. They will apply for and use permits for harvesting antlerless deer. They respond positively to the linkage between quality habitat and quality deer and are educable regarding need for lower deer density to produce an abundant and diverse understory of herbaceous vegetation, shrubs, and tree seedlings for food and cover for deer and other wildlife species. They are also receptive to the idea that to produce more and better food and habitat, landowners need to harvest trees as a source of income that helps pay for deer management.

These hunters respect and accept management strategies to benefit forest vegetation that may result in fewer (but healthier) deer, and accept professional resource managers and researchers as credible sources of information. They generally are computer- and smartphone-literate and will seek out and use information provided by multiple media sources, including websites, blogs, and Facebook. These hunters will attend workshops to learn more about deer and forest management and will volunteer to help with data collection for monitoring. They also will bring deer they harvest to checking station operations and will be credible proponents of deer management to benefit all forest resources on field trips held for other stakeholder groups, including natural resource administrators and legislators.

Managers can increase their outreach by using Twitter, Snapchat, and Instagram media to comment on deer management and send pictures of harvested deer at checking stations, trail cam pictures of trophy deer and other wildlife such as coyotes and bobcats on their DFMA, and other enticing images of their operation to their lists of followers who, in turn, will send this information on to other hunters who may be interested in hunting their DFMA.

If deer management programs include a guiding committee, such as the KQDC Demonstration Project (Chapter 34), managers should include a local alpha hunter or two to solicit helpful management suggestions and to enhance their credibility with local hunters and facilitate recruiting additional alpha/beta hunters to hunt deer on their DFMA.

22.7.2   HUNTERS: LOCAVORE

Because locavore hunters likely have not been interested in, or affected by, deer management issues and are neutral concerning the too many-too few deer controversy, they are not interested in pro and con arguments. Their main interest is in locating and obtaining permission to hunt on properties where they can hunt deer that are relatively close to where they live. Because they are novices with little to no practical hunting experience or even navigation within forested landscapes, they are receptive to information, including workshops, where they can obtain such information and experience. They will use permits to harvest antlerless deer and may need help in learning where and how to obtain such permits. Because there is no established locavore hunting organization with biases for or against the science of deer management as presented by professionals, they are receptive to information provided by professionals.

The biggest challenge to attracting and maintaining locavore hunters lies in connecting with them and making them aware of the opportunity to hunt deer on managers' DFMAs. Placing ads in local newspapers seeking locavore hunters, or advertising workshops held on the DFMA to educate/train hunters, may yield results. Locavore hunters likely search the internet for places to hunt, so websites and blogs are excellent ways to reach them. Experienced alpha hunters hunting managers' DFMAs may be solicited to mentor locavore hunters.

Once managers have attracted locavore hunters to their DFMA, they can recruit additional locavore hunters by using Facebook, Twitter, Snapchat, and Instagram to send visual images demonstrating successful hunts to locavore hunters, who can forward the information to their friends/followers.

22.7.3   HUNTERS: OMEGA

These hunters are primarily interested in hunting areas with abundant deer herds, where dozens of deer can be seen during hunting season and especially on opening day. If these hunters learn that local forested properties receive permits to hunt antlerless deer, they will obtain the permits and hunt the properties with the expectation of seeing many deer and harvesting one. In the initial year or two of listing forestlands for permit availability, these hunters will obtain the permit, hunt the property, and contribute to an initial high removal rate of deer. However, as deer density declines, these hunters will cease using the property as they no longer experience the high deer density they require.

Omega hunters are distrustful of information produced by scientists and agency biologists, preferring instead to rely upon information they receive in meetings and by word of mouth provided by leadership representing their culture. They will be receptive to print information provided by outdoor writers if those writers promote high deer density. They generally will not attend workshops conducted by professionals that promote deer density in line with healthy deer and healthy habitat. If deer density drops below some threshold higher than that required to sustain healthy deer and forests, omega hunters will stop hunting on DFMAs.

22.7.4   OWNERS OF FORESTED LANDSCAPES

Owners of large forested landscapes may hire consultants to manage their forest resources, including deer—often when maximizing deer abundance and quality supersedes management for sustainable timber and other nondeer forest resource outputs. Usually, the properties have been owned for generations of landowners whose wealthy and influential ancestors purchased the land to form exclusive clubs emphasizing recreational activities such as hunting and fishing. Many were formed in the late 1800s and early 1900s when deer populations were exploding under the relaxation of natural predation (removal of wolves, mountain lions, native Americans), heightened restrictions on antlerless hunting, and enormous increases in quality deer forage created by clearcutting or large natural disturbances (fires and large-scale tornadic events or ice storms) wherein overstory trees were removed and understory seedlings, shrubs, and herbs flourished.

The resulting high deer densities, coincident with production of trophy deer related to bountiful and nutritious forage, were perceived by these landowners and their descendants as the norm, representing the desired future condition. Concurrently, the desire to fund recreational activities by harvesting and selling timber, as had been done in the past, was stymied by lack of advance regeneration caused by browsing by overabundant deer herds. Eventually, they ran out of timber to harvest, which was not replaced by seedling regeneration because deer browsing had eliminated required advance regeneration of desired tree species at required stocking levels.

The difficulty for consultants attempting to produce the desired conditions (abundant and trophy deer and sustainable timber income) was that they were dealing with powerful and influential owners who were accustomed to success in their professional endeavors by demanding and obtaining results not bound by natural laws—by virtue of their power and influence, they were able to influence the "laws" regulating their particular enterprises to suit their needs. This was not the case when they tried to force natural systems to their will against natural laws, which are immune to political/economic pressure and cannot be altered to favor unsustainable management of forest resources. A prime example is the Adirondack League Club case history (Chapter 40) wherein landowners demanded production of sustained timber production and high abundance of quality deer from their forestry and wildlife consultants. Unable to obtain the impossible, the property owners oversaw a series of revolving-door dismissals of "unproductive" consultants in the search for ones who could produce the unattainable.

The message these stakeholders need to hear is that their desired conditions are not possible given natural laws regulating species and abundance of natural resources. They need to hear and accept the reality of what their resources can sustain given their expectations. They must be informed of how similar circumstances failed in the past, under circumstances similar to theirs, and they need to hear and see how realistic solutions were developed and pursued, which required lowered but realistic expectations and associated management. They need to hear the message from forestry/wildlife experts/consultants, and, more importantly, from peers credible to them from similar forested landscapes who succeeded by accepting management recommendations from wildlife and forestry professionals. Message delivery is simple—the landowners must see and experience the forest conditions on similar properties, under management based on scientific principles and findings that worked, as promoted by their peers.

22.7.5 NATURAL RESOURCE ADMINISTRATORS/LEGISLATORS: DEER ADMINISTRATIVE UNITS/STATEWIDE REGULATION OF DEER HARVEST

Natural resource administrators, whether in the top management echelon at state headquarters, or involved locally as supervisors of management units, affect the ability of managers to achieve and maintain desired deer density on DFMAs by regulating where, when, and how many deer may be harvested through season and bag limits, including permits for harvesting antlerless deer on DFMA and DAU units. They usually are lobbied heavily (to increase deer density) by deer hunter groups composed of many individuals and by landowner groups (to decrease deer density) composed of relatively few individuals whose property deer are negatively impacting.

Depending on the state, administrators may view deer hunters as primary clients and affected forestland managers as of less importance/influence. Some respond more positively to hunters touting culture, values, and tradition as reasons justifying high deer density than to managers and/

or forest/wildlife scientists seeking reduced deer density based on science and impact levels. These administrators are sensitive to demands of state legislators who may control/influence agency purse strings. Under pressure by large numbers of complaining hunters, such legislators might hold hostage funding levels for state natural resource agencies if they do not deliver regulations that favor increased rather than reduced deer density.

Natural resource administrators and legislators may be more heavily influenced by emotion and culture-based information presented by large numbers of constituents (primarily deer hunters) than by science and reason as promoted by smaller numbers of managers, biologists, farmers, and other groups negatively impacted by overabundant deer herds.

These same administrators and legislators are sensitive to the potential for legal action (class-action lawsuits as a "nuclear option") related to natural resource management brought by groups negatively affected by high deer density. They may also be sensitive to criticism of deer management leveled by respected forest certification groups such as the Forest Stewardship Council (FSC; https://us.fsc.org/en-us) and Sustainable Forestry Initiative (SFI; www.sfiprogram.org) performing certification assessments of forest resource management if such management does not address the joint issues of overabundant deer herds and high deer impact on forest resources.

In many cases, assessments of natural resource management by these credible sources identify overabundant deer herds as serious problems in sustainable forest and natural resource management. Such assessments may be performed on large, individual managers, such as commercial forests, public state forests, watersheds for municipal water providers; on smaller, individual private forestlands; or on groups of managers banded together in group certifications. Negative assessments may influence level of federal funding for state natural resource agencies, and these do carry significant weight. Third-party assessments are fairly expensive, and require additional, annual audits; they may be beyond the financial resources of smaller managers. However, they have been used by managers of larger entities, such as state forests, to educate natural resource administrators and influence how they deal with regulations concerning deer management.

Natural resource administrators and legislators are aware of and generally respect the value of science in driving management decisions. However, political pressure brought by large numbers of hunters urging state legislators and natural resource governing boards to increase deer density, even in the face of contradictory science, often trumps science as politicians and governing board members protect their job security by responding to the constituency group holding the greater voting power. Breaking this alliance among disgruntled deer hunters, natural resource administrators, and governing boards is one of the most difficult tasks facing promotion of science-based deer management to benefit all forest resources. Getting administrators, board members, and legislators to enact and promulgate regulations and management steps to manage deer density and impact that benefits all state residents education is a critical task and far beyond the ability of individual managers to accomplish on their own.

22.7.6 Natural Resource Administrators/Legislators: Deer–Forest Management Areas

Natural resource agencies in many states now implicitly recognize the need of managers who manage deer on DFMAs to reduce deer density to levels that permit sustainable management of forest resources. In these states, the agencies may provide permits to managers for hunters to harvest additional antlerless deer on specified DFMAs under programs generically called deer management assistance programs (DMAPs). This action is not designed to reduce deer density on DAUs/statewide specifically and thus avoids the appearance of a uniform attempt to reduce deer density, excepting on properties not under the direct jurisdiction of state natural resource agencies. However, in states (e.g., Pennsylvania) where large numbers of managers have used DMAP programs to reduce deer density, deer hunters have noticed the resulting reduction of deer density and have begun (successfully) to lobby state legislators/natural resource agencies to reduce or eliminate DMAP permits for reducing deer density on DFMAs.

22.7.7 OTHER LANDOWNERS AFFECTED NEGATIVELY BY DEER

Agricultural groups and affiliated farmers, homeowner associations in forested communities, managers of state parks, community watershed managers, environmentalists, and others represent a constituency exceeding the numbers of deer hunters. By combining forces and coordinating education/communication efforts with natural resource administrators, members of these groups can combine individual lobbying efforts with group efforts that will carry weight by virtue of sheer numbers. Setting up individual Facebook accounts that list members as "friends" allows them to exchange information and coordinate educational efforts, including testimony, field trips, and lobbying.

Small woodlot owner organizations represent such groups, where persons with similar situations can communicate and plan educational activities to inform natural resource administrators and legislators, as well as exchange useful management information.

Regardless of the stakeholder group(s) and education/communication for which efforts are targeted, it will be important to test them with these groups for feedback and evaluation: the programs must be revised and refined such that the message and delivery evolve—they cannot be static. Choice of media outlet(s) may be dictated by finances, stakeholders, and the existing communication network established with stakeholders is important. But managers should choose a system that is affordable and suits their needs—which basically are to identify, attract, and maintain two-way information flow between them and the stakeholders who can help them achieve their management goals for deer and forest resources.

22.7.8 DEER–FOREST MANAGEMENT AREA-SPECIFIC MESSAGES, MESSENGERS, AND DELIVERY SYSTEMS

The communication package developed by forestland managers is dictated by forestland size, availability of financial and human resources, and target stakeholder group(s).

22.7.8.1 Small Woodlot Owners

Small woodlot owners choosing to reduce deer density and impact by hunting usually depend on friends and family members to do the job. These forestland owners have limited resources for message preparation and delivery. The message is quite simple—that their forestlands offer a safe, uncrowded, and hassle-free area to harvest the number of deer the landowner(s) need removed, that a majority of deer harvested must be antlerless (does and fawns), and that the approximate number of deer to remove is identified. If possible, the landowners should advise hunters that they can provide antlerless permits to allow harvest of additional antlerless deer. The best way to recruit (local) hunters to hunt on these forestlands is probably word of mouth by friends/family who also hunt the property. Ads placed in newspapers do not allow for screening of potential hunters (and possible unpleasant interactions between hunters and landowners), and are costly and of short duration. Websites/blogs as carriers of the message suffer some of the same shortcomings (can't screen hunters, are costly to implement and maintain), and few would have the technical ability or staff to create websites or blogs, which are designed for larger and more disparate audiences.

22.7.8.2 Small Commercial Forestry Operations

Owners of small forestlands managed for sustainable timber production have concerns about the impact of overabundant deer on seedling regeneration needed for replacement of harvested timber. They produce abundant and high-value deer forage when they are able to thin or final harvest forest stands within their properties, and may provide hiding and thermal cover with other stands. Their audience is alpha, beta, and locavore hunters whom they recruit by word of mouth and possibly by interacting with local hunting clubs, including quality deer management groups. They rely upon repeat hunters from the local area whom they recruited from family members, friends, employees, and acquaintances. Their message is simple: by being able to harvest trees, they produce valuable deer forage. If deer density is maintained at the carrying capacity for production of regeneration of preferred seedling species, they will produce quality deer for hunters.

These forestland properties may be large enough to support lease hunting as a way to generate additional income and to manage groups of hunters who can be motivated to harvest required numbers of deer (or lose their hunting lease) to achieve target deer density. As with small woodlot owners, owners/managers of small commercial forestry operations will be the ones to deliver the message(s) and establish and maintain contact with their hunters, possibly through Facebook or Twitter. Also, the potential for sharing costs and communication programs with other small woodlot owners banded together in cooperatives involving numbers of adjoining small woodlot owners should be explored (Chapters 28 and 31).

22.7.8.3　Large Commercial Forestry Operations

The primary goal of large commercial forestry operations is sustainable production and harvest of commercially valuable forest products, chiefly for the production of sawtimber and paper products. These forestlands contain numerous forest stands representing all successional stages (possibly excepting old growth). When these properties are negatively impacted by overabundant deer herds, the number of deer to be removed to bring deer density and impact down to goal levels requires large numbers of hunters who will harvest antlerless as well as antlered deer. The number of hunters required may exceed the limited number of local hunters known to forestland owners. A major outreach effort to recruit hunters from local and distant communities may be necessary, requiring a significant investment in time and resources to develop and deliver the communication package. Messaging should express the requirement for replacement of harvested trees by a new cohort of trees represented by seedlings of desired species and that these activates create quality deer habitat that produce quality deer—if deer density is maintained at a level that permits regeneration by seedling trees. The message should emphasize the role of hunting to reduce deer density and impact to levels that allow continuous production of forest products through timber harvest. Because participating hunters may be from outside the local area, messaging should include information on local restaurants, sources of lodging, and stores selling equipment related to hunting. Owners of these forestlands may wish to create websites/blogs to recruit and interact with local/regional hunters. They should coordinate with local natural resource agencies to facilitate directing hunters to their properties.

22.7.8.4　Public Lands

Managers of public lands (e.g., state and national forests, state and national parks) may be operating under mandates for producing wood products or for enhancing and maintaining biodiversity while prohibiting harvest of wood products and/or prohibiting deer hunting. Some public lands may be of such size and mixed management goals that individual compartments contained within them may be under differing management mandates (e.g., some for timber production, some for protection of rare and endangered species and/or biodiversity, some for human recreation), which requires managing for different deer densities and carrying capacities, and the associated requirement to develop and deliver communication and education programs for each of the major stakeholders of different deer management mandates. These programs should be communicated to all stakeholder groups, not just the targeted ones.

REFERENCES

Abram, C. 2016. *Facebook for dummies*. Hoboken NJ. For Dummies.
Adams, K. 2017. The dangers of blending politics and deer management. https://www.qdma.com/dangers-blending-politics-deer-management/
Blair, A. L. 2016. *Blogging for dummies*. Hoboken NJ. John Wiley & Sons.

Cialdini, R. B. 1994. *Influence: The psychology of persuasion.* New York. HarperCollins Publishers.

Leopold, A. 1986 (revised edition). *Sand County almanac (outdoor essays & reflections).* New York. Ballantine Books.

Marquis, D. A., R. L. Ernst, and S. L. Stout. 1992. *Prescribing silvicultural treatments in hardwood stands of the Alleghenies (Revised).* U.S. Department of Agriculture Forest Service Northeast Forest Experiment Station General Technical Report NE-96. Radnor PA. Northeastern Research Station.

Pierson, T. G. and D. S. deCalesta. 2015. Methodology for estimating deer impact on forest resources. *Human–Wildlife Interactions* 9:67–77.

Poston, L. 2009. *Twitter for dummies.* Hoboken NJ. John Wiley & Sons.

Sabin-Wilson, L. 2017. *WordPress for dummies.* Hoboken NJ. John Wiley & Sons.

23 Managing Hunters
The Four Rs

David S. deCalesta

CONTENTS

> When some of my friends have asked me anxiously about their boys and whether they should let them hunt, I have answered yes—remembering that it was one of the best parts of my education—make them hunters.
>
> **Henry David Thoreau (1854)**

23.1 INTRODUCTION

Broad social changes, including urbanization, an aging population, and competing time commitments, have contributed to a nationwide trend in loss of hunters, resulting in decreased deer hunting and decreased recruitment and retention of deer hunters (Ryan and Shaw 2011).

As population demographics change, fewer people are likely to be initiated into hunting by their families, reducing the number of individuals recruited into hunting through traditional channels. As existing hunters age and cease hunting, the community that generates the camaraderie and support for new and existing hunters to start and continue hunting diminishes over time Duda et al. (1998). Larson et al. (2014) stated that "Decades of decline in the number of hunters have made hunter recruitment and retention a high priority within the North American wildlife management community."

Managers who use public hunting to reduce deer density and impact to tolerable levels face the double whammy of a declining population of hunters and difficulty in recruiting new hunters to help them control deer density and impact. The problem of recruiting and retaining hunters has been recognized by government agencies such as the U.S. Fish and Wildlife Service (2018), which promotes the R3 program (recruitment, retention, and reactivation of hunters) to address this need. This program is intended to create new participants and increase participation rates of current or lapsed outdoor recreationists. The resulting Outdoor Recreation Adoption Model promotes development of recruitment activities to generate awareness and interest in hunting, as well as providing opportunities for people to try it out. Retention activities are designed to provide the support necessary for novice recreationists to build their skills until they are able to participate independently. Reactivation activities are designed to help lapsed recreationists become active participants. The R3 program recognizes the problem of declining hunter numbers and describes three steps to address it. However, the program does not describe activities that might be developed to actually recruit, retain, and reactivate hunters. And there are an additional two steps necessary for success: (1) retraining hunters to educate them with the science and management steps of reducing deer density and impact so they will accept and participate in programs designed to reduce deer density and impact, and (2) rewarding hunters for participating in such programs and harvesting antlerless deer.

23.2 THE FOUR Rs OF HUNTER MANAGEMENT

The R3 program can be expended into an R4 program designed to optimize the potential for success in partnering with hunters to manage deer by controlling their density. R4 program components include: *recruiting* new hunters; *retraining* recruited hunters regarding how they view deer density, deer impact, and deer management for the benefit of all forest resources; and *rewarding and retaining* hunters with a package of incentives for hunting on specified deer–forest management areas (DFMAs—see Chapter 9) with an emphasis on harvesting antlerless deer. Rewarding and retaining successful hunters turns them into "repeat customers," the ones who keep coming back year after year and providing the service (harvesting deer) essential for attaining and maintaining deer density and impact goals.

Recruiting, retraining, rewarding, and retaining hunters can be accomplished on DFMAs if managers establish and maintain effective communications with hunters. Variability in hunting conditions and level of deer density and impact within large deer administrative units (DAUs—see Chapter 9) enclosing DFMAs, and difficulty of state wildlife agency administrators and wildlife/deer biologists in engaging in effective communications with hunters, renders them with few incentives to recruit, retrain, or reward hunters for hunting on DAUs. The only incentive such agencies have is to provide permits to harvest antlerless deer at the DAU level, and with the exception of alpha and beta hunters, most omega hunters either decline to apply for and use these permits, or, worse, obtain them with no intent of using them save to keep them out of the hands of hunters who would use them to reduce deer density, a management goal rejected by omega hunters.

23.2.1 RECRUITING HUNTERS

The Merriam-Webster definition of recruitment related to biology is, "The process of adding new individuals to a population or subpopulation (as of breeding or legally catchable individuals) by growth, reproduction, immigration, and stocking." In the terminology of deer management, recruitment refers to overwinter survival of fawns to become part of the breeding population. Here, we will expand the definition to define *hunter recruitment* as increasing the pool of hunters hunting a DFMA, be they youth hunters just beginning to hunt or hunters who have never hunted a particular DFMA but can be attracted to hunt that DFMA by a variety of management stratagems. The Wildlife Management Institute (2009) developed a website (http://www.huntingheritage.org) designed to share emerging information/strategies on hunter recruitment/retention with deer–forestland managers.

For managers of small woodlots in the hundreds of ha size, recruiting hunters, experienced or novice, is mostly a matter of encouraging family, friends, and nouveau hunters from both groups to hunt on their lands, with a given condition being that there would be no interference/safety concerns arising from uninvited or trespassing hunters hunting the property.

For managers of larger DFMAs, recruitment necessitates soliciting local and nonlocal hunters to hunt on their DFMA. The task is to identify potential hunters and attract them to DFMAs through word of mouth, advertising, and social media. The larger the property and the higher the need for large numbers of hunters, the less effective it is to rely on local hunters, families, and friends to harvest enough deer.

23.2.1.1 Recruiting Hunters by Communications

Media messages, including contact information for manager(s), should tout ease of access, condition of access roads, quality deer, quality hunting experience, and quality and abundance of other wildlife. Informational packages should be developed and made available to hunters, including travel directions to the DFMA, topographic maps of DFMA with access points, and contact information for local restaurants, motels, grocery stores, and sporting goods stores. Picture galleries of successful hunters with quality deer brought to checking stations, and trail cam pictures of quality deer from the DFMA, are powerful visual messages attractive to hunters. Testimonials from satisfied hunters extolling the benefits of hunting the DFMA are worth their weight in gold, and descriptions of rewards for hunters to hunt specific DFMAs, for example, prizes for the largest buck, largest doe, and drawings for hunting paraphernalia drawn from lists of hunters bringing harvested deer to DFMA checking stations, can be used to encourage hunters to hunt on DFMAs.

Hunters searching the internet using keywords such as *deer hunting area*, *quality deer hunting*, and *public deer hunting* can look for information on DFMAs for specific localities. An especially effective mass-media resource for attracting hunters to DFMAs is local and regional outdoor writers familiar with deer management on DFMAs who can promote hunting on them. Getting to know such writers is important, and having them serve on leadership teams specific to DFMAs (if they are large enough for such teams) provides the writers with comprehensive information about the hunting situation and quality of deer and habitat.

23.2.1.2 Mentoring as a Recruitment/Retention Tool

Attracting young and/or first time hunters is largely dependent on supportive parents who hunt and want to maintain a family hunting tradition. The best way to attract young hunters is to attract their hunting parents/relatives who will want them to hunt the same properties they hunt. Some conservation organizations (e.g., the Quality Deer Management Association) have experimented with initiating and recruiting new hunters through the development of mentored hunts and workshops. Some states offer mentoring programs where experienced hunters may mentor young hunters and help them learn to hunt and gain access to areas where they can hunt deer. Anecdotal evidence, however, suggests that many of these efforts recruit hunters who likely would have been socialized into hunting by their families or friends (Duda and Young 1998). And, like all other aspects of deer management, mentoring is not a one-time effort—rather, it should be an annually recurring event to ensure young/inexperienced hunters develop hunting skills and experience the rewards of positive hunting experiences that will keep them coming back to the specific DFMA where they were mentored.

23.2.2 RETRAINING HUNTERS

Most hunters lack formal education in deer–forest management, and the primary route to recruiting and retaining new hunters is through peers, family members, and mentors who initiate, train, and socialize their children into hunting tradition (Ryan and Shaw 2011). Much of this information is based on culture and tradition and may not relate to sustainable deer and habitat management (e.g., there may be a strong antidoe hunting philosophy and attitude that the more deer the better). Hunters with this kind of background will be resistant to management for quality deer and quantity habitat, favoring instead high deer densities. Manager interactions with these hunters should include training experiences that expose hunters to the concepts and outcomes of sustainable deer management and associated benefits to the hunters. Combinations of classroom-type presentations by peers hunters trust, with show-me field trips, can provide visual confirmation of how overabundant deer can impact habitat and demonstrate what habitat and deer look like under sustainable management practices. Participating hunters collect data on deer density and impact during such training sessions and can see how the data are analyzed to produce estimates of deer density and impact and be used to guide management. Chiefly by regulating deer density, hunters so educated are more apt to support deer management based on reducing deer density to improve deer and habitat quality.

Several of the case history DFMAs used such training sessions to retrain hunter thinking and perception initially: in the KQDC case history (Chapter 34), annual training sessions were part of overall management operations until all interested hunters and other stakeholders had received the retraining (about 7 years into the project), at which point the training sessions were suspended. Having hunters participate in other field data collection operations, such as daylight roadside counts to count antlered bucks, does, and fawns, and participation at deer checking stations during hunting seasons (either as checkers of deer, harvesters of deer, or both) invests them in management operations. These hunters are more receptive to the concept of managing for sustainable deer herds, including the need to harvest antlerless deer.

Checking stations are particularly effective opportunities for enhancing the values of sustainable deer management: hunters can see first-hand the results as evidenced by quality of deer harvested and brought to checking stations. Pictures of harvested quality deer posted on checking station walls reinforce the message of reducing deer density. Checking station personnel can interact directly with hunters, answer questions, and provide them with pamphlets and charts illustrating benefits of deer management.

The Quality Deer Management Association (Chapter 27) communicates with its members by a newsletter. The articles feature characterization of, and support for, sustainable deer management, including recommendations for managing deer density through harvesting antlerless deer.

23.2.3 Rewarding and Retaining Hunters

Once hunters have been recruited and presented with the science and practice of sustainable deer management, their hunting experiences should be enriched. Such enrichment can be provided by a system of incentives that will reinforce their decision to hunt the DFMA and encourage them to return on a recurring basis. The keys to retention include: (1) maintaining high-access road systems that provide hunters with good access to all parts of the DFMA; (2) maintaining a quality hunting experience, which includes hunters seeing quality deer and hunting in attractive habitats without legions of other hunters competing for hunting spots; (3) involving the hunters actively in the overall program through field opportunities such as participating in data collection and checking stations; (4) keeping hunters informed of deer–forest conditions; (5) providing them with information obtained from monitoring data collection and checking stations, including quality of habitat and deer; and (6) reinforcing their continued participation in hunting and harvesting deer through a system of incentives.

23.2.3.1 Rewarding Hunters with Incentives

Incentives for harvesting deer, especially antlerless deer, include: (1) enhancing opportunity for harvesting antlerless and antlered deer; (2) providing special recognition of hunters harvesting deer, such as providing hunter-recognition banquets where hunters are invited to attend the events; (3) providing cash/prizes awarded to hunters for harvesting deer; and (4) providing assistance to hunters for recovering harvested deer. Most incentives can only be applied at the DFMA level—state natural resource agencies do not offer monetary incentives and have too few resources to offer much in the way of incentives beyond attempts to increase antlerless deer harvest through "earn-a-buck" programs.

23.2.3.2 Earn-a-Buck Programs

Earn-a-buck programs are an attempt to increase antlerless deer harvest by requiring hunters to harvest an antlerless deer before they can harvest an antlered deer. Van Deelan et al. (2011) evaluated the effectiveness of an earn-a-buck program when paired with extending the length of the season for harvesting antlerless deer in Wisconsin. The earn-a-buck program resulted in a minimal increase in antlerless harvest (2 deer/km^2) and a small decrease (0.6 deer/km^2) in average antlered deer harvest. Extending the length of season to harvest antlerless deer resulted in similarly small reductions (<2 deer/km^2) and similarly small decreases in antlered deer harvest. The earn-a-buck part of the program was disliked by "many" hunters. Wisconsin fostered the earn-a-buck program as a way to facilitate an attempt to eradicate chronic wasting disease at the DAU level by greatly reducing the deer herd in specified areas within the state. The intent of the program may not have been described in a manner that won over hunters, whose resistance to the program resulted in the Wisconsin governor canceling the program. Effectiveness of earn-a-buck programs on DFMAs appears limited: Boulanger et al. (2012) experimentally applied the program on a small landscape surrounding the Cornell University campus, but, like the research of Van Deelan et al. (2011), did not compare increases in harvest of antlerless deer with changes in deer density and impact. Van Deelan et al. (2011) cited dislike of the program by hunters, and Boulanger et al. (2012) were concerned whether earn-a-buck programs could be maintained in the face of declining deer density.

Hunter distaste for earn-a-buck programs, and their limited effectiveness in reducing deer density, suggest that such programs will yield meager results. Pairing earn-a-buck programs with increasing length of antlerless seasons or providing late antlerless seasons could result in meaningful reductions in deer density and impact. Such pairings could be enhanced by additional opportunities for increasing antlerless harvest, such as with deer management assistance programs (DMAP—see Chapter 24 below) programs and special deer damage hunts.

23.2.3.3   Hunting Assistance Incentives

Incentives designed to improve hunter access and success include developing and maintaining good road access to all portions of the DFMA, providing safe hunting platforms (tree stands), controlling hunter numbers to prevent high hunter density (low-quality hunting), and helping hunters retrieve harvested deer. The West Branch Forest Preserve case history (Chapter 33) experimented with providing all-terrain vehicles for extracting harvested deer, but the program was little used, possibly because of the requirement that the vehicles be operated by preserve personnel and by the part-time presence of those personnel.

23.2.3.4   Material Incentives

Material incentives (providing credits for purchasing gasoline; giving away hunting caps, hunting knives, or other hunting paraphernalia) offered at the DFMA level (see Chapter 33) seemingly do not result in increased hunter participation or increased harvest and are perceived by hunters as being paid to kill does.

23.2.3.5   Recognition Incentives

Incentives based on recognizing hunter participation in DFMA-level management are few and of unknown effect: the Kinzua Quality Deer Cooperative (Chapter 34) held an annual hunter recognition banquet where successful hunters were invited to an annual banquet. At the banquet, hunters were thanked verbally for their participation, and a raffle provided material rewards to a limited number of hunters. The banquet was discontinued after the sponsoring organization (Sand County Foundation) ceased funding the banquet and hunter participation was declining (in the banquet—not in hunting), but hunter participation and effectiveness (harvesting enough deer to maintain the desired level of deer density and impact) did not decline in the years following cessation of the banquet.

Personnel and financial costs of some of the four Rs limit their application on some DFMAs, particularly small woodlots. And, as with all other management activities applied to achieve and maintain desired deer density and impact levels, they must be employed every year (as long as they are effective) for the duration of the deer–forest management program. However, as all but one of the case histories in Section V demonstrate, long-term, consistent application of some or all of the four Rs is effective, as improved through adaptive management.

REFERENCES

Boulanger, J. R., G. R. Goff, and P. D. Curtis. 2012. Use of "earn-a-buck" hunting to manage local deer overabundance. *Northeastern Journal* 19:159–172.

Duda, M. D., S. J. Bissell, and K. C. Young. 1998. *Wildlife and the American mind: Public opinion on and attitudes toward fish and wildlife management.* Harrisonburg VA: Responsive Management.

Duda, M. D. and K. C. Young. 1998. American attitudes toward scientific wildlife management and human use of fish and wildlife: Implications for effective public relations and communications strategies. *Transactions of the North American Wildlife and Natural Resources Conference* 63:589–603.

Larson, L. R., R. C. Stedman, D. J. Decker et al. 2014. Exploring the social habitat for hunting: Toward a comprehensive framework for understanding hunter recruitment and retention. *Human Dimensions of Wildlife* 19:105–122.

Ryan, E. L. and B. Shaw. 2011. Improving hunter recruitment and retention. *Human Dimensions of Wildlife* 16:311–317.

Thoreau, H. D. 1854. *Walden.* Boston. Tichner and Fields.

U.S. Fish and Wildlife Service. 2018. What is R3? Recruitment, retention and reactivation explained. https://www.fws.gov/midwest/news/WhatIsR3.html

Van Deelan, T. R., J. J. Dhuey, C. N. Jacques et al. 2011. Effects of earn-a-buck and special antlerless-only seasons on Wisconsin's deer harvests. *Journal of Wildlife Management* 74:1693–1700.

Wildlife Management Institute. 2009. Hunter recruitment and retention boosted by new website. *Outdoor News Bulletin* 63. http://www.huntingheritage.org

24 Managing Access and Antlerless Permits

David S. deCalesta

CONTENTS

> ...to curb the growth of our deer herd, hunters and landowners need to work cooperatively to get hunters to where the deer are.
>
> **Steve Merchant (2004)**

24.1 INTRODUCTION

If managers are to manage deer density and impact with hunting, it is essential to connect hunters with deer within deer–forest management areas (DFMAs—see Chapter 9) and to arm them with the wherewithal to reduce deer density. To do this, managers must facilitate hunter access to their DFMA and to permits for harvesting antlerless deer. And, they must communicate this information to hunters through multiple communication channels, as described in Chapter 22.

24.2 FACILITATING ACCESS

When I sought out stretches of streams for places to fish for trout, I kept walking along trails bordering the streams until the litter left by fishermen (candy and gum wrappers, cigarette butts, opened fishing equipment boxes) petered out—usually after about a half mile of walking. Those were the best stretches of little-fished water. Managers attempting to reduce deer density and impact on DFMAs have the opposite goal—to provide deer hunters with good access to all parts of their forestlands to reduce deer density and impact as uniformly as possible, hence the primary need to optimize and facilitate access by hunters to their DFMAs.

Providing vehicular (passenger cars, pick-up trucks, all-terrain vehicles [ATVs]) access to all parts of DFMAs is critical to reduce deer density and impact over entire DFMAs and achieve meaningful distribution of hunters and hunting. A road system that leaves no portion of the DFMAs farther than 600 m from vehicular access will provide the required distribution of hunting effort. In the case history of a cooperative with three major landowners (Chapter 34), the portion of the DFMA wherein the road system provided nearly 100% access within 800 m of graveled roads had the lowest deer density; that portion wherein access was less-well distributed had the highest deer density. Leaving pockets of forestland without vehicular access may result in "refugia" for deer to escape to during hunting season and avoid population reduction. Components of access include:

- Communication—the access components listed below must be facilitated through an effective communications program. Managers need to provide hunters with access details and should make this information to state natural resource regulatory agencies regarding where and when the antlerless permits may be used. Managers should make use of media outlets such as websites, blogs, Facebook, natural resource agency information programs, and local outdoor writers (Chapter 22).

• Maps—maps describing DFMA locations within deer administrative units (DAUs—see Chapter 9) should be provided to state natural resource agencies, which may note locations within DAUs in maps of hunting areas provided to hunters, greatly increasing potential for hunters to seek out identified DFMAs. Maps of DFMAs highlighting access roads should be made available and distributed to hunters through websites, blogs, Facebook, at checking stations, at kiosks, and at major entry points (stored in well-marked, weather-proof boxes such as realtor boxes for distributing information on homes for sale). Maps allow managers to identify "hot spots" where deer abundance is high and locations of recent timber harvests where deer may concentrate for foraging. Managers may also identify "refugia" (areas of dense vegetation where deer flee to when hunting begins—some hunters prefer to hunt refugia for trophy bucks). Maps displaying topographic features, such as waterways and steep vs. level hillsides, help hunters decide where they will hunt regarding terrain. Maps should provide information on how to travel to DFMAs along state highway systems. If managers can arrange to provide for the presence of personnel along major access roads of DFMAs (and at checking stations), especially during opening days, hunters can get directions from them.

• Access roads—access roads should be all-weather roads with a maintained graveled surface. Dirt roads tend to get sloppy and unstable during extended periods of precipitation and are hard to maintain at a level grade. Stability of access roads may be improved prior to hunting season by road-building equipment (an option available chiefly to large DFMAs and cooperatives); such stability includes plowing access roads after snowfalls sufficient to impede vehicular access.

• Parking areas—access roads should have pull-offs with hard surfaces where hunters can park their vehicles. The longer the access road, the more pull-offs there should be (there is no literature to provide guidance). Log landings make excellent pull-off sites.

• Gates—gates preventing entry onto access roads should be opened for the duration of hunting season and possibly in the months preceding hunting season for hunters wishing to "scout out" good hunting locations.

• Bridges—bridges over waterways, gullies, and other impediments to travel should be of study construction and well maintained. Skidders extracting timber from harvest sites may be able to ford streams or travel over "corduroy" bridges (logs placed in stream bottoms parallel to streamsides) but lack of established bridges will guarantee that hunter distribution will be uneven and that harvest sites accessed beyond streams and gullies will not be hunted, leaving advance regeneration of tree seedlings and other understory vegetation exposed to high deer density.

• Signing access—navigating forest road systems can be confusing, especially in low lighting conditions (when hunters may be moving to hunting sites or exiting DFMAs after hunting and/or when hunters do not possess maps of road systems). Failure to find their way within a confusing (and changing) road system was a cause of low hunter distribution in the cooperative case history (Chapter 34). Managers addressed this problem by clearly marking intersections with colored flagging that was reproduced on maps of the areas to help hunters navigate the area. On the national forest portion of this DFMA, access roads were graveled and well maintained and all intersections were marked with signs. Deer density and impact on this portion of the DFMA were lowest, and a disproportionately high number of deer brought to checking stations came from this DFMA.

• Checking station location—to optimize number of deer brought to DFMA checking stations, checking station location(s) should be at the intersection(s) of major roads within the DFMA and/or at main entry points to DFMAs.

- Use of ATVs—hunters may wish to use all-terrain vehicles to access DFMAs along trails and for use in hauling out harvested deer, but managers must determine whether state and local regulations permit use of such vehicles, especially on public lands such as national forests and state and county parks and forestlands.
- Contact information—managers should provide contact information (telephone numbers, e-mail addresses) so that hunters may obtain the information described above.

24.2.1 FACILITATING APPLICATIONS FOR ANTLERLESS PERMITS

Getting hunters in contact with deer is half the battle in reducing deer density. The other half is enabling them to harvest antlerless deer in addition to antlered deer harvested during regular deer seasons. Currently, many states have antlerless deer (doe) seasons, but many allow harvest of antlerless deer only during separate and short seasons (3 or so days). Most states allow for harvest of additional antlerless deer for hunters possessing special antlerless deer permits (e.g., deer depredation permits and deer management assistance program permits).

The annual process for obtaining and using antlerless permits includes: (1) landowners apply for the permits, generally after providing proof of unacceptable deer damage and stating how many permits they need; (2) the regulatory agency approves and makes the permits available, either by providing them to the landowners or by issuing them to hunters applying to the agency for the permits; (3) hunters apply for the permits, usually offered on a first-come, first-served basis through a process that restricts the number of permits they may receive; and (4) landowners provide the issuing agency with reports including the number of permits provided and number of deer harvested at the conclusion of hunting seasons, a process requiring record-keeping.

Generally, permits are issued on a year-by-year basis to participating landowners. Because hunting is the primary mortality factor for deer, and because hunters hunting during regular seasons are primarily hunting for antlered deer, achievement and maintenance of desired deer density usually requires annual use of antlerless deer permits, especially when managers attempt to keep deer at desired densities once they achieve those densities. Components of seeking and using antlerless permits include:

- Communications—managers need to communicate to hunters where and when the antlerless permits may be used and how to apply for them. As with communications about access, managers should make use of media outlets such as websites, blogs, Facebook, natural resource agency information programs, and local outdoor writers to provide the information presented below on antlerless permits.
- Process—managers need to obtain and communicate with potential hunters the process by which they apply for obtaining the permits, how they are to use the permits, and how they report the results of their use of the permits (harvest report).
- Monitoring—managers determine the need for antlerless permits after monitoring deer density and impact prior to deer hunting seasons. For managers to determine the number of permits to apply for, they need to monitor (annually) deer density and determine how far above the desired carrying capacity the deer herd is. Once they determine how many deer they need to harvest with antlerless permits, they need to approximate how many permits are required to harvest a single antlerless deer (generally three to five permits)—this provides them with justification for the number of permits they apply for. Because there is no hard and fast rule for number of permits required to reduce deer density and impact to desired levels, annual estimates of number of permits to seek are more of an art than a science. This adaptive management aspect of deer management requires managers to be flexible, to reduce the number of permits sought in one year if deer density seems to have been reduced too much or to increase the number sought in another year if density and impact remain above target levels. Record-keeping is essential for such posthunt assessments/predictions.

- Description—managers need to provide issuing agencies and hunters with a description of the area for which the permits are required. As with access requirements, providing maps of areas for which the permits are to be used, along with descriptions of locations and travel routes to the areas, makes it easier for hunters to locate the area(s) and allows issuing agencies to provide the same information to hunters on agency communication outlets.
- Hunter type—some hunters obtain antlerless permits with no intent to use them; rather, they aim to keep them out of the hands of hunters who would use them to harvest antlerless deer. If DFMA managers distribute the permits, they may be able to identify alpha and beta hunters and give them the permits and not make them available to omega hunters. On the other hand, if managers rely upon state natural resource agencies to distribute the permits, they have no way of controlling the type of hunter receiving the permits.
- Distribution of permits—managers should decide (if the option exists) whether to obtain permits from the issuing agency and distribute them directly to hunters or to rely upon the state agency to distribute permits. For smaller DFMAs, managers would have more control over who obtains the permits and may restrict their distribution to hunters they know. On the other hand, with larger DFMAs and larger numbers of permits, the task of distribution may overwhelm DFMA personnel. The staff at the DFMA cooperative case history (Chapter 34) was overwhelmed by the logistics of soliciting, granting, and distributing 3000 antlerless permits granted by the state natural resource agency and (gladly) accepted that agency's offer to take over the task of distributing the permits.
- Hunter feedback—as with other components of deer management, managers need to survey hunters using permits to determine whether there is any part of the process managers can improve upon to facilitate hunters using the permits to lower deer density.

REFERENCE

Merchant S. 2004. Quote https://www.messagem.edia.co/millelacs/record-number-of-antlerless-deer-permits-available/article_91564c08-8ffd-56ea-b279-8be15e5679f0.html

25 Integrating Ecological and Human Factors in Deer Management

David S. deCalesta

CONTENTS

> Management means...the substitution of thought for brawn and muscle, of knowledge for folklore and superstition, and of cooperation.
>
> **Peter Drucker (2013)**

> When deciding upon a deer management strategy for your property, don't forget to factor in people management. This alone will influence your success or failure.
>
> **Mississippi State University Deer Laboratory**

> Our goals can only be reached through a vehicle of a plan in which we must fervently believe and upon which we must vigorously act. There is no other route to success.
>
> **Pablo Picasso (2015)**

25.1 INTRODUCTION

Managing deer density and impact to secure goals for natural resources in forestlands is one of the most complex of wildlife management challenges. This is so because of the wide array of natural resources affected, number of stakeholder groups involved, political and cultural influences, and sheer number of intersecting biological and human dimension factors. Revisiting Figure 1.1 (Figure 25.1 in this chapter) from the introduction helps refresh one's appreciation for the degree of complexity and number of interacting factors and management actions to consider. Some factors weigh more heavily than others depending on biological and human dimension factors respective to individual deer–forest management areas (DFMAs—see Chapter 9). Not all managers will choose to address all biological and human dimensions aspects of deer management; of the nine case histories presented below, only one (Chapter 34) incorporated them all. And it is critical to restate the reality that addressing and integrating these factors can only be accomplished at the DFMA level where deer and habitat are actually manipulated.

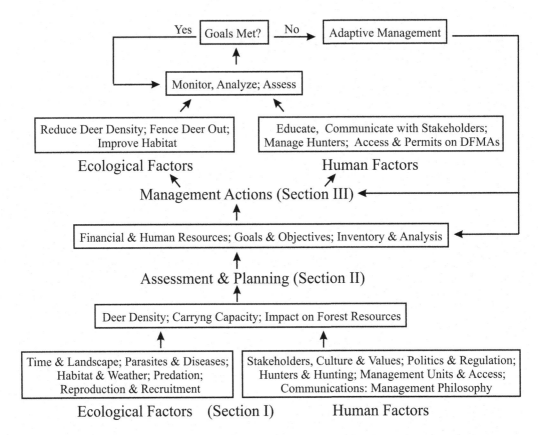

FIGURE 25.1 Factors and components involved in developing, delivering, and adjusting deer management actions to meet goals for forest resources as affected by deer.

At a minimum, seven major components should be included in any deer management plan. Managers should: (1) identify management goals within their respective DFMAs, including deer carrying capacity associated with the goal(s); (2) assess status of goals through baseline monitoring and analysis; (3) compare financial and human dimension resources available with those required to pursue their goals, which will dictate the extent of a developed management program as well as identifying whether to seek outside financial and human dimension resources; (4) develop a system for communicating with stakeholder groups, especially hunters; (5) optimize hunter access to DFMAs and distribution of permits for harvesting antlerless deer; (6) develop a program to address the four Rs of hunter management (see Chapter 23); and (7) monitor responses of deer density and impact to the developed management program to gauge success in achieving goals in anticipation of needed corrective action (adaptive management).

25.2 IDENTIFY AND INTEGRATE MANAGEMENT GOALS

Goals direct deer management and must include biological and human dimension components. Management on all of the case histories in Section V was shaped by goals, and all case histories integrated biological and human dimension goals in ways that made them compatible, excepting the last one (Chapter 40). In that case history, the goal of maximum sustained yield (MSY—designed to maximize deer density for hunting from a human dimension point of view) of the deer herd was incompatible with the goal of sustained production of timber products and optimal wildlife habitat. Deer density was not controlled to permit sustainable timber production, as restrictions on harvest of antlerless deer were in place. Both deer and forest habitat suffered: deer were of low quality and low abundance, seedling regeneration needed to replace harvested trees was either missing or replaced by undesirable seedling species, and understory vegetation, including susceptible herbaceous plants, was almost nonexistent.

Managers for one case history (Chapter 33) were unable to achieve desired hunter distribution and pressure on a section of the DFMA because it was so steep and poorly accessed by forest roads that the section was little hunted. In response, they altered the goal for that section, changing from sustained timber production to reservation of the section for development of unmanaged forest including old-growth characteristics.

25.3   CONDUCT BASELINE MONITORING

Managers cannot assess management needs without assessing (obtaining baseline information on) the status of deer density and health and the status of forest vegetation and other forest resources of management interest. In particular, assessment of the condition and landscape arrangement of deer habitat components can help assess deer movements, location of concentrated deer impact, and opportunities to improve the habitat in ways that benefit deer and other affected forest resources. Managers for all of the case histories conducted some form of baseline monitoring (for some it was obtained through the services of consulting specialists) of deer and forest vegetation to determine the need for corrective management. However, few assessed and detailed the presence and arrangement of deer habitat components. Managers in a number of case histories involved decision makers in initial assessments through field inventories wherein the administrators participated in data collection and were apprised of the analysis and interpretation required to assess the need for management action.

When managers of the cooperative case history (Chapter 34) were unable to obtain sufficient data to continue monitoring sex and age ratios of the deer herd through roadside counts, they ceased collecting that data. After collection of baseline data on deer density and impact, managers for most case histories scaled back subsequent monitoring of deer density and impact to infrequent assessments. The cooperative case history was the only one where deer density and impact were assessed annually (Chapter 34).

25.4   ASSESS FUNDING AND HUMAN RESOURCE NEEDS/OBTAIN OUTSIDE FUNDING

Financial and human resource costs of management were integral to management decisions in all case histories. Case histories for DFMAs less endowed with financial and human resources (small woodlot case histories; Chapters 38 and 39) planned management with reduced resources from the beginning. One case history wherein managers tried to enhance human resource components dropped some when they either were too costly or not used sufficiently to merit continuance (Chapter 33). One of the small woodlots in the case histories (Chapter 38) obtained outside funding from the USDA Forest Service for silvicultural practices integral to the management program, and the cooperative case history (Chapter 34) was initiated after a nongovernment organization (NGO) made a commitment to provide financial support for 10 years. Members of the club in the Quality Deer Management case history (Chapter 40) had sufficient funding from membership fees such that costs of management, including hiring consultants, were absorbed by members. Case histories using some form of permits for increasing harvest of antlerless deer benefitted from the support provided by state natural resource agencies.

25.5   DEVELOP A COMMUNICATION SYSTEM

All case histories used some form of communicating with stakeholders, chiefly hunters, with the exception of one of the two small woodlots (Chapter 38). Managers for this small woodlot declined to use hunting except for a small circle of friends and family, relying instead on various configurations of fencing and distribution of slash to reduce deer access to regenerating seedlings. For the other small woodlot (Chapter 39), word of mouth and telephone contacts with family and friends were the chief form of communication used by managers. Managers for the larger DFMAs used local newspaper outdoor writers to inform stakeholders, some used e-mail communications with stakeholders for whom they had contact information, and some (using state-provided permits for harvesting antlerless deer) had hunting information published by state natural resource agencies on annual season and bag information. The cooperative case history (Chapter 34) was the most comprehensive regarding communications, as it developed and used a website and blog. However, managers in this case history dropped use of the website and blog when the information transfer (IT) personnel working with the DFMA left, but adopted another form of communication with hunters (Facebook). Managers in the cooperative case history (Chapter 34) distributed written information to hunters at checking stations, and one (Chapter 33) provided maps and other hunting information from a kiosk within the DFMA.

25.6 OPTIMIZE HUNTER ACCESS AND ANTLERLESS PERMIT DISTRIBUTION

DFMAs in all case histories had well-developed and well-maintained access roads. Managers for the Forest Service portion of the cooperative DFMA (Chapter 34) opened roads during deer hunting seasons that normally were closed to public access at other times of the year. The DFMA for which access roads/trails to about half the area were limited and terrain was extremely steep changed its management goal for that section because hunting was unable to reduce deer density and potential for sustained output of timber was low.

All DFMAs using public hunting to reduce deer density utilized deer management assistance program (DMAP—see Chapter 24) programs to increase harvest of antlerless deer through availability of antlerless permits. Managers for the club DFMA (Chapter 40) allowed hunting for its members-only and discouraged them from harvesting antlerless deer.

25.7 ADDRESS THE FOUR Rs OF HUNTER MANAGEMENT

For any DFMA where managers use public hunting to manage deer density and impact, a key goal is to attract and retain alpha and beta hunters—the ones who prefer quality deer to high deer density—use permits to increase harvest of antlerless deer, and return every year to harvest deer. One of the small woodlot managers (Chapter 38) dropped hunters who refused to harvest antlerless deer; the manager for one of the lease properties described in Chapter 29 warned lease holders to harvest more antlerless deer or else, and when they did not, pulled their lease and awarded it to a different group of hunters who did use antlerless permits to reduce deer density. Gradually declining deer density led to attrition of omega hunters in the cooperative case history (Chapter 34). Although fewer hunters returned to hunt on that DFMA annually, by that time deer density was sufficiently low that these hunters were able to harvest enough deer to balance recruitment and maintain deer density (and impact) at desired levels.

In addition to rewarding and retaining repeat hunters by providing quality deer, DFMA managers should recruit new hunters and seek ways to increase incentives for new and repeat hunters to hunt and harvest deer on their DFMAs. Maintaining and improving vehicular access during hunting seasons and instituting/improving communications with hunters to inform them and maintain their interest in hunting DFMAs with optimal hunting conditions are critical components of management activities.

25.8 MONITOR RESULTS AND USE ADAPTIVE MANAGEMENT AS NEEDED

Managing deer is hardly an exact science. It combines scientifically obtained information on deer ecology with activities designed to address human dimension factors, but these factors are based more on the qualitative nature of culture and learning rather than on cold, hard quantitative data.

Thus, deer management is more art than science and progresses best when it moves in incremental steps, as illustrated by Chapter 36. Aldo Leopold described this process as intelligent tinkering, which retained all management components with little change unless proven wrong by experience— an apt depiction of the necessity of melding science and human nature in part by trial and error in what is defined as adaptive management.

However, there can be no adaptive management if there is no basis for comparing status of deer density and impact before and after application of designated management steps. And, initial steps taken to improve management based on an initial before-and-after comparison must likewise be evaluated after the passage of time to determine any need for adjustment. In other words, monitoring must be a recurring management activity to ensure progression toward goals continues and that goals, once achieved, are maintained. Yearly monitoring of deer density and impact allowed managers in the cooperative case history (Chapter 34) to make yearly, informed changes in number of permits for harvesting antlerless deer and maintain deer density and impact at desired levels.

REFERENCES

Drucker, P. 2013. Quoted by Swerdlow, R. In *Managing Skills for Teachers: Complete Teacher Program Academy* https://itunes.apple.com/us/book/complete-teacher/id630286227?mt=11.

Mississippi State University. No Date. Deer 101 Trophy Deer Management Starkville MS. MSU Extension Service Forest and Wildlife research center Division of agriculture, forestry and veterinary medicine. http://msudeerlab.com/mobile/trophydeer.asp

Picasso, P. 2015. https://opensource.com/open-organization/15/8/have-you-revised-your-goals-lately

26 Adaptive Management

David S. deCalesta

CONTENTS

> The challenge in using the adaptive management approach lies in finding the correct balance between gaining knowledge to improve management in the future and achieving the best short-term outcome based on current knowledge.
>
> **Catharine Allan and George Stankey (2009)**

> To keep every cog and wheel is the first precaution to intelligent tinkering.
>
> **Aldo Leopold (1949)**

26.1 INTRODUCTION

Adaptive management (AM), also known as adaptive resource management (ARM) or adaptive environmental assessment and management (AEAM), is a process developed in the early 1970s for adjusting management practices when original management operations do not produce the desired results: progress toward achieving goals is not reached (see Walters 2002). The impetus for the trial-and-error approach of adaptive management was simple: there was no established research available to help managers determine how to change management approaches that were not working. Managers with deer management failures were forced to alter what they were doing without guidelines.

Runge (2011) identified a principal reason for using adaptive management (in the context of management of rare and endangered species, but applicable to deer management): "recurrent management decisions plagued by uncertainty." He also identified key impediments to the application of adaptive management: (1) semantic confusion, (2) institutional inertia, (3) misperceptions about the suitability and utility, and (4) a lack of guiding examples. This chapter addresses the first three impediments. Section V, Case Histories, provides eight guiding examples of successful application of adaptive deer management at the DFMA level over a range of forestland size, ownership, and objectives for forest resources.

By the 1990s, wildlife professionals (Lancia et al. 1996) were describing adaptive management as, "the process of hypothesizing how ecosystems work, monitoring results, comparing them with expectations, and modifying management decisions to better achieve conservation objectives through improved understanding of ecological processes." Definitions for AM are rife and run the gamut of users from researchers to field practitioners. Because this book is designed to be used by practitioners, we will use the definition of Kremsater et al. (2003) as most apt: "adaptive management is a structured method for 'learning by doing' that includes establishing clear goals, defining practices to achieve those goals, implementing the practices, monitoring the outcomes of the practices, assessing how those practices are succeeding relative to the goals, and adjusting management in response to the assessments. It is designed to address questions such as: Where do we want to go? How do we get there? How do we know if we're there? If we're not there, how do we change to improve?" It is important to note that a key component of adaptive management is monitoring the outcomes of management practices, including deer density and status of affected forest resources.

The central requirement of AM (monitoring) helps determine whether management steps are working to achieve goals for forest resources. But it is critically important to understand that monitoring deer density and impact alone will not provide enough information to make management truly adaptive. For example, early definitions of AM did not include provision for, or monitoring of, values and culture of stakeholders. Irwin and Mickett (2008) noted that unlike other forms of management, AM includes all stakeholders in the management process and that many AM projects fail because of the lack of stakeholder identification, engagement, and continued involvement. Arnold et al. (2012) stated that adaptive, collaborative management emphasizes stakeholder engagement as a crucial component of resilient social-ecological systems (of which deer–forest management surely is one) and must be attuned to the knowledge, interests, and power of diverse stakeholder groups.

However, monitoring can only be effective if paired with a system for communicating with hunters that allows managers to solicit information from hunters to identify components of their AM program that are not effective and to apprise hunters of the status of deer hunting management components and changes to make hunting more attractive.

26.2   EXPANDING THE SCOPE OF ADAPTIVE MANAGEMENT

Monitoring stakeholder culture and values is an important component of AM. Moreover, as illustrated by Figure 1.1 (factors and components involved in developing, delivering, and adjusting deer management actions to meet goals for forest resources as affected by deer) in Chapter 1, deer management is a complex undertaking with multiple ecological and human dimensions components that affect management of deer density and impact. To make AM truly adaptive, practitioners should monitor and assesses effectiveness of *all* management activities often, usually annually, to determine which require adjustment(s) to make progress toward goals. And, as noted in Chapter 9, actual deer management is conducted at the deer–forest management area (DFMA—see Chapter 9) level, so it is at this level that AM activities influencing success must be conducted, monitored, analyzed, and adjusted.

26.3   ADAPTIVE MANAGEMENT ADJUSTMENTS: HUNTING

If there is no improvement in the status of forest resources as affected by deer, managers must determine which part(s) of their plan did not work and take corrective action. If the method chosen to reduce deer impact is hunting/culling, a reliable indicator of failure, which requires monitoring, is lack of reduction in deer density. Discounting the unlikely probability of significant immigration of deer onto affected DFMAs from adjacent landscapes, the question then becomes: What part of hunter management failed to produce reductions in deer density and impact?

26.3.1   FACTORS RELATED TO PUBLIC HUNTING

If public deer hunting is chosen as the means by which to reduce deer density/impact, and density/impact are not declining, managers may wish to revisit other components of their deer management program that may affect hunting, such as how they communicate with hunters, including the message(s) they convey, the kind(s) of hunters they are trying to attract, how they facilitate hunter access, how they obtain and distribute antlerless permits, and the kinds of incentives they provide for hunters to hunt on their DFMA. They may also wish to consider other lethal deer removal options, such as lease hunting or culling by contracted sharpshooters. If managers are unable to reduce deer density and impact by hunting/culling, they may wish to consider exclusion by deer fencing instead of hunting.

To separate out the potential causes of failure to reach goals for forest resources, managers must have information on: (1) weather conditions during hunting season, (2) number of hunters hunting the DFMA, (3) hunting access for the DFMA, (4) whether hunters obtained and used antlerless permits to enable them to harvest (additional) antlerless deer, (5) ability of managers to attract hunters to the DFMA, (6) perception by hunters that the DFMA was an unattractive place to hunt deer, (6) lack of incentives to attract hunters to the DFMA, and, (7) inability to attract alpha and beta hunters to hunt the DFMA and or educate novice hunters to make them more effective (locavore and mentoring/educating, especially for small woodlot owners). Evaluation of the contribution of these potential reasons for failure of hunting to reduce deer density and impact requires an effective communication system linking hunters to managers. Without hunter input, managers are left to guess which components of hunter management are not working. Managers *can* assess the effectiveness of some intuitively and take corrective action.

26.3.2   WEATHER

Weather factors (heavy precipitation, fog, temperature extremes) may depress hunting effort, especially if they occur on opening days of hunting season. Often, state natural resource agencies are aware of such weather factors and may extend hunting seasons appropriately. Managers must ensure they are aware of this condition and be prepared to use their communication system to make hunters aware of the availability of their DFMA for extended hunting season opportunities. Heavy snowfall can reduce hunter access to interior portions of DFMAs; if an option, managers should plow at least major access roads and communicate this action to hunters through their communication system.

26.3.3   ROAD SYSTEM/HUNTER ACCESS AND USE

Managers can assess the need for changes/improvements in hunter access by doing a bit of investigating prior to/during hunting season. Does the layout of their road system provide walking-distance access (<600 m) to most if not all of their DFMA? Is their road system well marked so hunters can reach all parts of the DFMA without getting lost in a maze of unmarked road intersections? Are their access roads in good shape, or are they rutted, slippery, partially under water, or covered with unplowed snow and ice? Are the only vehicles present during hunting season on roads and parking spaces four-wheel-drive pickup trucks, or are there passenger vehicles in the mix and all distributed throughout the road system?

If managers are unable to improve their road access, they have two choices: they can identify portions of their DFMA remote from access as quality deer hunting areas where hunters will have to travel on foot to areas with little competition with other hunters, a strategy employed in the cooperative case history (Chapter 34), or they can change their expectations and goals for the area of low/poor hunter access and concentrate their deer and forest management in more accessible areas, as in the case history in Chapter 33.

26.3.4   FAILURE OF HUNTERS TO USE ANTLERLESS PERMITS

Managers providing permits for harvesting antlerless deer should have a quantitative basis for estimating the number of permits they need to effect desired reductions in deer density and impact (Chapter 24). If many of the available permits are not requested by hunters, managers should embellish mechanisms for boosting numbers of permits requested by hunters, primarily by improving/expanding their communication system for reaching out to hunters (Chapter 22).

Some hunters obtain permits with the intent of preventing other hunters from obtaining them to reduce deer density on specified DFMAs. Managers may request survey information gathered by the issuing agency on numbers of permits obtained (if they do not distribute the permits themselves) and whether the permits were used to harvest an antlerless deer (if the agency collects such information and is willing to share it with managers of cooperating DFMAs).

If managers conduct checking stations, they can examine the age and sex composition of deer brought to the stations to determine whether antlerless permits were used to harvest antlerless deer. Based on their assessment of numbers of permits issued and used, managers may adjust numbers up or down in succeeding years, as was done in the cooperative case history (Chapter 34) when permit numbers were compared with annual estimates of deer density and impact.

26.3.5   HUNTER TYPE

In Chapter 10, "Hunters and Hunting," we promoted solicitation of alpha and beta hunters on the basis that they were interested in quality deer and deer hunting and could tolerate lower deer densities to promote healthy deer and a healthy forest understory, including wildlife habitat. However, on large DFMAs (thousands of ha) where deer density is at or above maximum sustained yield level (MSY—see Chapter 15), there probably aren't enough alpha and beta hunters to effect desired reductions in deer density. On these DFMAs, the original message to hunters may include the information that deer density is high, providing hunters with many opportunities to see large numbers of deer and to harvest a deer—a situation attractive to omega hunters, who require high deer density as a condition for hunting a DFMA. This was the case with the cooperative case history (Chapter 34): the original public information message noted that deer density was high. This message was buttressed by issuance of a large number of antlerless deer permits under the Pennsylvania deer management assistance program (DMAP), suggesting that hunters likely would see many deer with a good chance to harvest one on that DFMA. The original messaging also stressed that to produce quality deer and habitat, deer density should be reduced to lower levels. After the first 2 years under the DMAP program, deer density declined from $\sim$11 deer/km^2 to $\sim$4 deer/km^2, where it stabilized after $\sim$5 years.

Omega hunters began to complain about the scarcity of deer seen and harvested during later hunting seasons, and many ceased hunting the KQDC. But the remaining (assumed) alpha and beta hunters continued to hunt the KQDC, with many remarking that it was worth lower deer density to harvest deer of improving quality. Even though there were fewer hunters after the halving of deer density on the KQDC, deer density was sufficiently low that harvests provided by alpha and beta hunters, in conjunction with natural predation by bears and coyotes, resulted in stabilization of deer density at a level associated with successful establishment of seedlings of desirable tree species and observations of increased frequency and diversity of herbaceous vegetation.

Determining hunter type can only be assessed by contacting the participating hunters, either through written communications or by face-to-face exchanges as occur at deer checking stations. If the issue is hunter type, managers must work to enhance perceptions about quality hunting on their DFMAs through communications, such as provided by outdoor writers of local newspapers.

26.3.6   Hunter Perceptions

Managers should monitor perceptions of hunters hunting their DFMAs to determine whether hunting conditions, including quality and quantity of deer, are attractive enough to ensure repeat customers who may also promote the DFMA for hunting by friends and family. This can only be accomplished by surveying hunter attitudes as obtained by contacts at checking stations and/or through websites, blogs, and other social media communication systems.

26.3.7   Hunter Incentives

Providing hunters with incentives for harvesting deer, such as lottery/drawings prizes, cash awards, and recognition, has met with little if any improvement in reductions in deer density or impact (Chapters 33 and 34). The best hunter incentives are good road access, quality hunting experiences, and quality deer.

26.3.8   Temporal Framework for Hunting

Managers seeking to increase the level of deer harvest on DFMAs may seek special extra deer season permits from state natural resource agencies for hosting deer damage hunts held outside regular deer hunting season, if such permits are available. Because these hunts may represent a last chance to harvest a deer until next year's hunting seasons, they attract large numbers of hunters who can effect desired reductions in deer density and impact (deCalesta 1985).

26.3.9   Spatial Framework for Hunting

If hunting reduces deer density/impact satisfactorily on only part of a DFMA, managers may wish to partition the DFMA into separate, more uniform tracts, which may differ by topography (easier or harder to hunt), by access (amount and quality of road access; e.g., Chapter 33, where a portion of the DFMA was extremely steep and poorly roaded, making hunting access difficult), or by level of forage production through timber harvest (Chapter 34). Management on adjacent properties (or lack thereof) may make management of deer density and impact difficult or impossible if impact of deer moving from adjacent properties cannot be addressed/worked out with neighboring property managers/owners. Attempting to incorporate such neighboring properties in cooperatives (Chapter 31) could resolve the lack of consistent and effective deer management.

26.3.10   Communication System

26.3.10.1   Social Media

It has been stressed repeatedly that effective communications between hunters and managers are critical in effective deer management. Managers need to assess whether their communication system is working. A first consideration is how many hunters access the communication outlet, be it website, blog, or other social media such as Facebook. Websites and blogs can be adjusted to track number of contacts from outside persons (but do not differentiate between hunters using the DFMA and any other viewers), and they can be made interactive for two-way communication with hunters.

26.3.10.2   Education

Stakeholders are more likely to respond in helpful ways to communication efforts if they are actively involved in the educational process. If they invest their time in educational efforts, actually see in the field the problems faced by managers, and are forced to see actual field outcomes of deer impact and abundance and techniques used to combat them, they learn more. They will also have a better appreciation for, and understanding of, manager concerns, will be better able to absorb and understand the messages, and will be more receptive to the need for reducing deer density and impact in ways they can influence. Deer density workshops involving hunters, as described in Chapter 34, can provide for enhanced understanding of hunters with deer management realities and better communications with hunters.

Managers of the Ward Pound Ridge Reservation (Chapter 35) held a similar deer density and impact workshop for Park administrators and used the workshop, and subsequent analysis of deer density and impact, to convince the administrators to allow (bow) hunting on the park as a means of reducing deer impact on forest understory.

The consultant for the Hemlock Farms case history (Chapter 37) used comments by deer experts who toured the residential development and field visits by the board overseeing management on the development to convince the administrators to adapt a program of limited hunting by residents supplemented by culling by hired sharpshooters to successfully reduce deer density and impact.

26.3.10.3 Method of Communication

Managers relying upon word of mouth to attract hunters but failing to achieve desired reductions in deer density and impact may wish to try social media, including websites. Websites, while conveying comprehensive and useful information, require considerable effort in developing and maintaining, and skilled personnel for both. A downside of websites is they may take too much time and effort to maintain and be too cumbersome and difficult to navigate for hunters seeking a less frills/more hands-on method. Managers might consider blogs, a minimalist form of websites with the advantage of frequent message updating to convey their messages to hunters. User-friendly Facebook and Twitter social media methods may be more productive and easier to maintain than websites or blogs, as KQDC managers found.

26.3.11 CHANGE INCENTIVES

To recruit new hunters, and retain existing ones, managers often employ incentives, financial or otherwise. Because incentives can increase financial and personnel commitments of managers, their effectiveness should be evaluated and adapted as necessary. Some states have "earn-a-buck" programs, where harvesting an antlerless deer may be a prerequisite for obtaining a permit to harvest an antlered deer. Because most if not all such programs are applicable on deer administrative units (DAUs—see Chapter 9) by state regulation, rather than on DFMAs, the programs may not be DFMA-specific and difficult for private managers to apply.

26.4 ADAPTIVE MANAGEMENT ADJUSTMENTS: GOALS

If managers are not making satisfactory progress toward achieving goals by reducing deer density/impact, they may wish to adjust their goals, as exemplified by the case history in Chapter 33, wherein a portion of their DFMA was not amenable to public hunting (poor access, steep slopes) and managers produced a goal for that portion of the DFMA not reliant upon controlling deer density and impact. Chapter 34 provides an example of managing for higher deer density on one ownership within a cooperative, which was justified because of higher availability of forage created by forest management (and attracting hunters looking for areas with higher deer density).

REFERENCES

Allan, C. and G. H. Stankey. 2009. *Adaptive environmental management: A practitioner's guide.* New York. Springer Science & Business Media.

Arnold, J. S., M. Koro-Ljungberg, and W. Bartels. 2012. Power and conflict in adaptive management: Analyzing the discourse of riparian management on public lands. *Ecology and Society* 17:19. http://dx.doi.org/10.5751/ES-04636-170119

deCalesta, D. S. 1985. Influence of regulation on deer harvest. In *Symposium on game harvest management.* eds. S. L. Beasom and S. F. Roberson. Kingsville TX, Texas A & I University.

Irwin, E. R. and K. D. Mickett. 2008. Engaging stakeholders for adaptive management using structured decision analysis. Estes Park, Colorado. *The Third Interagency Conference on Research in the Watersheds,* September 2008.

Kremsater, L., F. Bunnell, D. Huggard et al. 2003. Indicators to assess biological diversity: Weyerhaeuser's coastal British Columbia forest project. *Forestry Chronicle* 79:590–601.

Lancia, R., C. Braun, M. Collopy et al. 1996. ARM! For the future: Adaptive resource management in the wildlife profession. *Wildlife Society Bulletin* 24:436–442.

Leopold, A. 1949. *A Sand County almanac and sketches here and there.* New York. Oxford University Press.

Runge, M. C. 2011. An introduction to adaptive management for threatened and endangered species. *Journal of Fish and Wildlife Management* 2:220–233.

Walters, C. 2002. *Adaptive management of renewable resources.* Caldwell NJ. The Blackburn Press.

Section IV

Special Cases

You can't solve a problem on the same level that it was created. You have to rise above it to the next level.

Albert Einstein

The ideal deer management situation occurs within homogeneous landscapes of a size, shape, ownership, and accessibility lending them to predictable outcomes based on management actions known to be effective at given deer density and impact levels. Many DFMAs do not fall within these prescribed confines, but can be grouped into categories, each with a set of specific and different adaptive management steps that have proven effective in respective, specified categories. The special cases addressed within this section are: (1) small woodlot properties, (2) forestlands managed under quality deer management guidelines, (3) forestlands leased for deer hunting, (4) forested public lands and residential communities not amenable to public hunting, and (5) cooperatives (aggregates of individual DFMAs) sometimes identified as compacts. The extreme variability of conditions within each of these special cases discourages research, with the result that specifics described in this book for these special cases are drawn from personal experience and observations as reflected by selected case histories presented in Section V. And, because there are no research-established guidelines for managing special cases, all (except one) are prime examples of the need for, and success of, adaptive management.

In a repetition of Sections II and III, I have not included Manager's Summaries for Section IV chapters. By reading these chapters in their entireties, managers will gain a comprehensive understanding of what is involved in deer management over a variety of different circumstances and be aware of the great variety of management activities used to succeed.

27 Special Case
Quality Deer Management

David S. deCalesta and Paul D. Curtis

CONTENTS

We are the next generation of hunters and the future of deer hunting is in our hands.

Quality Deer Management Association (2010)

27.1 INTRODUCTION

Managers searching for information to help them manage deer and other forest resources have expressed an interest in applying the principles and practices developed by the Quality Deer Management Association (QDMA). To address that need, this chapter provides a description of the QDMA program and how it addresses management needs of managers, and outlines how and whether managers of deer–forest management areas (DFMAs – see Chapter 9) might incorporate QDMA principles and practices into their management programs.

The QDM movement was stimulated by a problem faced by landowners trying to manage deer: how to incorporate scientific findings into management practices beneficial to their needs. A book,

Producing Quality Whitetails (Brothers and Ray 1975), addressed this need for research-driven management information on white-tailed deer (*Odocoileus virginianus*) in language the lay public could understand and use. This book organized deer research developed at Texas A&M University, the Caesar Kleberg Research Institute, and other research entities into a format detailing how to develop, deliver, and maintain deer management to produce quality deer.

A little more than a decade later (1988), the Quality Deer Management Association was established in South Carolina by a group of wildlife biologists led by Joe Hamilton following his observations of deer management in Australia. Philosophy, goals, principles, and practices of quality deer management (QDM) were published in a book (*Quality Whitetails: The Why and How of Quality Deer Management* by Miller and Marchinton 1995) that constitutes the QDMA bible for deer management. Original QDM objectives as espoused by founder Hamilton stressed quality—of deer, deer habitat, deer hunting, and deer hunters. Original QDM mandates were for herd management, habitat management, hunter management, and herd monitoring.

Later descriptions of QDMA, as offered by QDMA CEO Brian Murphy, expanded the definition of quality deer management to include "a management philosophy/practice that unites landowners, hunters, and managers in a common goal of producing biologically and socially balanced deer herds within existing environmental, social, and legal constraints."

Additionally, "QDM is about maximizing deer sighting rates for hunters while also maximizing deer health through balance with carrying capacity. Seeing the most deer, even if it means exceeding carrying capacity, sacrificing deer health, and causing damage to the habitat, is not compatible with QDM, nor is it an ethical or responsible goal."

QDM recommendations for forest vegetation management have also evolved to include emphasis on forage creation by forestry practices (silviculture), including the use of herbicides to change understory species composition. Timber harvest is another recommended practice for producing early-successional habitat that contains more nutritious deer forage than mature forests. QDMA Director of Education and Outreach Kip Adams recommends expanding the scope of monitoring to include quantitative assessments of deer density and deer impact on understory vegetation.

QDM was initially developed for private ranchers/managers who wanted to apply science to their land and deer management with an initial emphasis on quality deer. Managing for other forest resources, including commercial timber, was of secondary emphasis but promoted because of a side benefit of timber harvest—reverting maturing timber stands to early succession clearcuts or thinning to open the overstory and promoting development of understory forest vegetation containing abundant, high-quality deer forage.

27.2 QUALITY DEER MANAGEMENT COMPONENTS

27.2.1 Quality Habitat: Composition and Size

QDM is rooted in quality deer habitat, which contains foraging, fawning, escape/hiding, and thermal cover. Cover requirements are provided by vegetation within landscapes comprised of maturing timber, thickets of sapling/pole timber, densely vegetated forest openings and readily available water as found in riparian zones (creeks and rivers, edges of ponds and lakes, and springs). The recommended landscape size is generally 250–400 ha or more of contiguous forestland, containing all required habitat components. Such landscapes approximate the home range size of matriarchal groups of female deer.

If adjoining managers owning smaller areas (~10 ha or greater) combine their respective properties, the composite landscape could provide all required habitat components, including size, and qualify as a QDM Cooperative. A QDM Cooperative is "an informal agreement between multiple landowners, lease holders and hunters to practice QDM across property boundaries. As the average size of recreational hunting tracts has decreased in recent years, QDM Cooperatives have become more prevalent."

It should be noted that individual QDM properties, and QDM cooperatives, are managed under a set of guidelines developed for specific landscapes and incorporate the ecological and human factor components of management activities outlined for DFMAs in Chapter 25. Such activities may only be practiced at the DFMA level: ownerships, values, goals, and habitat and weather conditions of DFMAs within the larger, enclosing deer administrative units (DAUs—see Chapter 9) are so variable that no consistent, single management plan can be developed and delivered at the DAU level, setting aside the impossibility of providing required financial and human resource needs of areas as large and complex as DAUs.

27.2.2   QUALITY HABITAT: FORAGE CREATION

The quality habitat cornerstone of QDM emphasizes production of quality deer forage. Current QDMA doctrine for forage production focuses on silvicultural practices to increase quality and quantity of naturally occurring forage and on food plots to provide additional sources of high-quality deer food. Silvicultural practices (Chapters 13 and 21) leading to increased quantity and quality of deer forage are used by foresters to harvest timber products (e.g., clearcutting, single-tree or group-tree selection) or to enhance timber production (thinning, shelterwood seed cuts, crop tree release). All of these practices remove overstory trees, disturb the soil, and allow more sunlight to reach the forest floor and stimulate germination and development of tree and shrub seedlings. In some forest stands, competing plants such as ferns and seedlings of undesirable tree species such as American beech (*Fagus grandifolia*) and striped maple (*Acer pensylvanicum*) may dominate the understory, often a side effect of selective browsing by overabundant deer herds resulting in elimination of preferred tree and shrub species. Silvicultural practices used to reduce or eliminate completion from undesirable plant species include selective use of herbicides or repeated mowing. A side benefit of thinnings and timber harvest is resulting income that may be used to offset costs of QDM.

Creation and maintenance of food plots is another QDM practice for improving deer forage. Initial emphasis on developing quality forage featured food plots, and a branch of deer science evolved for testing mixes of forage species developed specifically for deer. Plant mixes including species maturing during different seasons (spring–fall) and identifying other factors required for development of such seedings (soil type, exposure to sunlight, soil drainage, and proper planting technique) are components of forage improvement efforts through food plots. Creation and maintenance of food plots requires expensive equipment (tractors, seeders, cultivators, etc.); costly materials (fertilizer, seeds, and herbicides); and large, seasonal time commitments (initial protection of the forages including erecting deer-proof fences, mowing, and other practices), which have dampened enthusiasm for food plots for some and encouraged other habitat enhancement practices. Food plots do attract deer during hunting seasons, aiding in deer harvest. However, if adequate numbers of female deer are not taken during open seasons, then food plots can sustain high numbers of deer, which will damage forest regeneration nearby. If food plots are included in a management program, then increased emphasis must be placed on harvesting the additional antlerless deer produced by the high-quality forage.

QDMA is promoting a new concept—old field management—as another way to increase quality and quantity of forage. Old fields are lands formerly cultivated for agriculture or grazing but later abandoned. Dominated by herbs and shrubs, old fields require far less investment of time and money to maintain as prime deer foraging areas.

27.2.3   QUALITY DEER

QDMA literature doesn't quantify physical attributes of a "quality deer," and for good reason—forage and other environmental conditions, including vegetation management, vary from locality to locality, making it impossible to provide a "one size fits all" quantitative description. Instead, emphasis is on herd management to develop characteristics of a quality deer herd: balanced age structure, proper doe:buck ratio (about 2:1), maintaining deer density at a level that provides and maintains quality habitat (as indicated by deer body weight), emphasis on harvesting antlerless deer to achieve desired deer density and buck:doe ratios, and refraining from harvesting the majority of young male deer (yearlings). If one must have a qualitative, generic description of a quality deer, the one provided in this book's preface serves.

A quality deer is defined as one that exhibits the best physical attributes (body weight and antler characteristics) obtainable from the nutrition afforded by local forage resources, including soil mineral content, when deer density is managed to allow full expression of understory vegetation in terms of species richness and abundance. When such deer are harvested in fall, they will exhibit high amounts of body fat (depth exceeding 1 cm) on the backstrap (muscle atop deer rump) and surrounding their kidneys, indicating sufficient quality and quantity of forage to carry them through severe winter weather in good condition.

27.2.4 QUALITY HUNTERS

QDM hunter management calls for educating hunters about deer ecology and behavior and identifying hunters as key participants in deer management, including control of abundance and sex and age ratios. They must be able to distinguish fawns, does, yearlings, and older bucks. The focus is more on the hunting experience than on harvesting a deer on every hunt and emphasizes forgoing harvest of immature bucks in favor of older, mature deer and on managing herd density by increasing harvest of antlerless deer. This approach is welcoming to alpha and beta hunters who are more receptive to science-based deer management than omega hunters whose primary interest is maximizing herd size and opportunities for seeing and harvesting deer. The QDMA provides seasonal training sessions on all aspects of deer biology and management, along with print magazine and online resources. Workshops are held by local QDMA chapters so that members can gain hands-on field experiences.

27.2.5 MONITORING

Initial emphasis of QDM monitoring featured antler characteristics, sex, age, lactation status, weight of harvested deer, and quality of forages in food plots. Deer data were collected at checking stations, with trail cameras, and by observations of deer during scouting and hunting trips. As quantitative information on deer monitoring became available through research, QDM monitoring incorporated additional techniques such as deer pellet group counts to estimate deer density. Evaluation of deer impact on forest vegetation and use of silvicultural treatments to open the overstory and improve native forage production were added to assessments of food plot success.

27.2.6 INHERENT QUALITY DEER MANAGEMENT CHARACTERISTICS

There are additional, implicit QDM components related to management of deer, forest vegetation, and hunters that align with chapters in Sections I through III of this book.

27.2.6.1 Quality Deer Management Preconditions

Before embarking on QDM, the QDMA recommends adherence to a checklist of preconditions, including: (1) minimum landscape size, generally >250 ha, either as individual property or as one of multiple properties in a cooperative (see below) formed by forestlands from a number of abutting landowners that aggregates to the minimum size; (2) commitment to follow QDM recommendations for at least 5 years to allow for vegetative and deer herd responses to management actions; and (3) adoption of a management philosophy including harvest of antlerless deer to achieve deer density goals and focusing buck harvest on mature, rather than immature (e.g., yearling), antlered deer.

27.2.6.2 Goals and Strategies

As identified in its information on establishing QDM cooperatives (see below) and contained in Chapter 25, QDMA advocates establishing deer management goals and strategies as first steps in developing a quality deer management program.

27.2.6.3   Carrying Capacity

QDM literature describes biological deer carrying capacity as "the maximum number of deer a landscape can support without detrimental effects." A herd at biological carrying capacity, also referred to as K-carrying capacity (see Figure 15.1) recruits no fawns into the population because density is too high and reproduction/recruitment are shut down. The definition is expanded to include *cultural carrying capacity* as the maximum number of deer that can coexist compatibly with local human populations as determined by their values and their tolerance for deer and deer-related issues. Such cultural or "social" carrying capacities include deer densities sufficiently high to represent unacceptable crop damage to farmers or number of deer/vehicle collisions to motorists. QDM philosophy seeks to avoid exceeding cultural as well as biological carrying capacity.

QDM further suggests that improvements in forage by forestry practices and/or by supplementing naturally occurring forages with food plots can increase deer carrying capacity due to increased forage abundance/quality.

These definitions do not relate to the full range of carrying capacities and values important to managers. As described in Chapter 15, there are at least four carrying capacities related to values of managers and hunters, each with its own threshold deer density. To address their values, managers need to manage for deer density that relates to their value-related carrying capacity. For the example drawn from the Kinzua Quality Deer Cooperative (Chapter 34) in the northern hardwood forest type in northwestern Pennsylvania, biological carrying capacity (related to inevitable population crash) occurs at about 27 deer/km^2; maximum sustainable yield (MSY—see Chapter 15) carrying capacity occurs at about 11 deer/km^2; carrying capacity required for successful regeneration of tree species desirable for harvest and promotion of quality deer occurs at about 7 deer/km^2; and carrying capacity required to foster optimum biodiversity, including wildlife species such as shrub-dependent songbirds and sensitive herbs such as Canada mayflower (*Maianthemum canadense*) and trillium (*Trillium* sp.), occurs at ≤ 5 deer/km^2. If QDM is to be practiced by managers whose values include the gamut of tree regeneration, quality deer, and biodiversity, desired deer density/abundance must be far below that identified for MSY.

The idea that increasing quality and quantity of deer forage will lead to an increase in carrying capacities associated with human values appears valid—at least for carrying capacity associated with successful tree regeneration, as suggested in Chapter 15. However, managing for carrying capacities associated with trophy deer characteristics and diversity of understory vegetation may not benefit from increasing amount and quantity of deer forage, unless performed judiciously. Field observations by the senior author of this chapter in landscapes surrounding food plots and winter deer feeding grounds indicated that deer concentrated in the landscape immediately surrounding the sources of forage rather that spreading out over the larger landscape, with resulting severe impact/ elimination of seedling regeneration and elimination of sensitive herbaceous species in forestland adjacent to the increased source of forage. Thus, increasing forage quality and quantity, including food plots, should be distributed uniformly and incrementally over the years to avoid untoward deer impact on understory vegetation.

27.2.6.4   Recruiting and Mentoring New Hunters

Learning opportunities for youth are emphasized as a way to educate, excite, and recruit younger hunters. QDMA volunteer branches introduce hundreds of young people to deer hunting. The QDMA National Office offers a National Youth Hunt program wherein young hunters who want to hunt deer but have no family or friend resources to draw on are taken on a "dream hunt." They experience the opportunity to harvest deer and hunting party camaraderie, and are taught how to age deer and learn hunting tactics. Attendees make lasting memories and friendships during the 4-day event on a property managed for quality deer and habitat. This program is supplemented by a mentored hunting program designed to educate, train, and retain new hunters on an extended basis.

The QDMA "mentored hunting program" is "designed to increase the number of youth and first-time hunters" and "builds a foundation for new hunters to become better stewards of natural resources."

Conducted over a period of several months, the training provides eight comprehensive packages: (1) background information on the QDMA program and hunting; 2) basic woodsman skills for becoming a successful deer hunter; 3) hunting ethics; 4) hunter safety; 5) development of hunting skills with small game hunting prior to deer hunting; 6) deer biology, behavior, ecology, and anatomy; 7) deer hunting safety, hunting technique, and management; and (8) emphasis on "celebration of the hunt" including social, cultural, and ethical dimensions of hunting rather than on harvesting (trophy or other) deer.

Volunteer mentors help new hunters develop the knowledge and skills to become competent hunters, learn about the hunter's roles in deer management, and become advocates for natural resources. Mentors are QDMA members selected and trained to provide experiences and continuing support for new hunters as they guide them through the eight steps of the mentoring program.

27.2.6.5 Stakeholders

Primary stakeholders are participating hunters who buy into QDMA philosophy and management practices. Managers of public forestlands court hunters to reduce deer density and cannot differentiate among alpha, beta, and omega hunters but rather must solicit all types for managing deer density. Conversely, QDM properties are hunted by alpha and beta hunters who are informed of and accept the need for controlling deer density, willingly harvest antlerless deer, and refrain from harvesting immature antlered deer. The typical science vs. values and tradition conflict between hunters and landowners for managing deer density, condition, and sex and age ratios is less of a factor—hunters (often heavily weighted to family and friends) are on the same relative page as the landowners and support science-based deer management. Abutting managers, another stakeholder group, benefit from, and are supportive of, QDMA management practices that control deer density and reduce negative impact on forest resources, including their own.

27.2.6.6 Communicating with and Educating Hunters and Other Stakeholders

The QDMA has a website (www.QDMA.com), blog (www.QDMA.com/homewidgets/blog-feed/), magazine (*Quality Whitetails*), and newsletter that provide two-way communication with hunters and other stakeholders, identify QDMA mission and objectives, provide biological information on deer, and announce opportunities for training and education. QDMA uses recognized deer experts (scientists, managers, and educators from university, federal and state research/management entities, consultants, and seasoned practitioners) in its educational and operational programs to provide current, qualified, and validated information on deer management.

27.2.6.7 Resources

The QDMA has 60,000+ members who, among others, donate >$2 million annually for QDMA programs. Individual forest landowner members have access to QDMA training and educational and management information as well as do volunteer QDMA members who can assist them in carrying out deer management programs.

The QDMA does not provide cash assistance to member landowners for habitat (including forage) improvement but does help them learn how to obtain financial assistance from federal and state aid programs, including: (1) the Conservation Reserve Program (CRP) where landowners are paid to take land out of agricultural production and plant permanent cover (such as trees and shrubs) and helps cover the cost of the initial habitat work as well as an annual rental payment for the life of the CRP contract (typically 10–15 years), and (2) the Environmental Quality Incentives Program (EQIP) for landowners with agricultural lands and nonindustrial private forestlands. EQIP provides financial and technical assistance to plan and implement conservation practices that address natural resource concerns and for opportunities to improve soil, water, plant, animal, air, and related resources.

27.2.6.8 Politics

Because QDM practices initially were applied on private forest/range lands, the political conflicts characteristic of deer management on public lands were avoided. As QDM concepts and management recommendations began to be applied on public lands, some state and federal wildlife management agencies embraced and promoted the QDM movement and its components. The primary objections to QDM have come from deer hunter groups that seek maximization of deer abundance without regard to environmental consequences that would ensue (e.g., loss of habitat and increased negative impacts on other wildlife species), and from state and federal wildlife agencies who prefer the traditional deer management approach.

27.2.6.9 Quality Deer Management Lands Certification

As an adjunct to training and education of hunters and landowners, the QDMA developed a program of standards and performance measures for certifying participating landowners/hunters. The 14 standards are augmented with respective performance measures and include: (1) commitment to QDMA ethics and conservation; (2) compliance with laws; (3) commitment to safety; (4) commitment to four cornerstones of QDM; (5) herd management activities, including the concept of, and commitment to, adaptive management; (6) forest vegetation management activities; (7) hunter management activities; (8) herd monitoring activities; (9) commitment to excellence and continued improvement; (10) conservation of biodiversity; (11) aesthetics; (12) special site identification and conservation; (13) air, water, and soil protection; and (14) long-term habitat protection and conservation.

Establishment of the standards and performance measures illustrates how the QDMA has evolved into an inclusive, ecologically aware, and representative organization by emphasizing factors in addition to hunting and deer. Now included in the QDMA core set of values and emphases is concern for the environment, biodiversity, preservation and protection of sensitive, plants, animals, and soils, and inclusion of environmental intangibles (e.g., commitment to excellence and improvement, and esthetics).

27.2.6.10 Adaptive Management

The QDMA has defined adaptive management as "A dynamic approach to deer management wherein the effects of treatments and decisions are continually monitored and used to modify management on a continuing basis to ensure that objectives are being met." Adherence to the concept of adaptive management is mandated as a performance measure in Standard 5 (Herd Management Activities) of the QDMA Land Certification Program and by performance measures identified for other Standards (4, 6, 7, and 8) for collecting and analyzing data on herd quality (e.g., adult sex ratio, deer density, antlerless deer management, antlered buck management) hunting quality, food plots, early successional management, tree/shrub plantings and water/pond/wetland management, hunter education efforts, deer harvest guidelines, deer sanctuary protection, and status of property access for hunting.

Further demonstration of QDMA adoption of adaptive management philosophies and practices is evidenced by its objective of "uniting landowners, hunters, and managers in a common goal of producing biologically and socially balanced deer herds within existing environmental, social, and legal constraints" by promotion of forage creation by silvicultural practices to change understory species composition, promotion of timber harvest to produce early successional habitat, and by expanding the scope of monitoring to include deer impact on understory vegetation.

27.2.6.11 Quality Deer Management Cooperatives

QDM Cooperatives are "groups of landowners and hunters working together to improve the quality of the deer herd and hunting experiences on their collective holdings. Cooperatives vary in size, number of participants, and structure depending on the needs and objectives of members. A primary benefit of cooperatives is that they enable hunters with small landholdings to participate in QDM. Cooperatives are voluntary and do not entitle neighboring hunters access to participating landowner's property or diminish the landowner's control. They are collections of landowners and hunters who agree to abide by similar deer management guidelines over a larger area."

QDM cooperatives address a problem common to many managers with small properties: how to manage deer, which occupy large landscapes, within holdings so small they are not conducive to manipulating deer habitat or density on a meaningful scale—deer from the resident herd spend most of their time on adjacent forest/farm lands. When managers combine their holdings with those of neighbors, they become much more able to manage deer density, age structure, sex ratio, and distribution, as well as impacts.

27.2.6.12 Benefits of Quality Deer Management Cooperatives

Cooperatives provide: (1) pooling of data on deer density, quality, harvest, forest vegetation, and impact at landscape levels and collection of comprehensive information useful for managing deer and forest vegetation at realistic management scales; (2) better management of density, age structure, sex ratio, distribution of a local deer herd, and deer impacts; (3) the opportunity to hunt a high-quality deer herd containing more mature bucks; (4) improved relationships with surrounding landowners and hunters who share common goals; (5) enhanced control of trespassing and poaching, as member hunters have an interest in preventing illegal access; and (6) reduced costs of management operations, including development and maintenance of food plots and sharing equipment and other resources related to silvicultural practices to manage vegetation, including seedling regeneration, reduction of competing vegetation, and creation of deer forage through timber harvest and thinnings.

27.3 RELATING DEER–FOREST MANAGEMENT AREA TO QUALITY DEER MANAGEMENT

Managers assimilating the underpinnings for deer management presented in Section I and incorporating elements of Sections II and III in their DFMA management plan are adhering to some, if not all, of the principles and practices espoused by the QDMA. For DFMAs ~200 ha or less, management options for deer and other resources will be limited unless managers can form a larger DFMA by combining forces and resources with abutting forestland owners. The publication describing formation of QDMA cooperatives (Quality Deer Management Association 2010) details how to form cooperatives.

In the event that management of specific DFMAs places higher priority on resources other than deer, such as timber for harvest or biodiversity, current QDMA management guidelines may need to be enhanced by greater emphasis on silvicultural practices designed specifically for managing forest vegetation. Because QDM emphasis for developing management plans for forest resources, including deer, is predicated on addressing landowner values, there is no impediment to developing a set of management activities designed to produce, for example, quality timber. Additional consultation with silviculturalists or forest ecologists may be required to develop comprehensive management plans to meet goals for forest resources of interest.

27.4 IMPACTS OF QUALITY DEER MANAGEMENT

The QDMA program has heightened public awareness of deer management, particularly for production of quality deer. Other, lesser-known components contribute to its comprehensive management philosophy and practices that promote management of all forest resources as directed by landowner values and goals.

Whether intentional or not, current deer management as practiced within the broad spectrum of conditions encountered by managers is influenced and improved by adoption of QDMA principles and practices. Case histories presented in Part V of this book subscribed to many if not all of the QDM components. Furthermore, the QDMA model for managing deer and habitats has been adopted and adapted by many types of managers of private and public land managers, including industrial forests, parks, state and national forests, and watersheds for public water supply.

However, we caution readers that simply adopting QDM practices may not be sufficient to achieve forest regeneration or biodiversity goals. Successful regeneration of northern hardwood forests requires not only successful deer management, but also incorporation of appropriate silvicultural practices, possibly including controlling native or invasive vegetation that would compete with seedling growth. Without addressing these interacting factors, at current deer densities common in many woodlands, successful regeneration of a diverse and productive forest is unlikely.

REFERENCES

Brothers, A. and M. Ray. 1975. *Producing quality whitetails*. Laredo TX. Fiesta Publishing.

Miller, K. V. and R. L. Marchinton eds. 1995. *Quality whitetails: The why and how of quality deer management*. Mechanicsville PA. Stackpole Books.

Quality Deer Management Association (QDMA). 2010. What is Quality Deer Management? http://wiredtohunt. com/2010/04/29/what-is-quality-deer-management/

28 Special Case
Small Woodlots

Jim Finley, Allyson Muth, and David S. deCalesta

CONTENTS

28.1 INTRODUCTION

By far the largest and most diffuse bloc of forestland in the northeastern and southeastern United States is owned and managed by small woodlot owners (Table 28.1), of whom <10% own forestlands large enough (>200 ha) to encompass a typical deer home range (Table 28.2). These owners have limited tenure on the land (generally less than 20 years) and may have goals that are not affected by deer impact, or they may not recognize that deer can affect their goals.

Because of truncated time frame, small landscape size, and management goals specific to small woodlot owners, many small woodlot owner/managers do not possess the wherewithal to address many of the seven important management components identified in Chapter 25 (identify goals for identified forest resources; obtain baseline monitoring data to determine whether management activities are called for; assess need for and availability of financial and human resources; develop a system for communicating with stakeholders, particularly hunters; manage road access and antlerless permits for harvesting deer; address needs of hunters; and monitor responses of resources of interest relative to application of management activities). This chapter is devoted to addressing deer management needs of small woodlot owners of limited financial and human resources and whether managing deer density and impact is required for them to meet their natural resource goals.

28.2 FACTORS DETERMINING NEED FOR, AND USE OF, DEER MANAGEMENT

28.2.1 CULTURE, VALUES, SCIENCE, AND GOALS

Hunters and resource managers are important influences for addressing economic, ecological, and social values held by small woodlot owners. The challenge is to move outside singular foci associated with individual interests and perspectives (e.g., hunting deer and harvesting trees) to appreciate and understand the values private managers hold. It is essential to learn about relationships between small

TABLE 28.1

Percent of Forestland in Northeastern and Southeastern States by Ownership Class

	Public Lands			Private Lands	
	Local	State	Federal	Corporations	Family
Northeastern states	4	13	8	20	55
Southeastern states	1	3	9	28	58

Source: Butler, B. J. 2008. Family forest owners of the United States, 2006. Department of Agriculture, Forest Service, Northern Research Station. General Technical Report NRS-27 Newtown Square PA.

TABLE 28.2

Percent of Forestland in Northeastern and Southeastern States by Size (ha) Class of Family Forest Owners

Size Class (ha)	<4	4–20	20–40	40–200	200–400	400–2000	>2000
East %/size class	7.9	24.5	18	32.8	6.5	7.8	2.6
East %/owners	60.8	27.6	6.5	4.7	0.3	0.1	0.0
North %/size class	9.7	29.0	20.3	31.7	4.4	3.6	1.3
North %/owners	62.4	27.3	6.2	3.9	0.2	0.1	<0.1

Source: United States Department of Agriculture Forest Service. 2008. *National woodland owner survey.* Amherst MA: Forest Inventory and Analysis Program National Woodland Owner Survey.

woodlot owners and the land, their vision for the future, and perceived constraints to their plans. Simple prescriptions and singular approaches may not lead to opportunities to hunt or manage deer. It is essential for managers to integrate scientific information with values.

The sheer number of private woodlot owners suggests there are myriad opportunities to guide management decisions across these landscapes; however, it is important to understand what drives the millions of small woodland owners to hold forestland. Objectives will vary by ownership, location on and of the property, management history, past activities, perceived and actual needs, and myriad other issues. We believe conversations with woodland owners ought to start by understanding what they value on their land and how they use it. Increasingly, resource managers are starting the conversation by seeking information from these perspectives. Knowing this, they can guide the landowner to make better-informed decisions in the context of how they value and use their woodlands. Professional foresters are taught to ask woodlot owners to articulate their objectives; many (mistakenly) believe that timber management and income are the primary reasons. This conclusion leads to conflicts as the forester learns that the landowner has other notions about other goals as well as how to cut the timber and how much income they might reap. The forester has little ability to direct the outcome to a sustainable well-managed forest—they develop management plans to achieve what the landowner dictates.

In the past 50 years, researchers interested in private forests have conducted studies to understand woodland owners. Repeatedly, these studies have found that most private managers seldom place timber, income, or investment high on their list. This is a common misconception of foresters and others about landowner motivations (Jones et al. 1995).

Owners of small woodlots emphasize and manage for goals other than producing harvestable timber, such as supporting wildlife, providing places to recreate, purifying air and water, reducing energy needs with windbreaks and shade, and providing (nontraditional) forest products such as mushrooms, firewood, and maple syrup (Ohio Department of Natural Resources, Division of Forestry).

A recent survey of reasons for owning small woodlots in northern and southern states (USDA Forest Service 2008) indicated that producing timber and other products of commercial/consumptive value was near the bottom of preferred goals (except on woodlots >200 ha) and that noncommercial cultural values such as beauty and scenery, protecting biodiversity, privacy, and family legacy, were ranked far higher (Table 28.3). Small woodlot owners favor amenity values (e.g., beauty/scenery, privacy) or have undefined uses for the woodlands (e.g., part of the home). Their woodlands also figure in their legacy and estate plans. Significantly, it is not until the land area becomes relatively large (>40 ha) that the importance of economic motivations (e.g., land investment, or timber, which moves to the top, or close to the top, of the ranking for the largest owners) gains ground. Interestingly, hunting and fishing are never ranked higher than fourth and decline as the ownerships become increasingly large. Clearly, hunting is a recognized use on many small woodlots, but other uses are held and valued by the owner for other objectives.

TABLE 28.3
Objectives of Small Woodlot Owners for Managing Forestlands by Rank of Value by Woodlot Size: 1 = Highest Rank; 12 = Lowest Rank

Objective	<10 ha	10–20 ha	20–40 ha	40–200 ha	200–400 ha	>400 ha
Northern states						
Beauty/scenery	1	1	1	1	2	3
Protect biodiversity	5	6	6	7	5	6
Land investment	8	7	7	2	1	2
Part of home	3	4	4	5	7	7
Part of farm	9	9	9	10	10	11
Privacy	2	2	2	6	8	9
Estate/legacy	4	3	3	3	4	4
Nontimber products	12	12	12	12	12	12
Firewood	11	11	11	11	11	10
Sawlogs, other products	10	10	10	4	3	1
Hunting/fishing	6	5	5	8	9	8
Other recreation	7	8	8	9	6	5
Southern states						
Beauty/scenery	2	3	3	5	5	5
Protect biodiversity	4	5	6	6	6	6
Land investment	3	2	2	1	1	1
Part of home	6	9	9	9	9	9
Part of farm	9	8	8	7	8	8
Privacy	5	7	7	8	7	7
Estate/legacy	1	1	1	2	2	2
Nontimber products	11	11	11	11	11	11
Firewood	12	12	12	12	12	12
Sawlogs, other products	8	4	4	3	3	3
Hunting/fishing	7	6	5	4	4	4
Other recreation	10	10	10	10	10	10

Source: United States Department of Agriculture Forest Service. 2008. *National woodland owner survey.* Amherst MA: Forest Inventory and Analysis Program National Woodland Owner Survey.

Understanding landowner values can provide opportunities to increase access for hunting, reduce wildlife and resource management conflicts, and lead to enhanced use of small woodlots. It requires learning about owner values and how to enhance them. Values may be embellished by including wildlife or aesthetics. For example, in their book *Bringing Nature Home*, Tallamy and Darke (2009) promoted the need to incorporate native plants to support native insect species that have very specific forage plants, making an argument linking insects to songbird populations.

Many woodland owners fail to understand the diverse linkages and interactions that affect management to achieve their stated values and objectives. Helping them understand the role of forest and deer management to attain or maintain desired future conditions is a process of education and communication that involves three steps: (1) creating awareness, (2) providing information, and (3) acquiring knowledge. Everyone involved in the management of privately held forestlands can help facilitate discussions about the role of forest and deer management. Much of the acceptance of the story hinges on conveying information in the context of the listener's values or needs. Consider the property owner focused on native plants or songbirds. Information on forest management will not find acceptance, but information on the need for native plants and desired structure and the role of deer in affecting the outcome might lead to opportunities to engage in management actions.

28.2.2 LANDSCAPE

Forestland ownership patterns extend across landscapes and include multiple ownerships. Relatively few small woodlot forest parcels are large enough to meet all the needs of wildlife species, especially deer whose home ranges and habitat requirements are of a size far larger than most small woodlots. The average small woodlot size barely exceeds 10 ha and about half of small woodlots are in ownerships smaller than 40 ha. The number of parcels larger than 100 ha is small, but the aggregated (and disparate) area of small woodlots represents about 41% of forestland in the East. Landscapes surrounding small woodlots present a matrix of interspersed properties of varying sizes and management goals. What this means is that small woodlot owners are at the mercy of larger, surrounding forestland ownerships because deer density and impact on small woodlot properties are dictated by deer density within the larger surrounding properties.

28.3 MONITORING

Estimating deer density is impossible on small woodlots because the area censused makes up only a fraction of the home ranges of impacting deer. Managers can obtain estimates of deer *abundance* (using pellet group counts or trail cameras) and deer impact on vegetation (and wildlife habitat, e.g., ground and shrub cover for forest birds) on their small woodlots, but because they cannot estimate effective deer density (density of deer in landscape including their woodlots), they cannot estimate number of deer to harvest to bring density down to levels not negatively impacting forest resources. What they can achieve by monitoring deer impact is determining whether and how to reduce deer impact.

28.4 REDUCING DEER IMPACT

If deer habitat(s) (seral stages providing foraging, fawning, hiding, and thermal cover) on small woodlots do not include foraging areas (because managers have not harvested timber or in other ways opened up the overstory to stimulate growth of deer forage) or areas scheduled for timber harvest (because they contain sufficient advance regeneration to progress from seedlings to fully stocked forest stands), there may be few reasons for managing deer density. One such reason could be an owner's goal of enhancing and maintaining diversity and abundance of understory forest vegetation, including uncommon wildflower and shrub species, in which case a management goal should be to reduce deer impact to almost nothing. With a goal like that, a management strategy employing hunting to eliminate deer family groups using the woodlot (the Rose Petal solution; Chapter 20) should be considered. Of the many hunting methods available (Chapter 20), perhaps the most effective on small woodlots is drive hunting, wherein moving lines of hunters drive deer to a stationary line of hunters waiting in ambush. Drives must be carefully arranged to prevent hunters from shooting at others within the drive. The only alternative to hunting is caging individual herb/shrub species or enclosing foraging areas such as timber harvest sites with expensive deer-proof fencing to prevent deer impact (Chapter 20).

28.5   ROLE OF HUNTING RELATIVE TO OWNER VALUES

Across landscapes populated with small woodlots, the mix of values and objectives, parcel sizes, and uses creates a matrix of conditions that affect wildlife and forest vegetation. Some owners permit hunting, others encourage it, and others post their properties against hunting or establish no-hunting safety zones that create refugia where deer are protected from hunting and may retreat to with the onset of hunting season. Managing forest vegetation and deer is difficult and complicated by the desire to manage forests for other/additional values such as species composition, regeneration, and habitat structure of forest vegetation—all generally requiring reductions of deer density and impact to meet goals.

28.6   FUNDING DEER–FOREST MANAGEMENT

Many private managers struggle with implementing management activities that lead to increased costs or risk changing forest conditions that go against their ownership goals. In much of the Northeast, sustainable harvesting for timber products depends on controlling competitive plants (e.g., native ferns and nonnative grasses and shrubs) before harvesting to establish adequate advanced hardwood regeneration. The costs are vastly increased by the need to erect deer-proof fencing if other solutions, such as hunting, are not applicable or are rejected by small woodlot owners. Research suggests that reducing deer densities (through hunting, if applicable) can provide opportunities for tree and other native species to outcompete competitive plants, enabling harvest plans that will lead to successful regeneration. Small woodlot owners can apply for federal funding (see Chapters 19, 37, and 38) to finance vegetation management designed to mitigate negative impacts of overabundant deer (negatively impacting species composition, abundance and structure of herbs, shrubs, and tree seedlings).

28.7   COMMUNICATIONS

Should small woodlot owners choose to manage deer impact with public hunting, they need to facilitate a hunting program by developing a communications network among themselves, participating hunters, and other stakeholders, such as local deer and forest biologists from state natural resource agencies (Chapter 22). Because they likely will not need a large number of hunters, and because they likely will feel more comfortable and in control if they use friends and family, their communication system may consist of no more than telephone and e-mail contact lists.

28.8   ADDRESSING THE FOUR Rs OF HUNTING

Should small woodlot managers decide to reduce deer density and impact with public hunting, they can easily address the needs for recruiting, retraining, rewarding, and retaining hunters as detailed in Chapter 23. This aspect of deer management will be simplified for small woodlot owners due to the number and nature of hunters, size of woodlot, and built-in advantages of addressing the needs of hunters with whom they are personally acquainted. Recruiting new hunters, and retaining existing hunters requires little more than maintaining friendly relationships with family and friends who will constitute most if not all of the hunters. Hunters will not need retraining because they will have been recruited based partly on their recognition of the need to reduce deer density and impact. And, recruited hunters will not need rewards beyond those provided by their relationships with the owners, familiarity with forestlands they hunt, and ability to harvest deer. As an added bonus, recruited hunters likely will be willing to perform some of the normal management tasks of the owners: posting and maintaining boundary lines, caring for and maintaining access roads and gates, and providing information on deer harvested as needed by the owners for maintaining and reporting harvest records as required for continued ability to obtain permits for harvesting antlerless deer.

28.9   ACCESS AND PERMITS

These management requirements also can be met with minimal investment by small woodlot owners, again because of the small size of their deer–forest management area (DFMA—see Chapter 9), minimal needs for access and access maintenance (most hunting areas may be reached simply by hiking from a central point, such as the woodlot owner's home if they reside on the property), and ease of providing permits for harvesting antlerless deer to their band of hunters and obtaining harvest reports from them.

28.10   ADAPTIVE MANAGEMENT

Monitoring, at least of deer impact on forest resources of management interest, should be an annual, springtime activity to identify the need for continued and/or heightened hunting and deer harvest. Landowners, by nature of their relationship with hunters, will obtain hunter feedback as hunters see the need. Depending on the size of the woodlot, increasing number of antlerless permits from issuing agencies can only be determined and defended by quantitative data on abundance and diversity of forestland resources of interest (given that managers have not already maxed out the number of permits allowable based on woodlot size).

REFERENCES

Butler, B. J. 2008. *Family forest owners of the United States, 2006*. Department of Agriculture, Forest Service, Northern Research Station. *General Technical Report NRS-27 Newtown Square PA*.

Jones, S. B., A. E. Luloff, and J. C. Finley. 1995. Another look at NIPFs: Facing our myths. *Journal of Forestry* 93:41–44.

United States Department of Agriculture Forest Service. 2008. *National woodland owner survey*. Amherst MA: Forest Inventory and Analysis Program National Woodland Owner Survey.

29 Special Case
Lease Hunting

Jeffrey Kochel, Michael McEntire, and Delores Costa

CONTENTS

Leasing forest and agricultural land can be a win-win situation for landowners and hunters.

Michael Jacobsen (2014)

29.1 INTRODUCTION

The information presented in this book is based on the premise that hunting by the general public on private and public forestlands is the best way to reduce and maintain deer density and impact at acceptable levels. However, managers of privately owned forestlands may choose to restrict hunting access to paying customers who lease hunting (and other) privileges on the forest owner's land(s). Successful managers of forestlands leased for hunting follow the same practices outlined in this book for forestlands open to public hunting, with a few notable exceptions. This chapter describes management practices we developed and implemented for lease hunting in the eastern and southeastern United States for 200+ properties on lands ranging in size from 40 to 10,000 ha.

With few exceptions, landowners who lease their forestlands for hunting have as a primary goal sustainable production and harvest of timber. For many, overabundant deer herds and resultant failure of advance regeneration required costly management practices (e.g., use of herbicides to reduce interference of ferns and undesirable woody plants resulting from removal by deer of preferred advance regeneration, cost of fencing to protect regeneration sites from deer browsing). When these forest owners attempted to reduce deer density and impact with public hunting, the hunters rarely reduced deer abundance sufficiently to permit success of advance regeneration. In large part, this was due to hunters preferentially harvesting antlered deer and declining to harvest does and fawns.

Lease hunting offers these forest owners the potential for reducing deer density and impact because of the control it exerts over hunter behavior—the hunters are motivated to harvest enough deer, including antlerless deer, to allow forest owners to practice sustainable forestry. That, or they are replaced by other groups of hunters who *are* willing and able to reduce deer density by aggressive harvest of antlered and antlerless deer.

29.2 BACKGROUND

Forest Investment Associates, a forest management firm with multiple holdings in the eastern United States, has a history of management issues related to overabundant white-tailed deer (*Odocoileus virginianus*). Clients were industrial forest owners with large forested landscapes. Properties were composed of multiple, single-age (50–90 years) hardwood stands with overstories of black cherry (*Prunus serotina*), birches (*Betula* sp.), sugar and red maples (*Acer saccharum* and *Acer rubrum*), eastern hemlock (*Tsuga canadensis*), and American beech (*Fagus grandifolia*). Long-term browsing impact by deer changed the dynamics of understory vegetation, including elimination of shrub and herbaceous species preferred by deer, domination by non–commercially valuable woody plant species less palatable to deer, and development of fern understories that choked out advance regeneration of desirable tree species. Our preferred approach to deer management on these lands was to attract and motivate deer hunters to reduce deer density and impact during deer hunting seasons. And, like many managers, we advertised in local and regional news outlets, offered free hunting access, provided maps of access roads, and tried other incentives to attract hunters to the properties. However, these efforts failed to reduce deer density to the point where regeneration was successful and biodiversity improved, forcing us to rely on protecting these forest resources with expensive 2.5-m-high woven-wire deer fencing exclosures constructed around regeneration sites. The problem with fencing was that only forest resources inside the fence were protected; those outside the fencing (generally over 80% of the area) remained unprotected. Leasing hunting rights on forestlands has a long history in the southeastern United States, and we wondered if by leasing hunting rights to the deer-impacted forestlands we managed we might succeed in reducing deer density and impact in northeastern as well as southeastern states.

29.3 WHAT IS LEASE HUNTING?

Lease hunting is an arrangement between landowners (lessors) and hunting associations (lessees) wherein hunting rights on the lessor's property are granted to the lessees. Leases are structured by a legally binding agreement between lessors and lessees wherein right of trespass is granted to lessees for the privilege of hunting on the leased forestland in exchange for an annual fee. The agreements stipulate that lessees will perform certain activities (e.g., post and patrol property boundaries during hunting seasons to keep out hunters not belonging to the lease group). Annual lease fees are identified and used to offset expenses related to administering the leases, educating lessees, and monitoring deer density and impact. In states requiring lessors to carry large liability insurance specifically covering the leases, that cost is passed on to lessees as part of the annual fee; in states with lower liability insurance costs, landowners absorb the insurance as part of timber management expenses. Maintenance costs of access roads, used for timber management purposes as well as for providing hunters access to interior portions of the leased property, are generally borne by lessors, unless lessees cause road damage related to their activities.

Lease agreements spell out the responsibilities of lessees and lessors and contain clauses specifying conditions under which the lease may be terminated. Members of the lease are named. Privileges granted to lessees generally include hunting for deer and other game species; fishing upon specified waterways; trapping for furbearers; and other recreational pursuits such as hiking, camping, and picnicking. Lessees are usually required to report annually all deer harvested by sex and approximate age—the approximate number to be harvested is stipulated by the forest manager for each property. Prohibited activities, such as cutting down trees or allowing use of the leased property by other than named lessees, are identified. Leases may be terminated at any time at the lessor's discretion for any reasons including failure by the lessees to harvest a specified number of deer. Generally, we find that there is high local demand for leases—groups want a controlled hunting environment and the ability to manage and hunt for quality deer.

Leases have size and time constraints. Landscapes under 40 ha are too small to attract lease groups and are not cost effective to administer. Most leases under 40 ha are hunted by family members only. On properties over 400 ha, it is difficult to maintain sufficiently large lease groups to guarantee harvesting enough deer to meet management goals. Such larger leases are divided into smaller ones with separate lessee groups. Lease agreements are generally renewable after 5 years; it takes lessees about 3 years to learn what it takes to be in compliance with lessor requirements. After 5 years, lessors can tell whether the lease group will perform according to the conditions of the lease. Some of our lease groups have been in existence for 20 or more years.

29.4   SETTING UP A LEASE

29.4.1   ASSESSMENT

We establish goals for forest management (e.g., successful seedling regeneration following timber harvest; diversity of overstory, understory trees, and shrub and herbaceous layers) with the primary focus on regeneration of commercial tree species. We assess levels of deer density and impact on these resources to determine whether existing public hunting has resulted in desirable levels of deer impact on properties we manage. If deer density and impact are too high, and the cost of fencing to exclude deer from regeneration sites is too high, we then ask a few questions to determine the feasibility of lease hunting to get us to goal.

- Is the property neither too small nor too large for lease hunting to be effective in reducing deer density and impact?
- Is the interior of the property accessible to hunters via a forest road system that is in good condition?
- Can we recruit local hunters to form lease groups and/or has there been interest by local groups to lease?
- Are there spokespersons within the lease we can communicate with effectively?
- Can we establish effective communications with lessees to use scientific findings in a way they will accept?
- Are lessees educable to the point they will accept lowered deer density to benefit forest and wildlife communities?
- Are hunting regulations such that we can expect that lease members will be able to obtain enough deer permits to achieve needed reductions in deer density?
- Are property boundaries clearly marked, and does the existing gating system ensure access to the property only for timber management activities and by lease members?
- Is our rapport with the state natural resource agency such that it will allow us to use/enhance existing regulations to reduce deer density and impact to target levels?

If we can answer these questions in the affirmative, we proceed through a number of steps to establish a hunting lease designed to reduce deer density and impact. In the process, we address a primary concern—that posted hunting leases might result in reduced hunting pressure, increased deer density, and a furthering rather than a reduction of deer impact.

29.4.2   INITIAL STEPS

We begin by determining whether a given property should serve as a single lease or be divided into smaller ones. Once we have determined the number of leases the property will support, we attempt to match up groups of local hunters with identified leases. We approach groups known to have hunted the property in the past, we evaluate requests from nonlocal groups seeking to lease the land(s), and in rare cases we may advertise locally the availability of the lease(s). We prefer that lessees be local residents because they will use the property year-round and be more protective of it (signing and patrolling boundaries, avoiding damaging access roads, policing the grounds to prevent trash deposition).

Prior to entering into lease agreements with lessees, we hold a meeting with core members of the lease to describe our primary goal for the property (successful advance regeneration of commercial tree species) and the negative impact overabundant deer have on that goal. We explain the lease group's responsibility to control deer density by harvesting deer, especially antlerless deer, and specify that these core members must enlist enough lease members to harvest the required number of deer.

We identify permitted uses (e.g., lessees may use campers on the property) and forbidden practices (e.g., cutting down trees, dumping trash). If spokespersons for the lessees agree to the conditions specified by the lease, we consummate the lease agreement.

We hold an initial educational meeting for all lessees and explain our forest management goals and describe the benefits of managing deer density to habitat and all wildlife species, their requirements to harvest a specified number of deer every year, and other responsibilities (e.g., posting and patrolling property boundaries, and allowed and disallowed uses of the leased property).

We acknowledge the importance to lessees of family hunting traditions and describe how we encourage that by allowing use of campers on leased property for multiple uses of the leased land. We also acknowledge that high deer density is a desired deer management goal for some lessees, but state that we use the science behind the relationship between deer density and negative impact to arrive at desirable deer density for the lease, and that a condition of the lease is that lessees harvest enough deer to get the herd and impact on the understory down to goal conditions.

We learned that deer management information presented by authority figures (e.g., wildlife conservation officers from natural resource agencies, university and agency scientists) was not credible with lessees, but that they would listen to, and seemed accepting of, such information when presented by our field foresters and others such as Quality Deer Management spokespersons.

At the initial meeting with club members, we hand out brochures and pamphlets containing deer management information. We do not allow supplemental winter feeding of deer and explain that if any lease members violate local hunting and fishing regulations, the lease will be terminated.

Within 3 years of the inception of the lease, we conduct a deer pellet group count to establish a quantitative estimate of deer density. We combine this count with observations by our managing forester of deer browsing damage on seedlings to set the required antlerless deer harvest for the lease. We require annual harvest reports from lessees that include the number, sex, and appropriate age of deer harvested from the lease property.

29.5   LEASE OPERATIONS

We conduct informal assessments of deer browsing impact annually to determine whether the lessees are harvesting enough deer to get the lease property to goal conditions for the deer herd and forest regeneration. If they do not harvest enough deer, we put the lessees on notice that if they do not harvest the number of deer specified in the lease agreement by the next deer hunting season, their lease will be terminated. Initially we conducted deer checking stations during hunting seasons on leased property to gather harvest information, but found that hunter distrust of checking station personnel (authority figures) resulted in so few deer being brought to checking stations that the information gathered was inadequate for characterizing harvest or deer health.

In the spring following notice of failure to harvest enough deer, we conduct a deer pellet group count to provide quantitative evidence of whether the lessees have harvested the specified number of deer. If they have not, we terminate the lease and offer it to groups that have expressed an interest (usually there is a waiting list—our leases are highly sought after).

Entrance and access roads to the leased property are maintained as needed for management activities related to timber management—associated costs are absorbed by the owner as part of timber management. Cost to repair damage to the roads caused by lessees is charged to the lease group(s): in our 20+ year experience with leasing, this rarely occurred. We charged an annual fee of $15–$40/ha of leased forestland (2014 data) to cover other costs of leasing.

From time to time, lessees requested permission to develop food plots for deer (about 10% of the leases). These were not discouraged, but lessees were required to clear locations and activities associated with candidate food plots with the managing forester. Lessees were not allowed to place food plots where they would interfere with ongoing or planned timber management activities.

Subsequent to the initial meeting with club officers and members, we communicated with lease members through their leadership core: annual meetings we held were poorly attended by rank-and-file lease members. By and large, our hunting leases resulted in reductions in deer density and impact sufficient to justify their continued use.

EDITOR'S NOTE

I conducted a deer density and impact assessment on a forestland leased by the chapter's authors who suspected that hunting pressure/antlerless harvest by members of the lease group was not reducing deer impact. The pellet group count and impact assessment I conducted confirmed the managers' concerns. They confronted lease representatives and required that more deer be harvested in the impending deer season. When it was apparent the lease club did not increase deer harvest over previous levels, the managers terminated the lease. They subsequently sought and obtained a different group of lease hunters who harvested enough deer to reduce deer density and impact. Informal surveys conducted by the managers revealed reduced impact on oak seedling regeneration and increase in abundance and vigor of previously heavily impacted herbaceous vegetation.

A drawback to leasing to reduce deer density and impact on deer–forest management areas (DFMAs—see Chapter 9) occurs when state natural resource agencies will not offer deer management assistance program (DMAP) permits for reducing deer density on leased forestlands. Managers must then rely upon lease hunters obtaining antlerless permits available within deer administrative units (DAUs – see Chapter 9) which does not guarantee hunters will use the permits on the leased DFMA.

David S. deCalesta, editor

REFERENCE

Jacobsen, M. 2014. *Forest finance 6: Leasing your land for hunting: Income and more.* State College Pennsylvania: College of Agricultural Sciences, Pennsylvania State University. Publication Code UH163.

30 Special Case
Landscapes Closed to Deer Hunting—Forested Public Parks and Residential Developments

David S. deCalesta and Marian Keegan

CONTENTS

> ...traditional hunting approaches may not be practical ... for some landscapes ... alternative methods such as limited, controlled hunts and sharpshooting have been developed.
>
> **Michelle Doerr, J. B. McvAinich, and E. P. Wiggers (2001)**

30.1 INTRODUCTION

People visit and/or live in wooded landscapes for a variety of reasons, including esthetic and lifestyle. Some meet their needs in visits to public forestlands such as parks; others choose to live in forested residential communities. When they are there, they assume that their safety and that of their property are assured. Most visitors/residents are unfamiliar with forest ecology and the composition of forest plant and animal communities, but it is safe to say that most expect—indeed, some look forward to—encountering deer. It is equally safe to say that prior to visiting/moving to the woods, most were unaware of the impact deer can have on vegetation and their health. And, sooner or later, many come into conflict with deer: some experience extensive browsing damage to shrubs and flowers on their residential woodland properties and increased storm water runoff from a lack of adequate understory vegetation, while others note severe impact on understory herbaceous vegetation in parks (Knight et al. 2000); some are involved in deer-vehicle collisions; and some contract deer tick-borne diseases (see Chapter 7) that require medical treatment and have the potential to be life threatening.

Woodland parks and residential communities share common properties that initially do not suit them for effective management of overabundant deer populations: regulations generally prohibit deer hunting within the forestlands. Parks and residential areas can become refugia for deer wherein deer densities and impacts are usually high; they are overseen by governing boards and staffs whose members generally are unfamiliar with managing deer density and impacts; stakeholders/homeowners hold conflicting views on how to manage deer and deer impact, including resistance to deer culling; and fencing to exclude deer may be prohibited by architectural committees and/or rejected as too expensive. Finally, managing deer density and impact under both situations is a long-term annual undertaking that requires awareness and preparedness by all parties to support and maintain.

Developing and delivering effective deer management programs is difficult. Responsible administrators must grapple with obtaining, evaluating, and implementing professional advice; educating themselves and stakeholders; resolving conflicts among stakeholders and outside special interest groups; and developing and delivering comprehensive management programs. Decker et al. (2004) provide an excellent framework for identifying and developing proactive programs to manage impacts of overabundant deer herds within human communities (which should work equally well in developing similar programs for public lands such as parks).

30.1.1 Program Development

30.1.1.1 Assess the Situation

Administrators of parks and residential communities become aware of the "deer problem" when they receive complaints from park users and homeowners. However, as Decker et al. (2004) note: "Agreement about the existence and nature of the deer problem must be sufficient to propel the issue toward resolution. If interest in the problem is not widespread or is held by those with little voice in the community, the issue may dissipate, regardless of whether the actual impacts of the concept are mitigated. Education and informative communication can be critical at this stage to minimize the probability of a rift among stakeholder groups in the community. The value of common community goals—which are essential for guiding discussion, analysis, and decisions—also becomes evident at this early stage."

Only when there is consensus among stakeholders regarding the need for action, and demand for such from the stakeholders, will administrators acknowledge and act upon the scope and nature of the problem(s) and seek solutions.

Identification and characterization of the problem are generally initiated by holding hearings open to the aggrieved stakeholders to define the depth and breadth of deer-related impacts. Administrators attempt to include key/influential stakeholder groups as participants to air their concerns. Because such input is subjective and poorly if at all quantified, administrators usually enlist the services of recognized regional deer–forestry experts for advice. These experts quantify the nature and scope of deer impacts through assessment/monitoring and begin to develop goals and identify requisite management activities, with associated requirements for staffing and operating budgets.

30.1.2 Initial Inventory and Monitoring

Contracted deer–forestry experts develop sampling programs (see Chapter 17, "Monitoring") to identify which resources are impacted by deer, the severity of the impact, distribution and timing of the impact across the park/residential landscape, and, if desired, deer abundance/density. The experts should also evaluate potential for impact originating from deer moving from abutting properties into the park/development landscape and course(s) of action if such movement(s) may occur.

30.1.3 IDENTIFY MANAGEMENT STRATEGY

30.1.3.1 Discard Ineffective Management Actions

With the help of experts, administrators should identify, explain, and discard proposed practices that have proven to be ineffective. The chemosterilant approach, wherein female deer are rendered sterile by injection of anticontraceptive drugs, is only effective within small landscapes totally closed off to immigration by deer from the enclosing landscape. Such closure occurs only on islands isolated from deer or within properties that are fenced to exclude deer. Fencing is successful by itself without the need for the additional expense of capturing and injecting does with a chemosterilant.

Live-trapping and removing deer is expensive, difficult to achieve, and the problem of where to release the deer is a real issue—some die in the process (which negates the argument of anticulling stakeholders who lobby against culling deer); potential recipient managers do not want the deer; and trapping programs must be practiced in perpetuity, as nonresident deer will immigrate from adjacent properties continually over time.

Introducing extirpated predators, such as mountain lions (*Felis concolor*) or wolves (*Canis lupus*), would be rejected by horrified stakeholders concerned about the safety of humans and pets as well as by testimony of experts who would point out that such introduced predators would likely either leave the introduction sites or be killed by motor vehicles. However, use of skilled deer hunters is one of a handful of potential and successful options.

Treating residential vegetation with chemical repellents has limited success, as it must be applied continually, is rendered ineffective when washed off by precipitation, and cannot be used to protect native forest vegetation such as in parks or woodlots in forested residential communities because of the enormous scope and cost of such action.

30.1.3.2 Identify and Promote Realistic Solutions

Reducing deer density and associated impacts on forest resources by public hunting or contracted sharpshooters is a proven effective method for reducing deer damage to vegetation and protection of human health and is discussed below.

Enclosing entire park and residential landscapes with fencing to exclude deer has the potential for success. However, such fencing is a major and costly option: (1) the perimeter of protected areas must be fenced in a way the prevents deer ingress, (2) the fence includes placement of deer-proof gates where breaks in fence lines are required to allow passage of motor vehicles (but not deer), (3) fence lines must be monitored on a regular basis to fix gaps created by trees falling on fences, and (4) it requires an extensive effort to drive deer out of fenced areas prior to complete closure.

Developing, delivering, and maintaining effective deer management programs on parks and residential areas requires contracting with wildlife/forestry experts or retaining them on staff to develop and implement management operations competently and on a recurring yearly basis.

30.1.3.3 Realistic Solution: Reduction in Deer Density

Reducing deer density to levels compatible with diverse animal and plant communities can be accomplished with public hunting and contracted sharpshooters in combination or singly. Public hunting using archery as the removal method is safer than with rifles/shotguns, but archery is less effective. When sharpshooters are contracted, the deer are removed by shooting at night with noise-suppressed rifles. Three case histories provide examples of successful application of hunting: two (Chapters 34 and 35) are from public lands and one (Chapter 36) is from a forested residential community. Other parks/residential communities successfully reducing deer density and impact by shooting deer include The Wickahisson and Pennypack Parks Forest in Chestnut Hill, Pennsylvania (http://www.chestnuthilllocal.com/2011/03/16/wissahickon-deer-kill-continues-as-opposition-wanes/); the Fox Chapel residential community near Pittsburgh, Pennsylvania (which, in addition to culling deer in the residential community, protected its famous Trillium Trail with a 2.5-m-tall deer fence; http://www.post-gazette.com/sports/outdoors/2015/03/15/

Finding-a-palatable-answer-to-deer-Mt-Lebanon-overpopulation/stories/201503150112); and Gettysburg National Park, where the National Park Service conducts annual (as needed) deer culling by qualified federal employees—public hunting is not allowed (https://www.nps.gov/gett/learn/news/deer-management.htm). Meat from culled deer is donated to Gettysburg local food banks.

30.1.3.4 Realistic Solution: Protection of Vegetation/Habitat with Fencing

Before commercial forestry operations were able to use expanded state-regulated hunting seasons to reduce deer density and impacts, managers relied on woven wire or electric wire deer fences to exclude deer from regeneration sites. On these sites, where overstory trees were removed or thinned to allow development of advance regeneration—seedlings of desirable tree species—deer exclusion fences enclosed regeneration sites to protect tree seedlings from browsing by overabundant deer herds. Forest vegetation and seedlings outside fenced areas were subjected to increased deer browsing pressure. In effect, fenced areas provided islands of overstory and understory plant species and diversity found nowhere else. Owners of one of the small woodlots in the case histories (Chapter 38) resorted to protecting regeneration sites with deer exclusion fencing because they were uncomfortable allowing deer hunting on their property by persons they did not know. To date, there are few known parks or forested residential areas that are protected by perimeter deer fencing—one such example is the Flat Rock Brook Nature Center in Englewood, New Jersey. Park officials discarded trapping and relocating deer, chemosterilization of does, and culling (park too close to homes in residential Englewood) and settled on fencing approximately 40 hectares of the 60 hectares that make up the park with a 2.5-m-high deer fence made of woven wire. Parks and large forested residential communities have yet to use deer-proof perimeter fencing to prevent deer impact. Building and maintaining such fences is described in Chapter 20.

30.1.4 Developing, Selling, and Funding the Deer Management Program

Assessing the severity of deer problems and settling on a management plan are the easy steps in deer management. Selling the program to a divergent group of stakeholders, often at cross-purposes, through an effective communications program and assembling the economic and human resources to deliver and maintain the program are far more difficult.

30.1.4.1 Identifying Common Ground

As discussed in Chapter 22, because of different values, cultures, and beliefs, information regarding deer management presented as factual to one stakeholder group may be disputed, along with its related conclusions, by another group. Arriving at consensus regarding existence and extent of deer impact, severity of such impacts, and potential solutions requires acknowledging differences among stakeholder groups and defining the issues of deer overabundance and impacts in terms meaningful to each group. The key is to acknowledge differences in beliefs held by stakeholder groups, find a common problem all can agree on and define, and develop a deer management program that addresses the common problem. For example, environmentally concerned stakeholders may contend that the issue of overabundant deer is reduction of diversity of plant and animal communities and opt for culling deer. Animal rights groups may say that they don't care about diversity and that the real issue is their objection to killing deer. A third group may claim that culling deer is too dangerous with residences so close to culling areas. A solution that resolves the concerns of all groups, while avoiding value-based issues, is to promote a deer exclusion fence that protects diversity, avoids killing deer, and poses no safety risk to humans living in the area.

30.1.4.2 Communicating with and Educating Stakeholders

Seeking input from stakeholder groups is indispensable for identifying deer impact problems and other deer-related concerns and is a first step. Once a deer management program is settled on, communicating the other way, from program administrators to stakeholders, as well as two-way communications between program directors and stakeholders, must be addressed with an effective, timely apparatus. Newsletters and other digital mailings posted at regular intervals, including updates/progress attached to e-mails sent to stakeholders, can be effective, especially if a means for obtaining input from stakeholders is included. Show-and-tell and dog-and-pony field trips, wherein different kinds of negative deer impacts and examples of how data are collected are on display, help bridge different perspectives among stakeholder groups. Including members from stakeholder groups on steering/organizing committees, wherein stakeholder concerns are represented, avoids misunderstandings and builds trust and confidence.

30.1.4.3 Funding the Program

Program administrators and stakeholders must be prepared for a long-term program—basically as long as they are responsible for managing natural resources on wooded properties. This includes long-term, recurring budgets for education and stakeholder engagement, program maintenance, supplies and materials, and staffing-up/retaining employees/consultants to run the program.

To determine progress toward goals, monitoring must continue so that any changes in the program needed to address unforeseen issues, or respond to lack of progress toward goals, can be anticipated and acted upon. Such monitoring may include deer density, deer impact, and stakeholder satisfaction.

REFERENCES

Decker, D. J., D. A. Raik, and W. F. Siemer. 2004. *Community-based deer management.* Ithaca NY: Northeast Wildlife Damage Management Research and Outreach Cooperative.

Doerr, M. L., J. B. McAninch, and E. P. Wiggers. 2001. Comparison of 4 methods to reduce white-tailed deer abundance in an urban community. *Wildlife Society Bulletin* 29:1105–1113.

Knight, T., S. Kalise, L. Smith et al. 2000. Effects of intense deer herbivory on the herbaceous understory at Trillium Trail. *Wildlife Damage Management Conference—Proceedings.* S2. http://digitalcommons.unl.edu/lcwdm_wdmconfproc/S2

31 Special Case
Deer Management Cooperatives

David S. deCalesta

CONTENTS

Co-ops are the future.

Ted Hansen, Mark Kenyon, Alex Robinson (2015)

When individuals join in a cooperative venture, the power generated far exceeds what they could have accomplished acting individually.

R. Buckminster Fuller

31.1 INTRODUCTION

Managers of deer–forest management areas (DFMAs—see Chapter 9) face a number of obstacles to managing deer density and impact to facilitate sustainable management of all forest resources. They may be unable to affect deer densities on abutting forestlands not managed to control deer density and impact. Their forestlands may be too small to allow them to use recommended management steps, including meaningful monitoring, or to attract enough hunters to reduce deer density and impact to desired levels. They may be limited in what they can provide by way of appropriately trained personnel and financial resources required to conduct deer management practices. And, they may lack the counsel of professional foresters or wildlife biologists required to develop and guide management. A solution to these deficiencies could be the formation of deer–forest management cooperatives among abutting DFMAs. The original impetus behind the formation of deer cooperatives was the desire of landowners who were deer hunters to produce quality deer for hunting.

In her study of deer cooperatives in Michigan, Mitterling (2013) found that hunter satisfaction was higher among hunters hunting cooperatives, and that in areas of high deer densities, cooperatives included a greater proportion of does within their deer harvest than elsewhere in Michigan. She speculated that cooperatives could be an effective way to contribute to deer population reduction and that hunter recruitment and retention could be increased by the formation of deer cooperatives. She also stated that cooperative leaders are relied upon by multiple members as resources for insight on deer management and that managers of properties within cooperatives can connect members with each other as they facilitate and organize distribution of information within cooperatives.

Cooperatives can erase deficiencies of individual DFMAs as well as providing additional benefits accruing from aggregating forestlands and pooling of financial and personnel resources. Sharing expenses and expertise among cooperators renders deer management more affordable on a per unit basis to monitor deer density and impact; to hire/share professional foresters/wildlife consultants/ staff; to create landscapes providing deer forage and hiding cover, to facilitate operation of deer checking stations for feedback on deer health; and to establish, sign, and monitor enclosing boundary lines. DFMAs within cooperatives no longer compete for attracting hunters; can attract a larger pool of hunters; and may be able to offer incentives, such as use of existing outbuildings for participating hunters and their family members, opportunities for other recreational uses of expanded DFMAs, such as hunting other game species, fishing, and hiking/picnicking, and a system of well-maintained and interconnecting access roads with multiple entry points. It is easier for cooperatives to obtain permits for harvesting antlerless deer from state natural resource agencies, and they may be better able to facilitate, and share expenses of, management of forest vegetation, including timber harvest and use of practices such as herbicides used to improve species richness and abundance of understory vegetation (thereby improving forage and cover characteristics for deer and other wildlife).

On the con side, joining cooperatives may reduce DFMA manager flexibility and/or increase financial and human resource outlays. Managers of DFMAs may be required to: (1) contribute (on a proportional basis) to financial and human resource costs borne by the cooperative, including road maintenance, monitoring of deer density and impact, posting of cooperative boundary lines if their property forms part of the cooperative boundary, and application of silvicultural treatments to improve habitat; (2) allow hunters they do not know to have access to their forestlands for hunting (unlike on small woodlots where hunters generally are family or friends of managers); (3) accept public hunting as the method for control of deer impact and remove fencing they may have erected to protect forest vegetation within their DFMA; and (4) serve on cooperative leadership committees and participate in developing and delivering communications with stakeholders, including deer hunters.

31.2 COOPERATIVES DEFINED/DESIGNED BY QUALITY DEER MANAGEMENT

The Quality Deer Management Association (QDMA) provided initial justifications/advantages and guidelines for forming cooperatives, most of which are described above, which include: opportunity for hunters with small forestlands to participate in Quality Deer Management (QDM). Cooperatives also enhanced: (1) ability to manage deer herd age and sex ratio; (2) distribution of local deer herds; (3) management on small incorporated individual properties where deer may spend only part of the time; (4) pooling of harvest and observational (trail cam and others) data; (5) relevance of harvest data with larger samples from larger areas; (6) ability to make more precise deer management recommendations based on larger area, larger herd, and better data; (7) opportunity to harvest quality deer with more mature bucks; (8) relationships among neighbors, including establishing trust and honesty among neighboring property owners involved in cooperatives; (9) costs of deer management by sharing equipment and supplies for food plots and facilities such as meeting places to discuss deer management; and (10) control of trespass and poaching.

31.2.1 Quality Deer Management Association Steps for Establishing Cooperatives

Universal steps recommended by the QDMA for establish cooperatives center on: (1) identifying potential properties for inclusion in a cooperative; (2) establishing a minimum size, generally identified as about 400 ha; (3) identifying potential participants among adjoining landowners; (4) holding meetings to describe cooperatives and solicit participation from adjoining landowners; and (5) holding subsequent meetings of interested participants to establish goals and begin to build the cooperative.

Recommended actions include: (1) obtaining services of qualified wildlife biologist to advise and evaluate the potential for success; (2) establishing and maintaining effective communications among participating landowners; (3) promoting restrictions on harvest to exclude (protect) young antlered bucks so they can survive to later seasons when their antlers will be more impressive; (4) promoting harvesting antlerless deer to maintain proper herd size; (5) describing the time frame for successful establishment of the cooperative as 2–5 years and expecting gradual rather than rapid improvement in deer quality; (6) ensuring goals and objectives are reasonable and based on habitat and management options; (7) dealing with nonsupportive adjacent landowners and associated problems; (8) compromising among cooperative participants on goals, objectives, and management operations; (9) collecting and maintaining records (data) on herd numbers/condition through use of trail cameras, deer check stations, and other means, to establish baseline conditions and progress toward goals; and (10) obtaining and using forest management practices from professional foresters to manage forest vegetation for production/promotion of deer forages and cover.

31.2.2 DIFFERENCES BETWEEN QUALITY DEER MANAGEMENT ASSOCIATION COOPERATIVES AND DEER–FOREST MANAGEMENT AREA COOPERATIVES

There are a number of differences of philosophy, goals, and management operations between QDM cooperatives and DFMA cooperatives: (1) QDM cooperatives were developed for the benefit of small forestland owners who also were deer hunters and had as a primary objective creation and maintenance of quality deer for quality hunting; (2) DFMA cooperatives are designed to promote sustainability of all forest resources by focusing on deer and vegetation management, not just quality of deer and deer hunting, although these latter two goals are integral to DFMA cooperatives; (3) DFMA forestlands exhibit a larger size range as opposed to QDM forestlands of smaller size (generally less than 50 ha)— DFMA landowners, in addition to small woodlot owners, include industrial timber companies with large forestland holdings, state and national public land administrators (e.g., state and national forest/ park lands, including national forests, national monuments, national parks); and (4) management practices for QDM cooperatives include enhancing deer forage with food plots, whereas DFMA cooperatives rely upon established forest management practices to create deer forage and cover as benefits of timber management practices designed primarily for sustainable production of timber.

31.3 PREREQUISITES FOR ESTABLISHING DEER–FOREST MANAGEMENT AREA COOPERATIVES

QDM suggestions for establishing cooperatives are a good starting point for establishing DFMA cooperatives and include: (1) identify potential properties, with a key ingredient being that cooperative landscapes should be integral and whole—if participating landscapes are fragmented by nonparticipating forestlands, everything, including monitoring, control of hunting/trespass, and poaching, will be more difficult; (2) identify and use managers of participating forestlands as key contacts among and within participating forestlands; (3) conduct an initial meeting to describe how cooperatives work with potential managers and describe typical goals and management practices of cooperatives; and (4) hold subsequent meetings to solidify participating forestlands and managers and develop goals and management strategies, including description of establishment and maintenance of cooperative boundaries. Persons attempting to form cooperatives should be prepared to address resistance to coordinated, cooperative deer management within the landscape of surrounding property owners, some of whom may have differing goals (including no hunting or no hunting of antlerless deer) or no goals for deer management.

Additional considerations for initiating DFMA cooperatives include: (1) identification of key stakeholder groups that will impact deer and forest management of the cooperative, such as adjacent, noncooperating landowners, local and state wildlife/forestry enforcement and management personnel, and representatives of local businesses affected by deer hunting; (2) development of a leadership team composed of representatives from each participating property and stakeholders from influential groups, such as state and local natural resource agencies, hunters, and local businesses affected by hunters and hunting—all designed to provide a stable administering group that identifies and sets goals, maintains monitoring, and develops adaptive management steps to make adjustments as needed to reach goal(s); (3) identification of a coordinator for the cooperative, and determination how that individual will be reimbursed, if needed; (4) establish how annual and recurring costs of operating cooperatives will be funded and shared among cooperators and how funds will be collected and disbursed; (5) establishing what will be monitored, who will collect and analyze data, and who will pay for monitoring and analysis, including deer density, deer impact on identified forest resources, and condition of ground, shrub, midstory, and overstory vegetation; (6) developing a process for obtaining feedback from hunters and other stakeholders regarding their concerns about management on the cooperative; (7) obtaining services of, and potential reimbursement for, experienced deer and forestry consultants who can provide dependable recommendations for management of deer, deer impact, and forest vegetation throughout the cooperative lifespan. Such professionals likely would be best equipped to develop and supervise collection of monitoring data, analyze the data, and make informed and professional management recommendations regarding management of deer, forest vegetation, and deer hunters; (8) developing a communication system for information exchanges among leadership team members and for communicating with/informing stakeholders of health and status of deer herd and habitat and key management information, including season and bag limits, access points, and road system(s) of the cooperative for hunters. Such communications systems should include plans for soliciting and informing hunters for the cooperative and should include building a communications network (e.g., telephone numbers, e-mail addresses) to provide effective and timely communications between stakeholders and the leadership team; (9) developing, implementing, and conducting coordinated forest vegetation management activities, including harvesting timber and application of chemicals (lime, fertilizer, herbicides) to enhance vegetation management; and (10) developing a plan for dealing with potential for chronic wasting disease (CWD) if it is detected among deer harvested from the cooperative as related to management of the deer herd and hunters.

Details describing how the above-itemized recommendations for developing and maintaining deer–forest cooperatives may be developed and conducted are found in Chapters 12 through 26.

Development and implementation of the above steps will vary among cooperatives, as no two are alike. The Kinzua Quality Deer Cooperative (Chapter 34) case history employed the above-suggested steps for organizing and running a deer–forest cooperative, and its continued success in managing deer density and impact for over a decade is a testimony to the effectiveness of such cooperatives to aggregate multiple properties into functioning entities.

31.4   PROBLEMS FACING COOPERATIVES

31.4.1   Buy-in of Managers of Properties within Cooperative Boundary

All property owners/stewards within proposed cooperative boundaries must agree to participate at all levels of deer management in the cooperative. Agreed-upon components include allowing hunters to hunt individual forestlands; allowing hunters to use all roads within individual forestlands; agreeing to the need for, and granting permission for, harvesting antlerless deer; allowing monitoring activities (chiefly deer density and impact on forest resources) to take place on their properties; and agreeing to contribute annually to funding on a prorated basis (based on area of individual forestland) for operational and maintenance components of the cooperative.

31.4.2 ROAD ACCESS

Road access for hunting among cooperating forestlands may be disparate and in need of development or enhancement. Difficulty in developing an integrated and interconnected system of access roads increases with the number and size of individual forestlands within the cooperative; some properties may not even have access roads. The costs of developing and maintaining access roads can be prohibitive unless cooperating managers can develop a system for sharing the costs associated with access among landowners/stewards.

31.4.3 HUMAN SAFETY FACTORS

Owners of forestlands within cooperatives may reside on their properties, making provision for absolute safety of them and their properties a key issue. Safety zones surrounding these areas must be established, of a size that absolutely guarantees human and property safety, and identified in a manner that prevents hunters from endangering humans and personal property. Hunters hunting within cooperatives including resident landowners must be educated sufficiently to guarantee they will not endanger human safety and private property via safety zone policing.

The number and distribution of hunters hunting on cooperatives may need to be restricted/controlled to prevent hunter density that negatively affects quality of hunting and/or increases danger to resident landowners. Two case histories (Chapters 35 and 36) describe systems for screening participating hunters and controlling their access and distribution within established boundaries.

31.4.4 COORDINATION OF MONITORING

An advantage of cooperatives is that respective managers can share in costs of monitoring and results will be more meaningful when collected over the larger landscapes included in cooperatives. A problem is in coordinating exactly where, how, and by whom monitoring will be conducted. Transects for monitoring should pass through every cooperating DFMA to provide representative coverage of the cooperative as well as for data specific to DFMAs. And, as much as possible, cooperating DFMAs should provide personnel for collecting monitoring data (data collection is not complicated and requires minimal training). Analysis and interpretation of monitoring data can be done by a single individual contracted to perform the service, or by appropriate personnel within the cooperative.

31.4.5 MAINTAINING COMMITMENT OF COOPERATING MANAGERS OVER TIME

Deer management is a long pull, and requires recurring actions and associated expenses, such as collection and analysis of data, maintenance of road access and other physical requirements, consistency of goal(s), communication among cooperative managers and stakeholders including hunters and local natural resource agency personnel, and commitment to adaptive management when and where needed. Cooperating managers need to know that results do not occur overnight but rather are incremental over 3–5 years. It took 3 years for the cooperative described in Chapter 34 to achieve target deer density, which was coincidental with a large number of antlerless permits issued every year under a deer management assistance program (DMAP—see Chapter 24) program.

31.4.6 RESISTANCE TO COOPERATIVES

Owners/managers of forestlands within the boundary of a proposed deer cooperative may disagree with goals, including deer density, proposed by other managers within the cooperative. Some who may reside on their forestland may be opposed to any form of deer hunting because of personal philosophy against hunting or may be concerned over hazards to their health and property. Placing such properties within large and well-marked safety zones may be enough to mollify recalcitrant landowners, but they may still be opposed to hunters hunting on their lands. In a somewhat similar situation, landowners bordering the boundaries of a county park in the Chapter 36 case history agreed to deer population reduction when conducted by bow hunting within the park: many allowed hunting on their property. The process of making landowners aware of the program, soliciting their input, and developing and delivering a program they supported took 7 years. Much time, planning, and involvement of stakeholders are required to launch deer cooperatives.

Resistance to coordinated, cooperative deer management within the landscape of surrounding/ included property owners should be anticipated and addressed early on in the development of deer cooperatives.

31.4.7 OBTAINING RECOGNITION FROM STATE NATURAL RESOURCE AGENCIES

If cooperatives can be recognized as DFMAs by state natural resource agencies, they should qualify for and request antlerless permits (if available) to aid in reducing deer density and impact. Such recognized DFMAs can then be identified on state natural resource websites and other forms of mass media as sites for which hunters may obtain permits for harvesting antlerless deer. Cooperative managers should develop long-term relationships with natural resource agencies and their local wildlife biologists, as these organizations and individuals can provide assistance in the form of equipment and volunteer personnel. Partnering with existing Quality Deer Management cooperatives, which are recognized by state natural resource agencies, should be pursued, if available, to increase the size of cooperatives and increase ability to share funding and other resources such as personnel and equipment.

REFERENCES

Hansen, T., M. Kenyon, and A. Robinson. 2015. The 6 new rules of whitetail deer management. *In Outdoor Life* March 2015 issue.
Mitterling, A. M. 2013. Understanding doe harvest behaviors in private deer cooperatives using social network analysis. Master's Thesis, Michigan State University, East Lansing MI.

Section V

Case Histories

In the field of game, however, it seems doubtful whether theories and plans alone, no matter how well supported by evidence, are nearly so useful as ...demonstrations of how those theories and plans work out in practice.

Aldo Leopold (1933)

The proof, it is said, lies in the pudding. This collection of nine case histories, from large and small landscapes, spanning public and private ownerships, with diverse goals for forest and deer management, provides examples of how applying the principles of deer–forest management, including adaptive management, can result in success. One case history (Chapter 33) is an example of typical DFMA management scenarios. The rest represent examples of special cases: one (Chapter 34) is of a cooperative, three (Chapters 34 through 36) represent residential/public forestlands without previous deer hunting, two (Chapters 37 and 38) are of small woodlot forestlands, and one (Chapter 40) is an example of a (failed) attempt at Quality Deer Management. This lone unsuccessful case history, wherein management ignored the advice of professional wildlife and forestry consultants, relying instead on culture and traditions of past deer management, serves as an example of how going it alone and ignoring lessons from the past may result in failure of management of deer and forest resources.

There are no Manager's Summaries provided for the Section V case histories. The benefits from the many insightful and varied approaches to deer–forest management employed far outweigh the savings in time accruing from merely reading a summary of case history stories. Additionally, it is good for managers to see which approaches failed to accomplish results, and, most importantly, to see how managers applied adaptive management to make deer management successful. Each case history concludes with an Editor's Comments section that notes innovative approaches of each case history contributing to its success/failure and/or makes observations of additional actions that could be taken to enhance management under similar management conditions.

REFERENCE

Leopold, A. 1933. Game Management. New York. C. Scribner's Sons,

32 Timberline Farms/Hyma Devore Lumber
The Power of Education

Jim Chapman

CONTENTS

32.1 INTRODUCTION

The private land holdings of Timberline Farms and Hyma Devore Lumber Mill (TFHD) have been managed for decades for forest products and agricultural crops. The property is approximately 1000 ha of farm and forestland in northwest Pennsylvania, Warren County, with most lying in Freehold Township. Forest management has been conducted by the forestry staff of Hyma Devore Lumber Mill, Inc., of Youngsville, Pennsylvania (Figure 32.1).

TFHD is primarily forestland with an inclusion of 97 ha of farmland in corn and soybean production with minor components of hay/native grass and wheat. The main forest management goal has been to establish and maintain a forest capable of sustainable long-term harvesting and economic future. Secondary goals were for enhancement of biodiversity and improvement in forest/ habitat health.

This mixture of forest and agricultural habitats on TFHD (hill, forest, farm fields and meadows, and swamp) has been difficult to manage in respect to deer. Deer carrying capacity in this mix of forest/agriculture habitat is extremely high. There has been ample food for deer throughout the year. Spring offers native grass and alfalfa sprouts in the agricultural fields and emerging tree shoots and germinating seedlings. Summer provides soybeans, alfalfa leaves, and developing seedling regeneration. Autumn provides apples, corn, soybeans, and alfalfa in addition to existing tree seedlings. For winter, deer have what remains of seedling regeneration until covered by snowfall. Thus, while deer browse on a variety of farm and forest vegetation, forest browse remains a year-round staple.

TFHD foresters had been unaware of the impact deer were having on forest regeneration for many years, assuming that the plethora of farm and forest forage available precluded unsustainable deer impact on forest resources. It took an assessment of forest regeneration following a shelter-wood seed cut in a previously thinned northern hardwood stand to make us aware of the deer problem. The shelterwood cut was textbook: advance regeneration (primarily sugar maple) was established prior

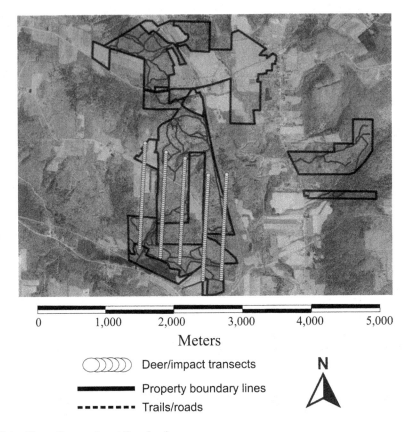

0   1,000   2,000   3,000   4,000   5,000

Meters

⬭⬭⬭⬭⬭ Deer/impact transects

▬▬▬▬▬ Property boundary lines

▬ ▬ ▬ ▬ Trails/roads

N

FIGURE 32.1   Hyma Devore forest/farmland.

to the cut sufficient to restock the stand. However, when we visited the stand the following spring after a relatively open winter with little snow, what we found was not a newly released future forest but a newly decimated regeneration site where almost all available advance regeneration had been severely browsed by deer. That was our wakeup call on deer impact.

32.2   PROGRAM DEVELOPMENT/MONITORING

Shortly after we realized the need to address deer overpopulation and impact, our forestry staff attended a deer density and carrying capacity workshop led by Dr. Tim Pierson, Penn State Extension Forester. The workshop focused on assessing and addressing deer density and impact on forest regeneration/wildlife habitat. At the same time, we began organizing a hunter base for reducing deer density and impact on TFHD by providing permission cards for invited hunters to harvest antlerless deer. An important part of this approach was conducting workshops/field exercises with the hunters to educate them regarding deer density and impact on forest regeneration and forest vegetation. Our goal was to gain support of the hunters in our management program to balance the deer herd with available habitat.

As part of our hunter education effort, we had Dr. Pierson conduct a deer density/impact workshop on THFD in spring 2001. Not only was it well attended, but it was a hit: most hunters enjoyed learning more about deer habitat, deer density, and deer impact on regeneration. This education of our key stakeholder group was the essential ingredient in our program's success. Most of the hunters took ownership of the program and were pleased to provide input into management planning.

We surveyed the main ownership in 2001 with two deer pellet counts collected from two "bow-tie" configurations of transect lines (each encompassing an area 2.6 km^2) and established that our overwintering density was 20 and 22 deer/km^2 respectively—a major problem! At this point we began to develop a written plan for deer management. We began sampling deer density and impact annually beginning in spring 2003 using a modification of the pellet group technique featuring five transect lines spaced 300 m apart in grids encompassing approximately 2.6 km^2.

Concurrently, we began collecting data and analyzing deer impact on forest regeneration using methodology developed by Tim Pierson for estimating deer impact on regeneration (Pierson and deCalesta 2015). After collecting and analyzing the data, we realized that we were losing ground by conventional hunting and needed to harvest more deer to reduce density and impact—deer density had actually increased from an average of 21 deer/km^2 overwinter in 2001 to 25 deer/km^2 in spring of 2003.

We now had the baseline data to use in crafting our management plan. We decided that our main indicator in achieving our goals was not deer density, but presence and health of forest regeneration/habitat—a directly measurable indicator of forest health and biodiversity. However, along with emphasizing forest health, we set a secondary target deer density of 10 deer/km^2—believing this would give us the forest vegetation/forest health goal we desired. At the time, that was considered carrying capacity deer density for forest/agricultural habitat. We had a long way to go.

32.3 PROGRAM FOR INCREASING DEER HARVEST

The Pennsylvania Game Commission (PGC) initiated a Deer Management Assistance Program (DMAP) in 2003 designed to allow landowners better control over deer density. The DMAP provided managers with unit-specific antlerless deer permits to increase deer harvest. The PGC required managers to submit a deer management plan prior to issuance of a specified number of DMAP permits if the owners requested more than one permit/20 ha. We requested 30 permits at the rate of 1 permit/7 ha for the 200 ha included in a proposed DFMA unit. We attempted to maximize our existing hunter base at 200 hunters focusing on a core target area of 200 ha and surrounding lands. We required hunters to report the results of their hunting efforts.

In the first year of DMAP, we removed a hunter-reported 96 deer from the 2 km^2 area. However, we discovered the following spring that we only reduced overwintering density to 17 deer/km^2, 7 more than the desired 10 deer/km^2. Accordingly, we increased number of requested DMAP permits to 110 and added four DMAP units for the 2004–2005 licensing year. Since 2007, we have maintained an overwinter average of 8 deer/km^2 over the combined five DMAP units with 55 DMAP permits annually. Hunter harvest of antlerless deer with DMAP tags, in addition to harvest of antlered and antlerless deer under regular hunting season regulations, has kept deer density (and associated levels of deer impact) where we want it.

32.4 PROGRAM COSTS

Program costs have been relatively low, as data collection and analysis have been incorporated into the workload of existing THFD foresters and use of unpaid volunteers. It takes about 18 hours to sample transects and tabulate data, and another 3 hours to update management plans and prepare DMAP applications. Based on prevailing technician salary rates, annual cost for the program would approximate $500.00 if we assigned salaried personnel to collect and use the data.

32.5 RESULTS

32.5.1 REDUCTION IN DEER DENSITY AND IMPACT

We achieved the goals of reducing deer impact to acceptable levels and reducing deer density to 10 or fewer deer/km^2 (Figure 32.2) by 2007 and have been able to maintain those levels. Hunter success rate with DMAP permits has been an impressive 45%–60%.

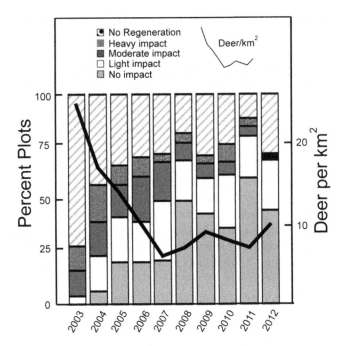

FIGURE 32.2 Reduction in deer density and change in impact levels on regeneration. [1]No impact: % plots with no browsing on seedlings; light impact = % plots with <50% twigs browsed; moderate impact = % plots with >50% stems browsed but seedlings not hedged; heavy impact = >50% seedlings browsed and plants hedged; no regeneration = % plots with no regeneration of any woody vegetation.

In addition to reduction of deer impact on regeneration generally, lower deer density has been associated with a shift in species composition of regenerating seedlings to favor acceptable and desirable species (Table 32.1). At the same time, however, prevalence of plots with regeneration of undesirable species, such as American beech and striped maple, has increased—undesirable as well as desirable seedlings benefitted from reduction in deer density. Nevertheless, stocking rates of acceptable and desirable tree seedlings fell within acceptable levels irrespective of increased proportion of undesired species, so the reduction in deer density has produced desirable results

TABLE 32.1
Deer Density and Impact on THFD, 2003–2012

Year	2003	2004	2005	2006	2007	2008	2009	2010	2011	2012
Deer/km²	25	17	14	10	6	7	9	8	7	10
Regeneration status					**Percent Plots**					
Undesirable[a]	12	26	25	22	26	40	34	32	35	24
Acceptable[b]	8	14	32	35	37	38	29	39	50	40
Preferred[c]	8	18	10	13	9	4	8	5	4	8
Acceptable/preferred[d]	16	32	42	48	46	42	37	44	54	48

[a] Undesirable regeneration = non–commercially valuable species such as striped maple.
[b] Acceptable regeneration = low commercial value species (e.g., beech).
[c] Preferred regeneration = high commercial value species (e.g., oaks, black cherry).
[d] Acceptable/preferred = percent plots with either acceptable or preferred species.

regarding tree regeneration. We have come to realize that the desired overwintering deer density for sustaining forest regeneration is approximately 6–7 deer/km^2.

32.5.2 ADDITIONAL BENEFITS

There have been additional benefits of the program. The greatest has been higher diversity and quality of all forest regeneration, seedlings and herbaceous vegetation alike. At the beginning of our program, it was almost impossible to achieve natural regeneration after timber harvest, most notably in stands dominated by oak and hickory. As changes progressed, areas that once were regenerated only by beech and other less palatable (to deer) species were displaying establishment of oak seedlings, providing the additional (future) benefit of mast production for deer and other wildlife species.

A second additional benefit has been the diversification of species within the management area. By effectively managing the deer population, we have seen an expansion of habitat diversity and wildlife diversity and abundance. In retrospect, the vertical structure of the forest has improved, creating habitat for a greater variety of songbirds as well as game birds. It has been easier to establish the quality early successional areas that are the building blocks for habitat. With this we have observed an increase in grouse on our lands, as well as the first woodcock we have seen in over 25 years.

A third benefit has been economical. Much of the regeneration sites would have to have been fenced to protect seedlings from deer over-browsing. Due to the steep and undulating topography, much of this area could not have been fenced without high cost. However, because of reduced deer density, fencing was not needed. Additionally, regeneration sites no longer are predominated by interfering woody seedlings or ferns and do not require expensive herbicide treatment normally required under high deer density.

A fourth benefit has been the change in attitudes of hunters. Involving and informing hunters in development of management plans and activities have been associated with a high level of support from these hunters. Initially, some hunters didn't like it when deer density was reduced precipitously. However, as time progressed, many skeptical hunters changed their minds as they saw improvement in quality of deer and habitat. Removing large numbers of antlerless deer created pockets of low deer density. These temporary voids were filled by an influx of juvenile bucks which, coupled with resulting higher buck:doe ratios, has enhanced reproduction and recruitment. These improvements in recruitment and proportion of antlered deer in the herd made believers out of most skeptics who enjoyed seeing more antlered deer even though they were seeing fewer overall deer. Hunters are now looking at the forest vegetation along with the animals they are hunting and valuing management and improvement of both.

Another positive outcome of our program has been incorporation of neighboring lands with those of TFHD for inclusion in deer–forest management. As neighbors saw improvements in regeneration, habitat, and deer quality resulting from proactive deer management, they included their lands in our program.

32.6 FINAL THOUGHTS

It takes time and patience to develop, implement, and adapt a dynamic deer management plan. We expanded our program by adding to the initial management area and going to five DMAP units with 110 permits requested in 2004. Over time, we combined these units into one DMAP unit and, because of reduced deer density, have reduced the need to 55 DMAP permits annually. It takes many years to work on a large landscape and improve it. Many factors must be reckoned with, including hunters, weather, wildlife food sources, predators, and the deer themselves—and all seem to be in a constant state of flux. We urge those embarking on a deer management plan to be patient and adapt to the changes that will be required to achieve their deer and forest management goals.

We learned that we had to reconcile two carrying capacities of deer. The first, associated with sustainable forest regeneration, is referred to as regeneration carrying capacity (deCalesta and Stout 1997) and requires 10 or fewer deer/km^2 on our landscape. The second relates to the requirements by hunters for some minimal number of deer regardless of sex or age that they encounter during hunting season, and that some minimal proportion of these deer are antlered with desirable antler characteristics (multiple antler points, width of spread of the antlers, and large diameter of antler bases). This carrying capacity is often referred to as the minimum sustained yield (MSY) of a deer herd and represents the maximum deer density the habitat can produce over time. We have communicated to our hunters that we cannot produce sustainable forest regeneration and remain in the timber business if we manage to produce MSY of deer. We have succeeded in changing our hunters' cultural carrying capacity to one that emphasizes healthy habitat and deer, which is in line with regeneration carrying capacity. Our educational efforts and transparency with our hunter base have allowed us to work together for what is now a shared goal for deer density and deer and habitat health.

Last, it is important to stick with the program. We have observed that heavily browsed regeneration usually does not recover without additional efforts at establishment. Further, establishment requires enough time for the seedlings to grow out of the reach of deer browsing, which takes time—a few years are not enough: it took us 6 years. The problem didn't occur overnight, and the correction will take longer than that. The key to success is recognizing that the upward pressure of deer reproduction requires that the herd be monitored and reduced as required on an annual basis, basically as long as the forest is being managed. Don't be afraid to alter your management plans—allow for changes to your deer management plan. Time and trial will make your plan balanced and effective, if you maintain the patience to see it through.

MANAGER SUMMARY

The case study for Timerline Farms/Hyma Devore Lumber is one of creating a deer management program that utilized and educated the stakeholders (hunters) of these lands. The hunters involved were given opportunity to: be educated in deer management, assist forest managers' efforts, offer input, and take some ownership on the deer management program. Though they had no definitive say in the mechanics of the program, they had ownership in the process. This is what generated our success.

EDITOR COMMENTS

With a minimal communication program, managers were able to attract and maintain a core of local alpha and beta hunters who, equipped with antlerless deer permits, were able to keep deer density and impact at desired levels. As with other successful case histories, monitoring was a central management practice. Over time, managers convinced abutting landowners to join in a loose, successful cooperative that was associated with being able to adjust the number of antlerless permits requested annually to maintain desired levels of deer density and impact. Even with a plethora of forage (farm and forest) available, the managers still could not increase carrying capacity for seedling regeneration or diversity above 6 deer/km^2.

David S. deCalesta, editor

REFERENCES

deCalesta, D. S. and S. L. Stout. 1997. Relative deer density and sustainability: a conceptual framework for integrating deer management with ecosystem management. *Wildlife Society Bulletin* 25:252–258.

Pierson, T. G. and D. S. deCalesta. 2015. Methodology for estimating deer impact on forest resources. *Human-Wildlife Interactions* 9:67–77.

33 The West Branch Forest Preserve

Whittling Away the Smorgasbord

Michael C. Eckley

CONTENTS

33.1   INTRODUCTION

Incorporated as a nonprofit organization in 1951, the Nature Conservancy (TNC; hereafter referred to as the Conservancy) has a mission to "conserve critical lands and waters on which all of life depends." The Conservancy has protected over 48 million ha of land and over 8000 km of waterways around the globe. Within the United States, the Conservancy manages 627,560 ha of forest and range lands within the range of the white-tailed deer (*Odocoileus virginianus*). Science-based quantitative analyses performed by the Conservancy consistently identified the white-tailed deer (*Odocoileus virginianus*) as being one of the top threats to biodiversity across its multistate landscapes. State chapters, particularly in the mid-Atlantic (NY, MD, NJ, and PA), report that their greatest ecological challenge is managing deer populations at levels compatible with sustaining diverse forests and rangelands. The Conservancy's Pennsylvania chapter owns 5666 ha of forestlands and has a controlling interest through conservation easements and other agreements on an additional 32,275 ha.

The Conservancy purchased a 1228-ha dry oak-mixed hardwood forest in the heavily forested wildernesslike setting of the Allegheny Plateau in north central Pennsylvania in 1999 from absentee landowners who had treated the land as a real estate investment. Their objective had been for the property to increase in value long-term while generating short-term income from natural gas exploration, timber extraction, and leasing for hunting and other forms of recreation. These activities occurred under loosely negotiated terms without professional guidance or oversight, and, coupled with high deer density, resulted in degraded forest conditions and low carrying capacity for wildlife species, including deer.

The property, named the West Branch Forest Preserve (hereinafter referred to as the preserve), is bounded on the east and west by a large public forest (the 121,408-ha Sproul State Forest) and on the north and south by private hunting camps (Figure 33.1). The eastern rectangular block of the preserve is hunter accessible and favorable to hunting: slopes range from 0% to 60% but average less than 25% for the majority of the block, and no place in the block is >400 m from a well-maintained road. The larger western block presents a challenge to hunters: it is steeply sloped (>20%, with a majority being from 40% to 60+%), and it is only accessed by trails that run along the tops of two ridgelines. Much of this block is >800 m from the one access road.

The region has a history of abundant natural resources and a strong hunting heritage. State-leased camps and private hunt clubs maintain family hunting traditions. However, decades of an out-of-balance deer herd led to significant degradation of the forests throughout the region (see Chapter 14), and the preserve is a prime example of how browsing by an overabundant deer herd combined with low deer harvest can devastate the health and productivity of a forest ecosystem.

Regional public land managers (Allegheny National Forest and Pennsylvania Bureau of Forestry state forests) used deer fencing to combat deer impact on forest resources, and some local private land owners planted and maintained food plots to create a surplus of forage to dilute deer impact, enhance deer nutrition, and maintain an abundant deer herd. The differences in values between private and public forest managers were too wide for development of a unified approach in deer management. The culture and traditions of private land managers were not accepting of science-based deer management when the solution (reducing deer density) to high deer impact was in conflict with local tradition regarding deer abundance (high). Although most of the preserve is bounded by the Sproul State Forest, a 526-ha privately owned hunt club borders the preserve on the north. The aging membership of this club struggled to recruit younger hunters to maintain its hunting and deer management tradition. Club members are well-educated and most understand the problem of local deer overabundance, but they have been unwilling to reduce and maintain deer densities by harvesting antlerless deer to bring the local herd in line with the diversity carrying capacity (Chapter 15) of the forest. To support high deer density, members relied upon food plots and winter feeding with shelled corn. They banned harvesting antlerless deer in an attempt to produce more fawns and maintain high deer density.

The previous owners of the preserve restricted access and hunting to authorized individuals and members of clubs. The property's 16.7-km boundary was posted and high tensile wire was strung between trees at well-known access points to emphasize the no-trespass message and to hamper access. As the

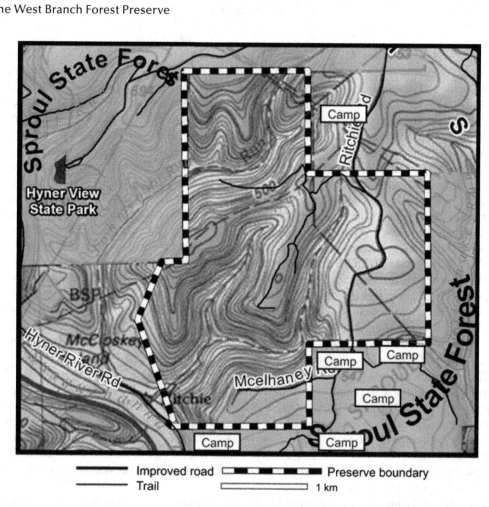

FIGURE 33.1 West Branch Forest Preserve.

majority of the surrounding landscape was state-owned and open to public hunting, there was a local perception that the preserve constituted a de facto sanctuary with more wildlife (and deer) and offered better hunting opportunities. When the Conservancy purchased the property, it came with a constituency of hunters worried about whether they would be able to hunt on the property. These concerns were eased when the Conservancy replaced No Hunting/No Trespassing posters with signage indicating the property was open for public hunting—the intent being to manage the deer population through public hunting.

Historically, many of the lands purchased by the Conservancy, including those in Pennsylvania, were designated as hands-off, no hunting preserves to protect rare, threatened, and endangered species that were thought to require unmanaged forest and range conditions. Outside of traditional stewardship activities that included maintaining signage, improving parking areas, and suppressing non-native invasive plants, Conservancy preserves were relatively unmanaged. The expectation was that such passive management would produce the desired environmental conditions.

Under this philosophy, deer management on the preserve progressed in two phases: an initial period when little was known about deer or forest management and initial estimates of deer density and impact were undergoing development, and a subsequent comprehensive period when staffing in forestry and deer management was added to develop and implement a strategy of proactive deer management, including obtaining and disseminating more comprehensive information on deer and habitat with an emphasis on outreach, collaboration, and communication with hunting and forestry communities.

33.1.1 INITIAL DEER MANAGEMENT

The early deer management program focused on learning about the deer and forest dynamics within the property and developing deer hunting as a forest management tool using Pennsylvania Game Commission (PGC) deer hunting seasons. In 2000, we registered the preserve in the Hunter Access Program, a term-lease agreement that creates a partnership between the PGC and managers to improve public hunting opportunities and wildlife habitat on enrolled property. The PGC provides a variety of benefits, including law enforcement patrols to deter visitors from breaking the law, for example, illegally using all-terrain vehicles and littering or dumping; free seedlings for wildlife food and habitat; and advice on soil conservation and habitat improvement. We hoped that this partnership would address social issues associated with various stakeholder interests in having access to hunt the property. We also hoped the partnership would enhance deer management by expanding the availability of the property for public hunting, reducing liability toward hunting-related incidents, increasing interaction with PGC Wildlife Conservation Officers, and improving access to PGC-sponsored habitat improvement cost share programs.

When Pennsylvania's Deer Management Assistance Program (DMAP) was established in 2003 to help landowners reduce deer impact by reducing deer density, the Conservancy enrolled its preserve property. In addition to providing a way to reduce deer density and impact on the preserve, DMAP was viewed as a communication and collaboration opportunity with local hunters who would likely make up the majority of participants in reducing deer density. In 2002, the Conservancy built six 20×20-m woven wire deer exclosures to demonstrate deer impact through comparison of paired fenced (enclosed) and unfenced plots.

The Conservancy staff had little experience with deer hunting, deer management, or the deer hunting community, so it relied upon the Western Clinton County Sportsman's Association, a neighboring club, to distribute DMAP permits allocated to the preserve. To gauge the success of this approach, the Conservancy developed a DMAP harvest report card that accompanied each antlerless permit. To encourage reporting, two hunters were selected through a random drawing of returned-mailed report cards each year to receive a $50 gift card. Data from the initial 2 years (2003–2004) indicated that hunters with DMAP permits were harvesting mostly antlered deer and harvesting few antlerless deer with DMAP permits.

Recognizing the need for a quantitative basis for developing forest management on the preserve, we contracted with Vision Air Research from Boise, Idaho, to provide aerial estimates of deer density with forward-looking infrared (FLIR) technology. On April 12, 2005, 97 deer were counted, for a density estimate of 7.9 deer/km^2; 116 deer were counted on March 21, 2006, for a density estimate of 9.5 deer/km^2. Similar FLIR surveys by the same company on PA Bureau of Forestry land (M. Benner, PA Bureau of Forestry biologist, personal communication) and on the KQDC case history land (deCalesta 2013) underestimated actual numbers by 20%–30%, so density estimates for the preserve likely were in the 7.9–10.5 deer/km^2 range in 2005 and 9.5–12.8 deer/km^2 in 2006. These densities fall within the range of deer density associated with negative impacts on northern hardwood forest regeneration and habitat (Chapter 8) and were demonstrated by comparisons between fenced and unfenced plots on the preserve in 2004.

On average, 47% of red maple (*Acer rubrum*) and oak (*Quercus* spp.) seedlings in unfenced plots were browsed by deer, whereas an average of 3% of seedlings were browsed in fenced plots (damage thought to be by rabbits/hares rather than deer). Also, diameters and heights of seedlings inside fenced plots were significantly larger than those in unfenced plots. These disturbing deer density and impact data led to the development of a comprehensive deer and habitat management plan on the preserve.

33.1.2 COMPREHENSIVE DEER AND HABITAT MANAGEMENT

The "no to low human impact" minimalist management philosophy, debated by the Conservancy's staff throughout the years, changed in 2007 when the Conservancy began development of its proactive deer management program. The goal was for the preserve to be managed for sustainable outputs of forest resources, which would be achieved by designing and implementing hunting strategies to reduce deer density and impact and improve forest regeneration and habitat. The ensuing strategies and activities formed a 5-year deer management program to balance the preserve deer population with diversity carrying capacity.

Conservancy administrators hired an ecologist, an outreach forester, a conservation planning specialist, and a forest technician intern in 2006–2007 to develop a comprehensive deer management program. In 2008, a Deer and Deer Hunter Management Plan was drafted and circulated for review among a cross-section of regional experts (foresters, ecologists, wildlife biologists, and avid hunters) and targeted stakeholders. The following year (2009), the Conservancy proceeded with Forest Stewardship Council (FSC) forest certification for some of TNC holdings in Pennsylvania: the Conservancy selected the preserve as an initial certification property. Certification required that natural resource management be based on quantitative analysis and a philosophy of sustainable outputs for all natural resources, including preserving and protecting areas of high conservation value (e.g., old-growth forests and other unique habitats required by threatened and endangered plant and animal species threatened by overabundant deer).

These investments were intended to inform and advance development of the Conservancy's Forest Conservation Program focused on private forestlands. Late in 2009, the Conservancy launched its Working Woodlands program (www.nature.org/workingwoodlands), designed to provide private managers FSC certification and access to ecosystem (forest carbon) markets at no cost in exchange for a perpetual easement that restricted commercial/residential development for nonforest uses. The result was the implementation of science-based land management activities and operations on select Conservancy properties (including the preserve) and easement lands enrolled in the Working Woodlands program.

Because other forestlands undergoing FSC certification in Pennsylvania almost universally were required to reduce deer impact as a condition of certification, and because the preserve is located in an area of high regional deer impact, administrators for the preserve realized that they needed to make deer management a major component of their operations, at least in part to satisfy FSC standards.

33.1.3 DEER MANAGEMENT PROGRAM

We compiled feedback from the Deer and Deer Hunter Management Plan and integrated it with the Working Woodlands program and requirements for FSC certification to produce a comprehensive deer management program based on integration of hunter management, herd management, habitat management, and monitoring. These themes are cornerstones of the Quality Deer Management (QDM) program and were used to guide development of a deer, habitat, and hunter management plan specific for the preserve while complying with FSC standards for sustainable forest management. We proposed the following management activities, based in part on QDMA cornerstones, to achieve our goals for deer management:

- Incorporate state and federal forestland management assistance program opportunities for managing deer and forests and educating stakeholders
- Educate hunters on deer biology and management with workshops
- Communicate with hunters, using a blog, e-mail, annual reports, and updated hunter information signage at key locations within the property
- Use PGC programs to increase antlerless (and thereby overall) harvest via additional antlerless permits
- Develop and distribute harvest report cards for hunters issued DMAP permits to gather information about deer harvest
- Recruit effective hunters (alpha and beta hunters) to harvest antlerless deer
- Provide monetary incentives for obtaining antlerless harvest permits, for bringing harvested deer to checking stations, and for hunters to return harvest report cards

- Operate a deer checking station and gather information on successful hunters
  - Provide recovery assistance for hunters harvesting deer
  - Encourage early season antlerless harvests, particularly in early muzzleloader season
  - Enhance hunting access on the preserve
- Coordinate deer drives to increase deer harvest
- Quantify deer density with pellet group counts
- Conduct an inventory for characterizing wildlife habitat, regeneration, and deer impact including seedling counts, seedling development, deer browse impact ratings, presence of non-native invasive species, and percent cover of fern and woody shrubs
- Treat up to 40 ha annually for the following 10 years with silvicultural practices to eliminate undesirable competing vegetation; enhance regeneration of desired tree seedling species; increase tree, shrub, and herbaceous diversity; improve wildlife habitat; and improve deer carrying capacity
- Create a number of small ($\leq$0.2-ha) food plots from old log landings and abandoned gas wellhead sites
- Quantify and assess deer impact

33.1.4 Landscape Management: Habitat, Hunters, and Deer

We integrated the components of the deer management program across the preserve landscape, partly to comply with the requirement for development of a FSC-certified landscape forest management plan. Our objective was to learn whether an integrated deer management program could succeed at a local landscape level, and if so, which components would work and which would be impractical/ unproductive, given the constraints of annually variable funding and administrative support.

33.1.5 Allocation of Personnel and Financial Resources

The Conservancy operated on a budget controlled by one or two administrators with secondary oversight by upper level leadership. Midlevel managers and on-the-ground personnel were not provided with an annual budget and rarely were involved with fiscal planning. Staff requests for funding for add-on stewardship-related activities were given to the immediate supervisor, who forwarded them to a budgetary manager for review and approval. Prior to 2008, the Conservancy received a 5%–8% rate of return on its Stewardship Endowments established for each of its 30 preserves within Pennsylvania. Annual expenses for stewardship-related activities seldom exceeded a few thousand dollars. But as the Forest Conservation and Working Woodlands programs became established, sustainable forest management plans were being developed and green certification was obtained, funding required for additional staffing and operational costs rose.

Although initial commitments for the elevated approach to deer and deer hunter management were honored for the preserve in 2008, by 2009 and later, funding and staffing levels were severely reduced. As one consequence, the deer checking station was not operated in 2009, disappointing local hunters, as many reported their harvests by e-mail and voice messages while inquiring about the status of the checking station operation.

33.1.6 Reduction in Scope of Management Program

As we began to implement the deer and forest management programs, our limited financial and personnel resources, combined with the difficulty in gaining access to the steeply sloped western block of the preserve, dictated that we restrict deer and forest management to the eastern block with its much more favorable access and gentler slopes. We left the western block as a nonmanaged, hands-off de facto natural area wherein nature would take its course, potentially reverting the block to an old-growth seral state. We realized that this created a natural sanctuary within the preserve wherein deer could escape from hunters during hunting seasons, and whence deer could replenish deer harvested from the managed eastern block. In reality, this condition is no different from one in which abutting, unhunted forestland owned/managed by others provides sanctuary to deer on neighboring hunted lands. The hope is that the Rose Petal effect (McNulty et al. 1997) promotes at least temporary reduction in deer density on hunted lands and that movement of deer from neighboring lands to repopulate takes several years. This may have been the case: after deer density was reduced on the eastern block by hunting, it remained low for 6 years, possibly kept in check by fairly constant, year-to-year hunting by experienced hunters who harvested remaining resident and immigrating deer.

33.2 MANAGEMENT ACTIONS

Based on our analysis, the array of management steps/activities identified above was developed for consideration. Some actions were conducted on an irregular but continuing basis. Others were discontinued after 2008, in part because of a precipitous drop in funding from TNC, and in part because some were difficult to implement/continue or did not produce useful results. Some, after initial consideration, were never attempted. Allocation of staffing and funding initially planned for development and implementation of the deer management program were diluted by the additional demands of preparing for the 2009 FSC certification assessment and the 2009 development and launch of the Working Woodlands Program. Unexpected staff departures, additional declines in Conservancy budgets, and continued lack of desirable regeneration at the preserve, requiring additional unanticipated funding to mitigate, further depleted resources available for the deer management program. This "perfect storm" of competing programs for reduced funding and staffing resulted in scaling back or elimination of many of the deer management program funds and staff time. These reductions, in turn, led to curtailment of some ongoing components of the deer management program and failure to implement others, as described below.

33.2.1 USE STATE/FEDERAL/UNIVERSITY ASSISTANCE PROGRAMS

We incorporated PGC programs (Hunter Access and DMAP), which we continue to use; a Pennsylvania State University Extension deer density workshop (conducted for 1 year); and FSC forest management assessment evaluations (initial assessment followed by yearly audits) to add outside expertise and management recommendations to help guide deer and forest management and to improve our ability to educate stakeholders about deer and forest management.

33.2.2 MANAGE HUNTERS AND THE DEER HERD

33.2.2.1 Educate Hunters

In 2008, Conservancy staff assisted the Pennsylvania State University Cooperative Extension with its Deer Density and Carrying Capacity Workshop. The 1-day course included a review of deer biology, deer health, and deer habitat requirements, put attendees in the field for a half day collecting data on deer impact and deer density on targeted properties, and analyzed the data to estimate deer density and impact. These data were then used to form recommendations for managing deer harvest to achieve forest goals. The preserve manager incorporated the workshop's data collection protocols as part of the Conservancy's continuing deer and habitat monitoring procedures. Conservancy staff, volunteers, and local West Branch hunters were encouraged to attend this workshop. Unfortunately, as a result of cuts in staffing and funding, we were able to conduct the deer density/impact workshops for only 1 year (2008).

33.2.2.2 Communicate with Hunters

Prior to 2008, the primary means of communicating with hunters was through a formal letter that accompanied DMAP permits for individuals who had sought antlerless permits for the preserve.

Building on this initial communications attempt, we developed two-way communications and relationship-building with local hunt clubs, sportsman's groups, and local hunters. Our intentions were to gain an understanding of the history of hunting on the property and the dynamics of current hunters and their effectiveness in harvesting deer, to obtain the trust and collaboration of local hunters for managing the preserve's deer, and to facilitate hunter access and success hunting deer on the preserve.

We improved and expanded signage on the preserve gated roads and parking areas to provide additional information for hunters. Signage included recreational use rules for the preserve, Conservancy contact information, PGC Public Access Cooperator information, and Quality Deer Management Association posters espousing the principles of Quality Deer Management. We placed a wooden kiosk (Figure 33.2) at a primary entry point into the preserve adjoining a primary parking area and gated road that provided access into a core area of the property. Built by preserve staff with the help of local hunters, the kiosk served as a field communication center where hunters could get updated deer management information and view maps of the preserve. The kiosk has been viewed by an array of forest users, including hikers, berry pickers, gas workers, and families that stop by to explore the preserve. Frequent updating of the kiosk with relevant and timely information kept it attractive and useful.

We developed a blog for providing basic information about the deer management program and hunting/deer updates to improve communications among hunters and preserve staff. We created a contact list for the blog from information on hunters obtained from preserve harvest report cards and the e-mail list. Although the blog was developed for hunters, it was also used by other stakeholders (e.g., hikers, adjoining landowners, fellow TNC staff members). Some hunters posted pictures, videos, and comments.

Integration of trail camera pictures with the blog enhanced communication with hunters. Hunters remarked that they enjoyed viewing pictures of wildlife and successful hunters posing with harvested deer. Trail camera photos taken at the preserve depicting deer, bear, turkey, grouse, and other wildlife species created an uptick in "hits" by blog visitors and an increase in e-mail inquiries, primarily from hunters asking specific questions or making comments about hunting.

Game camera posts seemed to improve slumping morale among hunters who complained of no sign or sightings of deer, as the posts provided evidence that trophy bucks, mature doe, bears and other game and nongame species were, in fact, present on the preserve. A few hunters posted trail

FIGURE 33.2   Kiosk for hunter information.

cam pictures they took on the blog. An additional benefit of the blog was that it provided compliance with FSC certification standards regarding stakeholder engagement.

The blog was discontinued in 2011. In 2010, the program manager was tasked with additional responsibilities for programs that took priority over administering the blog. The Conservancy couldn't justify the time and expense required of the program manager for producing the detail of information on the blog that had been available from 2008 through 2010.

The annual reports regarding deer management on the preserve that we included with packets of information sent to hunters requesting DMAP permits were discontinued in 2010 when we turned distribution of the permits over to the PGC.

33.2.2.3 Use Pennsylvania Game Commission Programs to Increase Antlerless Deer Harvest with Special Permits

The preserve is within Wildlife Management Unit (WMU) 2G, a top destination for deer hunting in north-central Pennsylvania. The PGC allocates permits for harvesting antlerless deer in an attempt to reduce deer density in WMUs where deer impacts endanger forest resources; antlerless permits have been available for harvesting antlerless deer in WMU 2G since the 1990s. Because the preserve makes up a small proportion of WMU 2G, few WMU 2G permits have been used on the preserve, with the result that deer density and impact were not noticeably reduced by the program.

To rectify this situation, which occurred on many forested properties within the state, the PGC launched a Deer Management Assistance Program in 2003. Under the DMAP, managers with unacceptable deer impact on forest resources could obtain additional permits for harvesting antlerless deer on their properties and distribute them to hunters. We used hunter interest in obtaining antlerless permits as a way to enhance and maintain communications with current hunters, to recruit and develop communications with new hunters, and to assist in identifying hunters most likely to harvest deer, especially antlerless deer. The PGC's Deer Management Assistance Program enabled us to secure additional, property-specific antlerless permits and control allocation of the permits. The Conservancy enrolled the preserve in DMAP in 2003. From 2003 to 2007, the Conservancy obtained the base rate allocation of 1 antlerless permit per 20 ha of forestland, equating to 60 antlerless permits.

After deer browse impact ratings were collected and analyzed with deer density estimates from FLIR flights and pellet group counts, it became apparent that deer impact was still too high and that we could justify applying for an increase in antlerless DMAP permits. Hunter harvest data derived from the Conservancy's custom report cards and PGC DMAP harvest reports 2005–2007 indicated a DMAP-hunter success rate of 10% per year on the preserve. Based on this estimated harvest rate for hunters with DMAP permits, and our estimates of deer density and impact, we asked for and obtained an increase in permits from 60 to 150 for 2008 and a further increase to 200 permits for 2009. We recruited more hunters to purchase and use multiple DMAP permits 2008–2010, assuming we would obtain an additional harvest of 25 deer (20 antlerless, 5 antlered deer) from the preserve in 2008 and 35 additional deer in 2009. In 2010, we reduced costs by transferring administration of DMAP permits to the PGC when its automated system became available. Enrollment in, and administration of, the DMAP program was easy to manage and seemingly effective in recruiting new hunters, as evidenced by the increase in use of DMAP permits since the inception of DMAP in 2003 (Figure 33.3).

The relationship between number of DMAP permits available and harvest of antlerless deer was positive and significant: as the number of available permits increased, so did the number of antlerless deer harvested ($r^2 = 0.77$, $P < 0.001$, Figure 33.4); when the number of permits declined, so did the reported antlerless harvest. An initial large number of antlerless deer harvested in 2003 reflected a common phenomenon: introduction of large increases in numbers of antlerless permits is associated with a spike in number of antlerless deer harvested the first year the permits are available, followed by reductions in number of antlerless deer harvested in subsequent years with continued availability of permits.

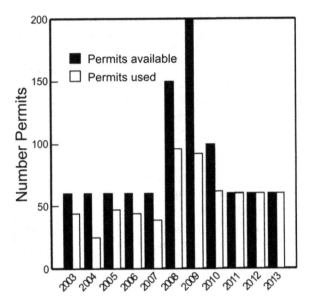

FIGURE 33.3   Numbers of DMAP permits available and used.

All DMAP permits offered were taken when they were distributed by the PGC, when they were available to all hunters, rather than by the Conservancy, which had provided them only to local hunters. Seemingly, offering DMAP permits for the preserve to a much wider hunter pool (via PGC) resulted in complete use and possibly higher antlerless harvest rates. This belies the notion, expressed by repeat hunters of the preserve, that turning allocation of permits over to the PGC would result in omega hunters obtaining tags and not using them so as to deny them to alpha and beta hunters who would use them to harvest antlerless deer, resulting in lower harvest of antlerless deer.

Nonreporting rates for whether hunters harvested deer with permits were higher when permits were administered by the PGC: possibly hunters were not as fearful of losing future opportunities

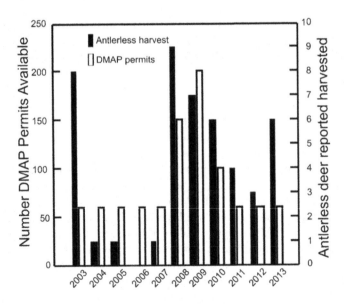

FIGURE 33.4   Relationship between number of DMAP permits and reported antlerless harvest.

for obtaining DMAP permits by not responding than local hunters when permits were administered by the Conservancy.

Examination of antlerless harvest rates (number of antlerless permits offered to produce a reported antlerless harvest) is also instructive. When 60 permits were offered per year by the TNC, an average of 46.9 was required to harvest an antlerless deer. When permits offered ranged from 85 to 200 and were obtained from TNC, 19.9 were required to harvest an antlerless deer. But when 60 permits were offered per year and distributed by the PGC, only 15 on average were required to harvest an antlerless deer. These data suggest that: (1) monetary incentives had little if any effect on antlerless harvest; (2) increasing numbers of antlerless permits results in higher numbers of antlerless harvest relative to numbers of permits offered; and (3) familiarity of hunters with the DMAP program (by the time allocation was turned over to the PGC, the program had been in place for 8 years), in the face of likely reduced deer density, may have favored disproportionately higher use of permits by alpha and beta hunters and associated higher hunting prowess and harvest rates.

Higher nonresponding rates for 2011–2013 may have biased estimates of the number of permits required to harvest an antlerless deer if unsuccessful hunters in 2011–2013 reported at a lower rate that successful hunters in preceding years when permits were administered by TNC. This could be the case if the ratio of successful hunters responding to unsuccessful hunters responding was higher during 2011–2013.

33.2.2.4 Harvest Report Cards

We compiled hunter harvest information from TNC harvest report cards (sent out with DMAP permits) and PGC DMAP hunter reports. Data included number of days hunting on the preserve, number of deer sighted, number of deer harvested, and numbers of permits required to harvest a deer. Reported deer harvest, including antlerless deer, was low from both sources, excepting the first year DMAP permits were offered, when all hunters using DMAP permits sent reports to the PGC.

The number of deer reported harvested was highest in 2008–2010, which coincided with monetary incentives offered by TNC (2008) for reporting harvests and with the highest number of DMAP permits available (85–200) (Table 33.1). Nonresponses increased over time as hunters learned that penalties for nonresponding rarely were levied. High response rates may produce underestimates of harvest, unless all successful hunters using DMAP permits report, an unlikely scenario. When TNC distributed DMAP permits, not all were used, but all were taken when issued by the PGC.

Large increases in hunters requesting and using DMAP permits in 2008–2011 may have been related to an increase in nonlocal hunters applying for permits specific to the preserve, as they likely

TABLE 33.1

Antlerless Harvest Reports Submitted by Hunters, TNC and PGC

Year	Permits Available	Permits Used	Deer Harvested (PGC)	Deer Harvested (TNC)	Nonresponse % (PGC)	Nonresponse % (TNC)
2003	60	41	8	n/a	0%	n/a
2004	60	24	1	n/a	58.3%	n/a
2005	60	46	1	2	37.0%	26.1%
2006	60	42	0	2	33.3%	42.9%
2007	60	35	1	0	40.0%	35.0%
2008	150	99	9	12	38.4%	59.6%
2009	200	92	7	8	62.0%	67.4%
2010	85	64	8	n/a	51.6%	n/a
2011	60	60	4	n/a	46.7%	n/a
2012	60	60	3	n/a	70.0%	n/a
2013	60	60	6	n/a	70.0%	n/a

were aware of the large increase in availability of permits. Also, the number of local hunters applying for permits never exceeded 46 prior to the large increases in availability and may have represented an upper level of locally available hunters inclined to apply for and use antlerless permits.

The fewest permits were required to harvest a reported antlerless deer (mean = 10.7) in years when the highest number of permits were offered (2008–2010). For years when 60 permits were offered, hunters reported a lower ratio of deer harvested to permits offered (mean = 15.0) when PGC distributed permits; when TNC distributed permits, it required an average of 27.5 permits on average to harvest one antlerless deer. Seemingly more antlerless deer were harvested when permits were distributed by the PGC. And, all permits offered were taken when distributed by the PGC; on average, 62.6% were applied for when distributed by TNC.

The response rate for TNC report cards was higher when 60 permits were available rather than when 150 and 200 were offered in 2008 and 2009 (Table 33.2) Even so, hunters returning report cards reported harvesting more antlered and antlerless deer when the higher numbers of permits were available; an increase in DMAP permit availability was associated with higher reported harvests of antlered and antlerless deer.

The number of hunters who reported hunting from TNC reports remained fairly constant even as the number of DMAP permits more than doubled and tripled. However, the mean number of days hunted and deer seen increased when DMAP permits increased, as did deer harvest, suggesting that hunters afield during increased availability of DMAP permits were more effective in seeing and harvesting deer, in light of likely reduced deer density.

We achieved target deer density on the portion of the preserve readily accessible to hunting by 2010. Beginning in 2011, we reduced the requested number of DMAP permits to the 2003–2007 level because we thought that number (60) would provide enough permits for experienced hunters to maintain reduced deer density with resulting reduced recruitment requiring smaller harvests to maintain deer density at goal. This strategy seems to have worked: the number of deer harvested under reduced numbers of DMAP permits for 2011–2013 was higher than during initial years of the DMAP program when the same number of permits was available annually (2003–2007; Table 33.1).

33.2.2.5   Recruit Alpha and Beta Hunters

The Conservancy inherited alpha, beta, and omega hunters when it acquired the preserve. The most beneficial to reducing deer density and impact were assumed to be local, nonlease alpha and beta hunters who were familiar with the preserve landscape and had spent many years hunting and recreating there for multiple generations. These hunters were dedicated to spending time in the field, were familiar with the terrain, had a sense for deer movements within the preserve, and demonstrated a willingness and ability to harvest deer on an annual basis. The second most valuable hunter category was nearby lease camp holders or those who had private deer camps that enabled them to maintain their family outdoor heritage with hunting and other recreational activities.

TABLE 33.2
Hunter Report Card Responses TNC

Year	Hunters Hunting	Days Hunted[a]	Deer Seen[a]	Antlered Harvest	Doe Harvest	Fawn Harvest	Total Harvest
2005	30	3.5	2.7	1	2	0	3
2006	20	3.5	1.7	0	2	0	2
2007	20	4.0	4.0	2	0	0	2
2008	40	4.8	3.5	6	10	2	18
2009	30	4.4	3.3	7	5	3	15

[a] Numbers are averages.

We developed a hunter spreadsheet with data from TNC's harvest report card and PGC's DMAP reporting system. Hunters were categorized as alpha, beta, or omega partially based on this harvest data and partially on interactions with 50 repeat hunters. Eight of these hunters were identified as alpha hunters (16%), 10 as beta hunters (20%), and the remaining 32 (64%) as omega hunters. We believed that the alpha or beta hunters were more inclined to harvest antlerless deer and support density reduction. The preserve manager ensured that these hunters were given first priority in receiving DMAP permits and provided them with information on areas of high deer impact and/or higher availability of forage and deer movement patterns to help locate and harvest deer.

Communications with preserve hunters were directed at alpha and beta hunters to provide specialized opportunities (e.g., permission for use of their all-terrain vehicles [ATVs] for deer recovery, identifying locations where deer impacts were high to increase hunting pressure). In the initial year (2008) of the deer management program, the preserve manager was a constant presence on the preserve during the hunting season to collect additional hunting information by talking directly with hunters and becoming familiar with those who hunted the preserve heavily. The conversations generated additional information on hunter history, views, and experiences related to deer hunting on the preserve, and generated contact information for facilitating follow-up communications.

An emphasis on recruiting and supporting alpha and beta hunters was associated with a higher number of days spent hunting, higher numbers of deer seen, and higher harvest rates (Table 33.2), suggesting that our efforts to attract and promote alpha and beta hunters were successful in producing higher deer harvests.

33.2.2.6   Provide Monetary Incentives for Hunting and Harvesting Deer

We provided incentives in 2008 to increase doe harvest, reduce deer density, and improve deer health to counter the notion, particularly among older hunters, that harvesting antlerless deer is counterproductive to deer health. A second goal of the incentive program was to blunt opposition to lowering deer density among younger hunters who are likely more aware of, and inclined to support, science-based deer management as a counterbalance to generational hunting lore and values promoting high deer density. Last, the incentive program was designed to enhance meeting and developing relationships with participating hunters. The incentives were advertised on the blog and spread by word of mouth within the local hunting community.

We raffled off two $50 gift cards at the end of the 2008 deer season for hunters who obtained and used DMAP permits. As another incentive, we gave a $200 gift card and a $10 reimbursement for the purchase price of their DMAP permit to the hunters who brought to the checking station the heaviest doe and the oldest doe. Additionally, all hunters who brought harvested deer to the checking station were given a $25 gas card and $10 as reimbursement for the cost of the DMAP permit.

The incentives were advertised on the blog and spread by word of mouth within the local hunting community. The monetary incentives for hunters using DMAP permits and for the biggest/oldest doe were given only in 2008. Funding for these incentives was exhausted at year's end due to higher than anticipated hunter participation and harvest success rates. Significantly reduced operating budgets in 2009 resulted in the elimination of the checking station/incentive program.

The biggest doe and oldest doe contests were associated with high participation rates at the deer checking station, increased hunter interest, and success as indicated by an increase in antlerless deer harvesting (Figure 33.4). Most of the visitors were participants in the Biggest Doe/Oldest Doe Harvest Contest and had been issued a DMAP permit for use at the preserve.

The monetary incentive programs (the biggest doe and oldest doe contests, gas cards and reimbursement for DMAP permits for hunters harvesting antlerless deer) instituted in 2008 coincided with large increases in DMAP permits made available 2008–2010; these actions were associated with high participation rates at the deer checking station, increased hunter interest, and success (Tables 33.1 and 33.2). However, we cannot ascribe the degree to which monetary incentives or increased DMAP permits individually contributed to the increase in antlerless harvests on the preserve.

33.2.2.7 Operate a Checking Station/Collect Deer Harvest Data

Harvest data reported by DMAP hunters to the PGC was combined with information from the preserve hunter survey from 2005–2010 and hunter report cards to provide quantitative information on hunters and deer harvest for adapting management steps to increase deer harvest and reduce deer density and impact. The checking station was well staffed for the first 3 days of the 2008 rifle season, which is when conventional wisdom and practical experience tell us that the majority of the deer harvest occurs.

Hunter participation exceeded staff expectations, with 14 hunters visiting the checking station over the first 3 days of the 2008 deer rifle season. Most of these hunters were participants in the Biggest Doe/Oldest Doe Harvest Contest and/or had been issued a DMAP permit for the preserve. We operated the checking station for only 1 year—the significantly reduced budget in 2009 resulted in the elimination of the checking station.

The number of deer reported harvested increased sevenfold in 2005–2010, and hunter success increased 2.5–3×. However, after concentrating on harvesting antlerless deer the first few years, hunters reverted to harvesting antlered deer the last 2 years, which was a common and continuing trend in north-central Pennsylvania (Table 33.2).

33.2.2.8 Deer Recovery Assistance Program

Evidence suggesting that preserve hunters were unwilling to harvest deer because of limited motor vehicle access and the steep and rugged terrain characterizing more than half the preserve prompted the staff to offer a deer recovery assistance program. Hunters were advised that an all-terrain vehicle would be manned at the checking station by staff for use on roads and trails to aid in recovering harvested deer from difficult terrain. Hunters expressed an interest in this service, but it was not staffed for the entire 2-week rifle season and used too little to determine whether it contributed to deer harvest and/or an increase in number of deer brought to the checking station—it too was dropped after the initial year.

33.2.2.9 Emphasize Early/Late Season Deer Hunts

We promoted early and late season (archery and black powder) deer seasons to increase numbers of deer harvested above those from rifle season. Some of the monetary incentives were designed to encourage early muzzleloader season hunting on the preserve for 2 consecutive years, 2008–2009. Hunter participation rates did not meet expectations and harvest success rates were low, resulting in few additional deer harvested, so the emphasis on early season hunts was abandoned. Informal conversations with preserve hunters indicated that not all hunters embraced the early seasons because some did not agree with the of use modern muzzleloading equipment.

33.2.2.10 Enhance Hunter Access

In addition to maintaining the all-weather road bisecting the eastern block of the preserve, we improved hunter access by expanding and hardening primary parking areas and opening up portions of remnant logging roads that led into more remote portions of the preserve. There are approximately 17 km of logging trails throughout the preserve; many were overgrown with dense thickets of saplings and brambles. Our efforts to encourage hunters to use logging trails were enhanced by a group of local hikers (the PA Trail Dogs). This group adopted 12 of the existing 17 km of logging trails and improved and maintained them. Although the trails weren't open for vehicular traffic, they did provide improved access for foot travel, promote still hunting, and work well for harvest recovery with game carts. We felt that that these trails promoted increased hunter pressure and increased deer harvests within areas of the preserve considered unhuntable due to rugged terrain and limited access.

A select group of trusted alpha hunters who consistently harvest deer on an annual basis were given permission to use their motor vehicles to recover deer that they harvested in remote locations. They were advised that this privilege was confidential, not to be advertised to other hunters, and required that they act responsibly and operate equipment at their own risk.

33.2.2.11 Deer Drives

We organized and supervised two deer drives on the preserve (2008 and 2009) with members of adjacent hunt clubs in an attempt to harvest deer in more remote sections of the property that seldom received hunting pressure. However, hunter unfamiliarity with the terrain and lack of organization resulted in few deer sighted (12 in 2008 and 2 in 2009) and none harvested.

33.2.2.12 Quantify Deer Density with Pellet Group Counts, Aerial Counts, and Trail Cam Counts

We estimated deer density on the preserve in 2008–2010 with the pellet group technique (deCalesta 2013). Because the access road on the preserve ran through the eastern block, making it easy to run transects, we obtained density estimates for 2008–2010 for the eastern block. Density declined from 10.4 deer/km^2 in 2008 to 4.6 deer/km^2 in 2009 and 3.9 deer/km^2 in 2010. We estimated density on the larger western block only in 2009 (because there was virtually no road access) at 12.7 deer/km^2, which was higher than estimates obtained in 2005 (7.9 deer/km^2) and 2006 (9.5 deer/km^2) from FLIR flights. However, when FLIR estimates are adjusted for the approximate 80% detection rate typical of the technique, the estimates correspond to density (9.9 and 11.9 deer/km^2) similar to overall estimated densities at the project's initiation. We were unable to use trail camera recordings to produce estimates of deer density.

33.2.3 Habitat Management: Silvicultural Treatments

33.2.3.1 Initial Inventory

We developed an inventory methodology in 2007 based on Silviculture of Allegheny Hardwoods (SILVAH; Marquis et al. 1992) and Treatment Unit Sustainability Assessment Form (TUSAF; Finley et al. 2007) guidelines to collect and characterize wildlife habitat and regeneration data (seedling counts, seedling development, deer browse impact ratings, presence of non-native invasive species, and percent cover of fern and woody shrubs). Approximately 60% of the property is operationally feasible, based on slopes ≤40%; a goal was to treat up to 40.5 ha annually on this portion of the preserve (primarily the eastern block) with silvicultural prescriptions.

Seedling establishment and advanced regeneration were severely impacted by deer browsing. Approximately 60% of plots had no advance regeneration (large seedlings and saplings) and nearly 30% were not stocked with seedlings of any kind. Regeneration on plots was dominated by striped maple (*Acer pensylvaticum*), red maple (*Acer rubrum*), and black birch (*Betula lenta*). Striped maple is resistant to deer browsing, and red maple and birches often dominate regeneration on sites because they thrive under enhanced lighting conditions on such sites. The few plots that were stocked with acceptable species (mixed oak [*Quercus* sp.], yellow poplar [*Liriodendron tulipfera*], black cherry [*Prunus serotina*], sugar maple [*Acer saccharum*], and hickory [*Carya* sp.]) were often lacking in stem densities, root collar caliper, and/or height indicative of successful regeneration, rendering them unlikely to progress to produce harvestable timber and useful wildlife habitat. Deer browse impact was rated high on 70% of the plots. With deer impact this high, advance regeneration will fail, there will be no replacement cohort of trees to replace those harvested in logging operations, and degraded wildlife habitat will support fewer wildlife species and lower abundance of species that persist.

33.2.3.2 Silvicultural Treatments

We conducted silvicultural treatments to increase tree, shrub, and plant diversity; improve wildlife habitat; increase carrying capacity for deer; and create a healthier, more ecologically and economically productive forest. Treatments included mechanical operations (i.e., mowing), and/or prescribed fire in conjunction with or followed by two or three shelterwood harvests and elimination of undesirable competing vegetation through herbicide application (Table 33.3).

TABLE 33.3
Silvicultural Treatments to Improve Habitat

Year	Treatment	Ha
2006	Mowing of undesirable competing vegetation (mountain laurel)	40.4
2007	Prescribed fire within mowed areas	20.2
2008	Prescribed fire within mowed areas	20.2
2008	Mowing of undesirable competing vegetation (mountain laurel)	52.52
2009	Prescribed fire within mowed areas	20.2
2012	Shelterwood harvest (whole tree chip)	70.7
2012	Herbicide treatment targeting hay-scented fern	8.08
2013	Fencing constructed to protect herbicide treatment area	8.08

About 80 ha of degraded forest in the eastern block of the preserve was treated with a combination of prescribed fire and cuttings (mowing) of undesirable vegetation in an effort to promote seedling regeneration, deer browse, and improved wildlife habitat.

Unfortunately, even with reduction in deer abundance to target density in the area treated with silvicultural applications, by 2011, competing vegetation recovered and expanded, while seedling establishment and advancement remained low (Tables 33.4 and 33.5)—creation of additional forage was not sufficient to achieve desired understory conditions. It is likely, given the chronic paucity of deer forage on the preserve, that target deer density was too high and should have been lower. Deer browse impact remained high likely because of low carrying capacity within the local

TABLE 33.4
Average Seedling Density and Percent Cover of Competing Vegetation after Silvicultural Treatments

Year	2008	2010	2014
Seedling density	$3.46/m^2$ plot[a]	$2.74/m^2$ plot	$2.4/m^2$ plot
Percent cover, competing vegetation	44%[b]	61%	96%

[a] With low deer impact, need 3.75 seedlings/m^2 plot; with high deer impact, need 12.5 seedlings/m^2 plot for successful regeneration.

[b] If ground cover of competing vegetation is >30%, resulting interference will cause failure of advanced regeneration.

TABLE 33.5
Response of Seedling Species Diversity to Silvicultural Treatments

Treatment	Average Seedling Species Diversity		
	2008	2010	2014
Mowed	4	7	6
Mowed and Burned	6	6	7
Mowed and Thinned	7	8	9
Thinned, Sprayed, Fenced	N/A	N/A	17

landscape—the size of treatment areas was too small a proportion of the local, forage-depleted landscape and contributed little to overall forage availability.

There were nearly twice as many seedling species recorded in areas protected by deer-proof fencing after thinning and spraying, a clear indication of the potential for silvicultural treatments to improve regeneration without heavy deer browsing pressure.

33.2.4 HABITAT MANAGEMENT: FOOD PLOTS

Checking station surveys and discussions with local hunters during the 2008 deer rifle season indicated broad interest and support for the establishment of wildlife food plots. At the conclusion of a deer density and carrying capacity workshop conducted in spring 2010, we surveyed participants regarding their recommendations for managing an understocked dry oak heath forest stand. There was virtually no available browse in an understory dominated by hay-scented fern, blueberry thickets, and interspersed mountain laurel where deer density was estimated at less than $4/km^2$. We were disappointed when hunters suggested that establishing food plots would correct the situation.

Although we did not advocate artificially increasing carrying capacity by converting forested areas into food plots, we recognized that establishing food plots might increase hunter morale, deer sightings, and deer harvest, which might help to achieve specific management objectives. Additionally, food plots were of growing interest to private managers, particularly hunt clubs and family groups, which are landowner types targeted for enrollment in the Working Woodlands program.

To test these ideas and gain a better understanding of the costs and benefits of food plots, we arranged a spring volunteer workday wherein volunteer hunters built two small ($\sim$0.2 ha) food plots, one on a log landing and another on an abandoned gas well site. The plots become established, but the planted oats, wheat, and clover were quickly consumed by deer well in advance of the deer rifle season. The first plot was close to a major road, and evidence suggested that it was associated with an illegal buck harvest. The second plot was more remote and there were unconfirmed reports that archery and muzzleloader hunters frequented the plots, but we recorded no confirmed harvests. We did not derive an estimate of how these plots contributed to early season harvest rates, as hunters who helped install the larger food plot preferred that we not advertise its location to other hunters. We did not attempt to develop additional food plots because local hunters viewed their contributions to food plot construction as entitlement to hunt the food plots exclusively by restricting knowledge of food plot locations.

33.3 LESSONS LEARNED

33.3.1 ACQUIRING AND SUSTAINING COMMITMENT OF RESOURCES

The deer management paradigm we chose (using public hunting to achieve and maintain deer density below levels that result in negative impact to forest resources) requires annual commitment of funding and personnel to insure that sufficient numbers of deer are harvested every year. We developed an aggressive program for managing deer that was unsustainable, partly due to undertaking additional, unrelated projects without securing additional funding and partly due to reductions in permanent and temporary personnel. We were forced to abandon most of the deer management components we designed, including incentivizing deer harvest, conducting checking stations, providing hunter assistance in gaining access to the preserve and in extracting harvested deer, providing educational programs including workshops and a blog, surrendering administration of DMAP permits to the PGC, and surveying hunter harvest information.

Having a defined and committed budget devoted to facilitating deer management activities is imperative to the overall success and credibility of a deer and deer hunter management program. At a minimum, it would have benefitted deer management on the preserve if we had been able to maintain annual hunter workshops to educate hunters and provide enough personnel to conduct annual data collection for deer density and impact and to continue deer checking stations, the blog, annual reports, and annual surveys of hunter participation and success on the preserve.

33.3.2 HUNTER RESISTANCE TO EDUCATION

The perception among preserve staff was that hunters participating in preserve deer hunting were uninterested in expanding their understanding of deer management or the science behind deer management and that few viewed themselves as a contributing to managing the local deer herd through hunting. Their values seemed to be associated with hunting to maintain family traditions and a hunting experience characterized by seeing many deer during hunting season (with the associated requirement for an overabundant deer herd) and harvesting only antlered deer. We discerned little hunter awareness of their role in deer management and little appreciation for healthy and diverse forest conditions required to support a productive and healthy deer herd. Future funding and efforts should be dedicated to working with/educating/recruiting/keeping alpha and beta hunters.

EDITOR COMMENTS

The manager of this case history was well-versed in deer ecology and human dimension factors. In an all-inclusive attempt at creating a deer management program, many factors thought to affect hunter participation, retention, and success were instituted, requiring considerable time and energy on the part of the manager. Over a period of a few years, additional responsibilities placed on this manager, shrinking administrative budgets, and unsuccessful attempts to increase hunter participation and success resulted in an adaptive management winnowing of management activities. This case history, in marked contrast to the Mianus River Gorge Park case history (Chapter 36), began deer management with a comprehensive and ambitious shotgun approach that had to be scaled back due to habitat/terrain features, hunter and administration compliance, finances, feasibility, and staff reductions. The result was a slimmed-down management system and resulting hunter disappointment at reductions and reduced interactions with hunters. The ensuing ad hoc process for winnowing down management to affordable levels serves as an example of how to use adaptive management to experiment with and develop a deer management program with little guidance from the literature. It appears that the program was held together after major management reductions by the continued and successful DMAP program and continued access of the preserve to hunting. Lack of monitoring of deer density and impact left preserve managers to hope that initial reductions in deer density in 2003–2008 would be maintained by a core of alpha and beta hunters who needed little encouragement or information save access to DMAP permits and to the preserve via existing road and trail systems.

David S. deCalesta, editor

REFERENCES

deCalesta, D. S. 2013. Reliability and precision of pellet-group counts for estimating landscape-level deer density. *Human-Wildlife Interactions* 7:60–68.

Finley, J. C., S. L. Stout, T. G. Pierson et al. 2007. *Managing timber to promote sustainable forests: A second-level course for the sustainable forestry initiative of Pennsylvania. General Technical Report NRS-11.* Radnor PA. Northern Research Station, USDA Forest Service.

Marquis, D. A., R. L. Ernst, and S. L. Stout. 1992. *Prescribing silvicultural treatments in hardwood stands of the Allegheny (Revised). General Technical Report NE-96.* Radnor PA. Northern Research Station, USDA Forest Service.

McNulty, S. A., W. F. Porter, N. E. Matthews et al. 1997. Localized management for reducing white-tailed deer populations. *Wildlife Society Bulletin* 25:265–271.

34 The Kinzua Quality Deer Cooperative
Integrating Ownerships and Goals

David S. deCalesta

CONTENTS

34.1 INTRODUCTION

In 2001, the Kinzua Quality Deer Cooperative (KQDC) was formed in northwest Pennsylvania to test the hypothesis that public hunting could reduce local white-tailed deer (*Odocoileus virginianus*) density to the point where deer impact did not threaten forest sustainability. A cooperative of public and private forest landowners, foresters, hunters, wildlife managers, scientists, and local business interests developed and implemented a 10-year demonstration program on a ~30,000-ha (73,250-acre) northern hardwood forest located in northwestern Pennsylvania (Figure 34.1).

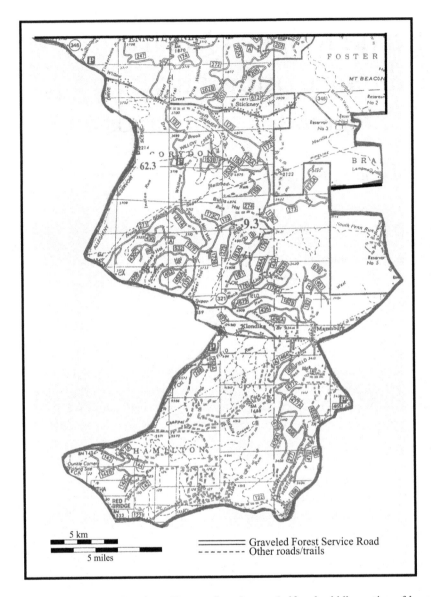

FIGURE 34.1 Deer Demonstration Area. East section of upper half and middle section of lower half less well roaded than surrounding ANF Forest. ANF = Allegheny National Forest; BWA = Bradford Watershed Authority; CP = Collins Pine; FIA = Forest Investment Associates.

This deer demonstration area (DDA) is composed of 24,342 ha of public land (19,567 ha Allegheny National Forest [ANF] and 4775 ha Bradford City Watershed [BWA]) and 5302 ha of private land (3695 ha Forest Investment Associates [FIA]; 1214 ha Collins Pine [CP]), and 393 ha Ram Forest Products [RFP]). The DDA was managed for sustainable production of multiple forest products (timber, hunting, fishing and other forms of recreation, public water supply, and wildlife habitat). Although road condition and maintenance were similar among ownerships, differences in hunter access to the DDA varied among ownerships: ANF roads were many, longstanding, well-signed, and familiar to hunters. The private portions were not as well-maintained or signed-hunters had much less familiarity with them-nor were the areas as densely roaded as on the ANF. These differences translated into differential hunting pressure and associated harvest and required proactive steps to increase hunter awareness and use of FIA, RFP, BWA, and CP road systems.

The genesis of the KQDC project was twofold: the 100-year history of the impact of an overpopulated deer herd on forest ecosystems in Pennsylvania, and the difficulty the Pennsylvania Game Commission (PGC) had in reducing deer density to the point where it was in balance with resources and no longer threatened sustainability of forest resources.

34.2 THE MANAGEMENT SITUATION

In the aftermath of the near elimination of deer in Pennsylvania by the late 1800s (caused primarily by unregulated market hunting), the PGC limited deer harvest to antlered deer and imported 1200 deer from other states to reestablish the herd in the early 1900s (Kosak 1995). Concurrently, widespread clearcutting resulted in production of vast quantities of prime deer forage statewide. In response, the deer population skyrocketed during the early 1900s, especially in northwestern Pennsylvania. Reacting to the deer overpopulation and resulting habitat degradation that occurred over the next two decades, the PGC held an antlerless-only deer season in 1938, which coincided with a severe winter with deep snow persisting into late spring. These two factors resulted in a massive deer reduction, and hunters blamed the resulting low deer abundance on doe hunting. Responding to pressure from hunters, the PGC reverted to antlered deer–only harvests, resulting in deer density climbing into the 20+ deer/km^2 (50 deer/mile^2) range and remaining at that level over the next five decades, much to the detriment of the forest ecosystem as a whole and to ecosystem components individually (deCalesta 2012).

Emerging deer science identified a set of environmental consequences when deer density in northern hardwood forests in the eastern United States exceeds 6–10 deer/km^2 (15–20 deer/mile^2): tree regeneration is stifled or altered, diversity and abundance of affected natural resources (e.g., native bird species and sensitive herbaceous species) are reduced, and deer health suffers (deCalesta and Stout 1997). PGC deer mangers used this information to set goal deer densities of ~7.7 deer/km^2 range (20 deer/mile^2), to reinstitute regulations permitting harvest of antlerless deer, and to initiate a "bonus" antlerless deer program that allowed hunters to harvest additional antlerless deer.

However, the antlerless harvest programs did not result in reduction of deer density sufficient to reduce impacts to desired levels. Hunters resisted harvesting antlerless deer, and PGC commissioners began reducing the number of antlerless permits because of pressure from hunters as deer abundance declined. Deer management pitted hunters who wanted more deer against an increasing number and diversity of stakeholders who wanted fewer deer and lower deer impact.

The situation became untenable with the rising tensions and conflicts among the various stakeholder groups, state and federal agencies, the PGC, the PGC commissioners, legislators, and lobbyists. In 1999, PGC administrators restructured deer management in Pennsylvania based on the premises that habitat health influenced deer health and antler size, and that changes in hunting regulations to effect reductions in the deer herd would improve habitat and deer health. The PGC established three goals for deer management: a healthy and sustainable deer herd, deer-human conflicts reduced to safe and acceptable levels, and healthy and sustainable forest habitat. These goals were to be addressed by the following regulations:

- In 2001, hunters were permitted to harvest antlered and antlerless deer concurrently.
- In 2002, harvest of antlered deer was restricted to bucks with at least three antler points on at least one side to reduce harvest of yearling bucks and increase the proportion of mature bucks in the population.
- In 2003, a Deer Management Assistance Program (DMAP) was established to increase harvest of antlerless deer by allowing landowners experiencing high deer impact to obtain permits and distribute them to hunters to harvest antlerless deer.

This proactive deer management program was headed by Dr. Gary Alt, who had led the successful black bear (*Ursus americanus*) management program in Pennsylvania. In addition to the new hunting regulations and programs, Alt developed and delivered a comprehensive educational program aimed at hunters and other stakeholders, including politicians and administrators. Recognizing that hunters did not relate to the message that deer density greater than 7 deer/km^2 is ecologically unsustainable, harming deer, their habitat, and other wildlife, Alt instead touted reasons appealing to deer hunters: that only by reducing deer could the habitat recover and produce quality (trophy) deer. Because of his reputation as an internationally recognized bear biologist and excellent rapport with hunters, Alt was the ideal messenger of this new concept tying reducing deer density to improved wildlife habitat and improved deer health.

These management changes occurred at the same time the KQDC was forming, providing an opportunity to determine whether they would work when incorporated into a program designed to return deer population densities to levels compatible with sustainable use of forest resources using public hunting as the primary tool.

34.3   THE KINZUA QUALITY DEER COOPERATIVE PROGRAM

Initiation and coordination of the KQDC were orchestrated by the Sand County Foundation (SCF), a private nonprofit organization dedicated to working with landholders to improve natural habitats on their land. The SCF served as a conduit and facilitator for sharing ecological and management information between and among private individuals, scientists, wildlife and wildland managers and landowners, hunters, and local communities. SFC also provided financial support for implementation and maintenance of the program.

34.3.1   Integrating Cultures, Traditions, Values, and Science

The KQDC leadership team understood that a sound scientific basis for the program, featuring reducing deer density to levels leading to improved forest health, would not be sufficient to obtain PGC or hunter support for proactive deer management. The overarching influences of values and traditions on deer management were learned through many years of providing testimony at PGC annual hearings on deer management, The PGC commissioners were far more heavily influenced by, and interested in addressing, values and traditions of the many hunters testifying at these meetings than in the scientific arguments voiced by a much smaller number of other stakeholders. For the KQDC program to succeed, local stakeholders in forest management (forest landowners, hunters, natural resource management agencies, forest ecologists, local business and industries tied to forest management) had to be engaged and their values, traditions, and interests acknowledged, represented, and integrated into the program.

The KQDC leadership team set target deer density at 6–8 deer/km^2. It was realized that by promoting a deer density lower than required by traditional deer hunters, a significant number of these hunters would choose not to hunt on the DDA, especially as deer density declined over time. The strategy was to seek out and retain progressive deer hunters and attempt with education to convert traditional hunters into progressive hunters who could accept lower deer density.

A primary given was that the program had to appeal to, and actively engage, progressive hunters. Their acceptance of and participation in the program were of paramount importance. Accordingly, a kick-off information event was held at a local community college in 2001 with Dr. Alt as the main (and credible to hunters) speaker to provide background for the KQDC program and repeat his message of reducing deer density to improve habitat and deer health. The target audience was hunters; emphases were quality deer, especially bucks with desirable antler characteristics, quality habitat for deer and other game species, and a quality hunting experience. Hundreds of hunters attended.

34.3.2   ADAPTIVE MANAGEMENT

The leadership team developed and implemented an adaptive management program patterned after that described in Chapter 26, which was:

1. Establish quantitative deer density and impact goals by integrating values of stakeholder groups with science-based deer density and impact levels related to sustainable forest management.
2. Compare goal levels for deer density and impact with existing conditions on the DDA with a monitoring system to identify needed changes in deer density and impact.
3. Identify and implement steps to bring deer density and impact to goal levels by:
   a. Gaining support and trust of hunters for the adaptive management program by soliciting their input, recognizing and validating their values, providing them with the scientific basis for managing deer density and impact, and describing the adaptive management program. Three common requests of hunters for deer management (use smaller management units than currently in practice, use checking station operations to gather information on deer condition, and provide valid estimates of deer density at the management unit level) were included as components of the adaptive management program.
   b. Refining and implementing emerging management tools to increase deer harvest and reduce deer density and impact (concurrent antlerless and antlered deer seasons, restrictions on harvest of antlered deer to increase proportion of mature bucks in the deer herd, and DMAP and other antlerless harvest permit opportunities).
   c. Motivating hunters to harvest antlerless deer by providing incentives (rewards for bringing harvested deer to checking stations, big doe and big buck contests) and including them in monitoring activities.
   d. Enhancing hunter success by improving, maintaining, signing, and advertising road access to the vast and well-developed DDA road system.
   e. Communicating with hunters on a regular basis through local media (articles in local newspapers by cooperating landowners and outdoor writers), annual KQDC reports showing improvements in deer and habitat health, and development of a website that explained the scientific basis for the program and provided hunters with information (maps of the road system on the DDA, information on how to obtain permits for harvesting antlerless deer, information on lodging and restaurants, improvements in herd and habitat condition).
   f. Developing, implementing, and reporting annually on a system for monitoring deer density, deer health, deer impact, deer sex and age ratios, and hunter satisfaction as the basis for making adaptive changes in the program to continue progress towards goals, and, once at goals, to remain there.
   g. Developing and implementing a process for identifying and dividing the DDA into smaller, more individual management units to better integrate stakeholder goals, habitat (including deer forage) and deer density conditions, and making site-specific adaptive management changes.
   h. Developing and maintaining a trusting relationship with the PGC so administrators were aware of KQDC support for PGC deer management and appreciation for management tools made available.
   i. Soliciting and implementing review and recommendations of an outside review team to assess the adaptive management program and make suggestions for improvements/continuity of the KQDC program.

34.3.3 Kinzua Quality Deer Cooperative Goals

Goals were developed jointly by a team of scientists, forest and wildlife managers, SCF administrators, local hunters, foresters, PGC wildlife conservation officers, and the local recreational bureau (Allegheny National Forest Vacation Bureau). They were:

- Reduce deer density to 6–8 deer/km^2 (15–19 deer/mile^2), later amended to 4–6 deer/km^2 (10–15 deer/mile^2).
- Reduce/eliminate need for fencing to protect tree regeneration from deer.
- Improve deer health (body weight, antler characteristics).
- Monitor and improve hunter satisfaction.
- Improve species richness and structural diversity of herbaceous and shrub layers.

34.3.4 Monitoring

We established baseline data through a monitoring program that was conducted annually to measure progress/identify areas needing adjustment. The pellet group technique (deCalesta 2013) was used to provide landscape estimates of deer density. Subgroups within the leadership team developed two systems for evaluating impact of deer on forest resources: one for tree seedlings as indicators of overall deer impact (Pierson and deCalesta 2015) and one for evaluating deer impact on herbaceous vegetation (Royo et al. 2010).

Recruitment and sex and age ratios of the deer herd were assessed annually by counting deer of various categories (fawn, adult doe, antlered bucks, unidentified) during late summer roadside counts. Herd health was assessed by collecting data (body weights and antler characteristics) from deer harvested within the DDA and brought to checking stations. Deer harvest locations, obtained from hunters bringing deer to checking stations, were plotted yearly and aggregated over time.

Hunter satisfaction was assessed by questionnaires filled out by hunters bringing deer to checking stations.

34.3.5 Hunters: Involvement/Communication/Motivation

The leadership team included local hunters whose input and suggestions were sought and implemented. One-day deer density and impact training workshops were held in spring to educate and involve local hunters in the science, technology, and monitoring of deer management as applied directly to the KQDC program. Optimizing deer and habitat health were stressed as principal factors, with reducing impact on tree regeneration and need for fencing as secondary goals. Deer roadside counts were conducted by local hunters and Allegheny National Forest personnel. Hunters were encouraged to bring harvested deer to KQDC checking stations and participate in data collection.

Allegheny National Forest and Allegheny Visitor Bureau personnel provided timely news releases to local newspapers regarding information on the KQDC program and on hunting seasons. One of the leadership team members (an established outdoor writer) wrote stories supportive of deer management on the KQDC for local and regional newspapers. Detailed (and condensed-version) annual KQDC progress reports were made available to hunters. The leadership team developed an interactive website that provided hunters with information on the KQDC program, the science behind deer management, downloadable hunting maps for the KQDC, instructions for obtaining various deer licenses and permits, and lodging and restaurant information.

Hunters bringing harvested deer to checking stations were invited to attend a hunter appreciation banquet, and tickets were issued to these hunters (one for antlered deer, two for antlerless deer) for prizes at a raffle held at the banquet, and renowned (e.g., credible to deer hunters) guest speakers provided entertaining and informative discussions of deer management related to deer and habitat health.

34.3.6 HUNTER ACCESS

The DDA was well roaded and roads were well maintained. Road density and placement were such that most areas of the DDA were within 800 m or less of a paved or improved road. Maps were provided to hunters to help them identify access points to the DDA, and a color-coding system was incorporated to help hunters locate major access roads on private landowner parts of the DDA. Roads normally closed to recreational activities were opened for deer season and plowed when possible if snow hindered vehicle access.

34.3.7 PENNSYLVANIA GAME COMMISSION DEER MANAGEMENT INITIATIVES

Each of the four PGC initiatives (concurrent antlered and antlerless hunting season, three-point antler restriction on buck harvest, provision of antlerless permits on a management unit basis, and the DMAP program) was used and effect(s) monitored through changes in deer density, deer sex and age ratios, and deer impact. The number of DMAP permits requested was adjusted annually based on desired reduction in deer density and number (derived empirically from KQDC data) of permits required to harvest desired number of antlerless deer.

34.3.8 COST–BENEFIT

A dollar value could not be assigned to some of the benefits/outputs (e.g., improvement in deer and habitat health, hunter satisfaction), even though they were tangible and quantitative. Costs of monitoring and preparation of reports and recommendations were quantifiable and were estimated on an annual/km^2 basis and compared with economic benefits such as elimination of need (and costs) of fencing to protect regenerating seedlings from deer and elimination/reduction of need to apply herbicides to offset increases in interfering plants (ferns and undesirable seedlings) fostered by high deer densities. Monitoring, reporting, and development of adaptive management decisions were considered recurring annual costs and defrayed against annually estimated savings from silvicultural techniques not employed.

34.4 RESULTS

34.4.1 DEER DENSITY

In spring 2002, deer density for the DDA was 10.4 deer/km^2 (27 deer/mile^2), roughly twice the desired density, and the impact level was assessed as severe. A program of increased opportunity for hunters to harvest antlerless deer was developed. Initially, the only incentives available to motivate hunters to harvest more antlerless deer and reduce overall deer density were the science stating that lower deer density produced quality deer and the checking station operation, which gave hunters who brought harvested antlerless deer two tickets to the raffle/banquet. This situation changed dramatically when the PGC began its concurrent antlered/antlerless seasons and the DMAP program (2001–2003).

Deer density responded differently to the different PGC management initiatives (Figure 34.2). There were no baseline, preconcurrent season density estimates for the DDA, so the role of concurrent antlered and antlerless seasons in affecting deer density is unclear.

Deer density increased slightly in 2002–2003, reflecting the initial impact of the three-point antler restriction regulation: hunters were still primarily hunting for antlered deer in 2002, and the three-point antler restriction greatly reduced the number of antlered deer legal for harvest. In 2002, hunters likely harvested many fewer buck deer while refraining from harvesting antlerless deer, resulting in a higher deer density the following spring.

The large reduction in deer density from 2003 to 2004 reflects the impact of the DMAP program and the changeover to issuing antlerless permits for Deer Management Unit 2F rather than only

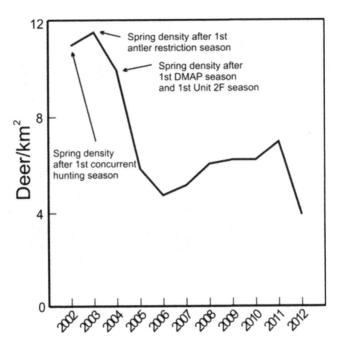

FIGURE 34.2   Response of deer density to PGC management initiatives.

for McKean County, with concomitant large increases in number of antlerless permits. The steep decline in deer density in 2004–2006 reflects the continuing influence of the DMAP and Unit 2F antlerless permit programs, and possibly the concurrent antlered/antlerless seasons on the DDA. Annual fluctuations in deer density closely followed fluctuations in numbers of DMAP and Unit 2F antlerless permits (Figure 34.3).

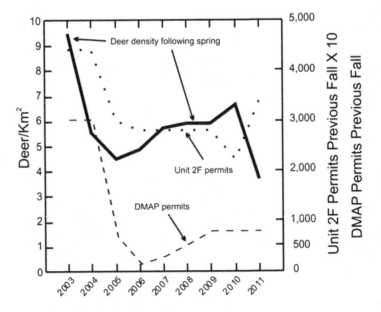

FIGURE 34.3   Trends in deer density and numbers of DMAP and Unit 2F permits.

An initial decline in deer density in 2003–2006 was followed by a gradual climb in density in 2006–2011. When deer density was increasing (2005–2010), the number of DMAP and Unit 2F antlerless permits declined/remained low, reflecting a reduction in hunting pressure on antlerless deer. Deer density peaked in 2011 following a 21% reduction in number of Unit 2F antlerless permits and a leveling off of number of DMAP permits. Deer density dropped 44% in 2011–2012 when the number of Unit 2F permits increased by 53.5% and DMAP permits remained stable.

Individual contributions of the concurrent antlerless/antlered seasons and DMAP and Unit 2F programs to the variation in deer density cannot be separated, and it is impossible to determine whether either had more or less influence on deer density. Prudence dictated retaining all three PGC incentives, as the combination may have been more than simply additive.

The KQDC leadership team was able to manipulate one of the three PGC initiatives (number of DMAP permits) in an adaptive management manner—by decreasing the number after deer density reached goal in order to avoid reducing the herd below goal, and increasing the number when deer density began to creep above goal. The KQDC leadership team aggressively promoted use of DMAP and Unit 2F permits through the website, local news releases, workshops, and testimony at annual PGC season and bag limit hearings.

34.4.2 Deer Impact

Percent plots with no deer impact on seedlings, as a coarse measure of deer impact, reflected an inverse relationship with deer density. When deer density dropped, percent plots no impact climbed and vice-versa (Figure 34.4). Percent no impact was highest when deer density was lowest, and highest when deer density was lowest. The same trend was observed when deer density was graphed against impact levels on individual indicator seedling species.

34.4.2.1 Red Maple

Red maple is a highly preferred deer browse species, and prior to reductions in deer density after 2003, few plots had red maple seedlings, and on a measurable proportion of those plots, the impact level was heavy-severe (more than 50% of seedlings were browsed and hedged, with some being

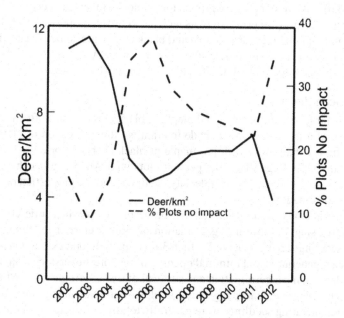

FIGURE 34.4 Comparison of deer density and percent plots no impact.

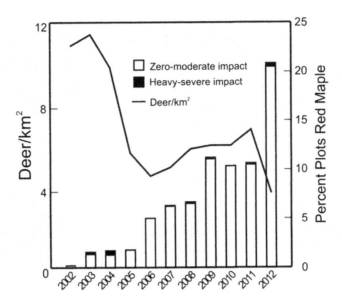

FIGURE 34.5 Comparison of deer density with impact level and abundance of red maple.

browsed to less than 15 cm height) (Figure 34.5). After the steep decline in to its lowest point in 2006, deer density climbed slightly, then leveled off after 2008.

In the same time interval, percent plots with red maple seedlings climbed, and the proportion of plots with heavy-severe impact dropped to almost zero. As with the coarse impact indicator (percent plots no impact), percent plots with red maple seedlings varied inversely with deer density—reductions in deer density associated with increased harvest of antlerless deer resulted in a significant increase in abundance and condition of red maple seedlings.

34.4.2.2 Black Cherry

Black cherry, like red maple, was observed on less than 5% of plots prior to the large decline in deer density (Figure 34.6). Like red maple, the proportion of plots with black cherry seedlings doubled as deer density plummeted, and proportion of plots with heavy-severe impact was much reduced as deer density declined and remained low. Red maple and black cherry are somewhat shade intolerant, and because timber harvest occurred on only a small portion of DDA areas, there was little stimulation for large increases in germination and development of these two species.

34.4.2.3 American Beech

American beech is far less preferred than red maple or black cherry—it is often referred to as being "deer-resistant"—and it is also far more shade tolerant, so the proportion of plots with American beech was much higher than those with red maple or black cherry prior to the large reductions in deer density (Figure 34.7). Regardless, the percent plots with American beech, as with red maple and black cherry, greatly increased as deer density declined, and the proportion of plots with heavy-severe impact also declined.

American beech is considered a "starvation" forage in winter when little else is available due to past deer overbrowsing and minimal germination of more preferred seedling species. But even American beech seedlings are present on less than 30% of plots, a joint expression of decades of deer suppression of tree regeneration and minimal opening of the forest overstory from timber harvest. It should be noted also that most germination and survival of tree seedlings prior to the large reduction in deer density occurred within fenced regeneration sites—without fencing following timber harvest, germinating and regenerating seedlings were generally totally eliminated by the overabundant deer herd on the DDA.

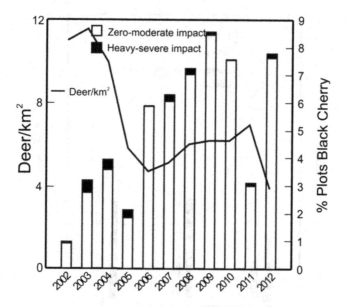

FIGURE 34.6   Comparison of deer density with impact level and abundance of black cherry.

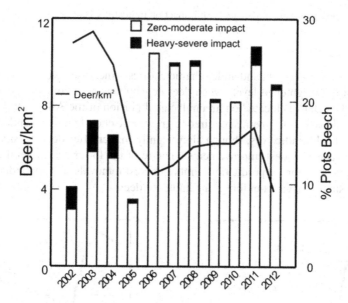

FIGURE 34.7   Comparison of deer density with impact level and abundance of American beech.

34.4.2.4 Wildflowers

Although data are only available for 3 years in the 2002–2012 span, the response of selected wildflower species to changes in deer density is similar to that of woody indicator species (Figure 34.8, from Royo et al. 2010). Prior to reduction in deer density, representative wildflowers were suppressed, likely from decades of overbrowsing. By the time deer density had been reduced to goal levels (2006–2011), impact levels on wildflowers had been significantly reduced and had remained at reduced levels for 4 years (one indicator, % cucumberroot flowering, continued to increase from 2006–2011). Thus, reduction to, and maintenance of, deer density to goal levels was associated with significant and lasting improvements in woody and herbaceous vegetation.

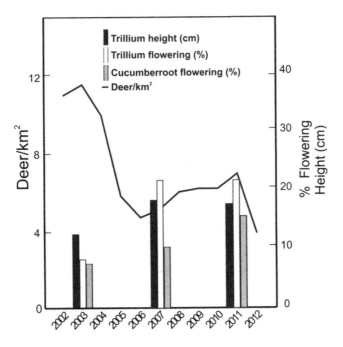

FIGURE 34.8   Comparison of deer density with impact on indicator wildflowers.

34.4.3   Deer Health

The indicators of health—weight and antler characteristics—increased when deer density declined, then stabilized at the heightened levels when deer density was maintained at goal levels (Figures 34.9 and 34.10). The DDA is located on an unglaciated portion of the Allegheny Plateau. Soils are old, and relatively depleted of minerals required to grow large deer (bones and antler characteristics). Increases in deer health indicators, though significantly higher after deer density was brought to goal than before, reflect modest improvements because of the lower soil mineral content and lack of significant increase in forage production: timber harvest removals were light during 2002–2012, resulting in only small increases in forage available for deer.

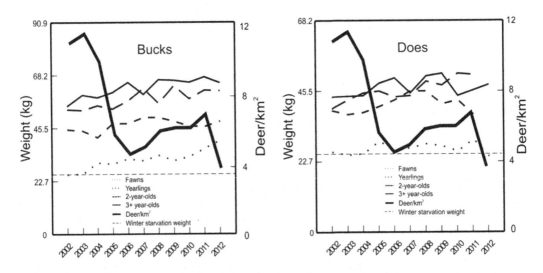

FIGURE 34.9   Weights of bucks and does brought to checking stations by age groups.

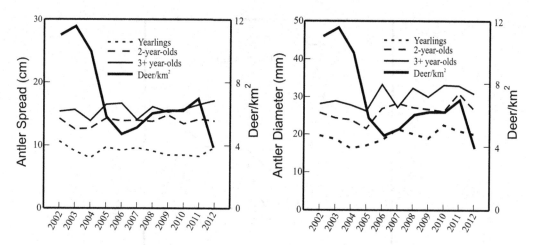

FIGURE 34.10   Antler characteristics of bucks brought to checking stations by age groups.

34.4.3.1   Body Weight

Body weights of bucks and does in four age groups (fawns, yearlings, 2½-year-olds, and 3+ year-olds) increased as deer density declined, then leveled off when deer density stabilized (Figure 34.9). Body weights of all classes for bucks and does were significantly higher after 2006 than in 2002–2003 when deer density was highest. The average increase in weights from 2002–2003 to 2004–2012 for female deer was 4.6%; for bucks, it was 9.4%. Fawns entering winter with field-dressed body weight of less than 25 kg are at risk of mortality during severe winters: body weights of buck and doe fawns were below that weight prior to reductions in deer density but rose above that and remained there once deer density was stabilized in the goal density range.

34.4.3.2   Antler Characteristics

Spread between the main antler beams and antler diameter increased slightly after deer density declined to low levels after 2004 and then remained fairly stable as deer density remained within the goal interval (Figure 34.10).

34.4.4   Recruitment and Sex and Age Ratios

Deer density/abundance is affected by fawn recruitment: the herd increases as the number of fawns surviving to winter increases and decreases as the number of fawns surviving to winter decreases. Because the DDA is such a large landscape, changes in deer density resulting from immigration and emigration of deer are much less likely than from changes in fawn recruitment. Fawn recruitment, and resulting changes in deer herd abundance, are affected by reproduction and fawn mortality.

Fawn mortality prior to winter is affected by predation. A study in Pennsylvania (Vreeland et al. 2004) indicated that nearly half the fawns born may be killed by predators, primarily coyotes and bears. Fawn predation is limited to the first few weeks of life—fawns are large and quick enough to elude predators by fall. Predators do not rely upon fawns as a primary food source, so their abundance is neither limited nor enhanced by fawn availability. As Figure 34.11 indicates, numbers of predators counted on roadside counts fluctuated but overall remained stable while deer density was halved from 2002 to 2006: the absolute number of fawns removed from the population by predation over this time span likely was stable but constituted a higher proportion of deer as overall deer density declined.

The impact of predators on fawn recruitment would thus be highest when deer density was lowest and the number of fawns removed by predators would constitute a higher proportion of the total number of deer in the herd.

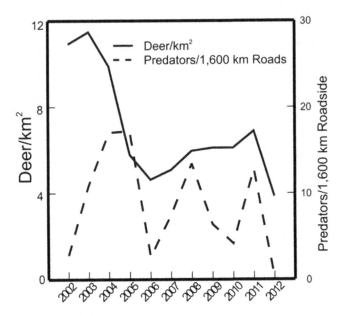

FIGURE 34.11   Fluctuations in predator counts and deer density.

Reproduction is affected by doe condition and breeding rate. The ratio of fawns birthed per doe increases as quality and quantity of forage available per doe increase and decreases when forage quality and quantity decline. Quality and quantity of forage per doe is inversely related to deer density. At high deer density, forage is limited and of lesser quality—does will not be in optimal breeding condition and the ratio of fawns birthed per doe decreases. Conversely, as deer density declines, quality and quantity of forage increase and the ratio of fawns birthed per doe will increase. As deer density declined on the DDA, forage quality and quantity should have increased, resulting in absolute increases in recruitment. Absolute recruitment did increase after 2007 (Figure 34.12) when deer density had been reduced to low levels for a number of years, allowing forage quality and quantity to improve. However, absolute recruitment after 2008 began to decline, while deer density remained low (4–6 deer/km^2).

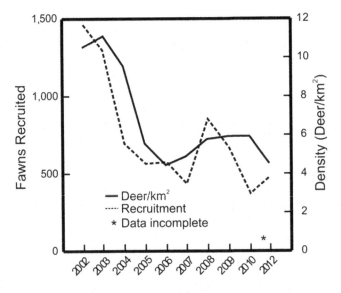

FIGURE 34.12   Relationship between absolute recruitment and deer density.

Another factor seeming to affect recruitment (absolute was well as rate) is ratio of does to breeding-age bucks (yearling and older). The recruitment rate in 2002–2012 was inversely proportional to the ratio of adult does to antlered bucks. When the ratio of does:bucks in fall was low and a higher proportion of does would have been bred, recruitment the following spring was high and vice versa (Figure 34.13). The implication for deer herd management is clear: if hunters are primarily hunting antlered deer and avoiding harvesting antlerless deer, as occurred on the DDA after 2007, the recruitment rate will be down, absolute numbers of fawns recruited will be down, and the deer herd will stabilize at lower density. This scenario (hunters avoiding harvesting antlerless deer when deer density is low) facilitates retaining deer density at a level compatible with forest sustainability.

The number of fawns recruited will be high when deer density is high, even if the doe:buck ratio is not optimal, because there are many does in a large deer population. Absolute recruitment was highest on the DDA when deer density and abundance were highest. As deer density declined because of hunting mortality in 2002–2006, absolute numbers of fawns recruited declined because there were fewer does. The doe:buck ratio was ≥4 during this same period, so reproduction was less than optimal, as likely not all does were bred. After 2006, the doe:buck ratio declined to <4 while deer density began to increase, and, as a result, absolute numbers of fawns recruited rose. By 2008, even though deer density had increased, the doe:buck ratio had again increased and so recruitment was low.

Thus, attaining and maintaining deer density at a goal level on the DDA has been a combined function of high initial harvest of antlerless deer fostered by the concurrent antlerless and antlered seasons, DMAP and Unit 2 permit programs, sustained predation on fawns, and hunter preference for harvesting antlered deer with resultant low recruitment. Deer density on the DDA stabilized in 2007–2012 at the goal level, and can likely be maintained by the continuing and consistent pressure by predators, concurrent antlered and antlerless seasons, availability of DMAP and Unit 2F permits, and removal of ~600 deer per year by hunting to offset the approximately 600 fawns recruited annually. The number of Unit 2F permits issued by the PGC had been monitored annually, and declines or increases were countered by offsetting increases or decreases in DMAP permits, as determined by data inspection and trial and error.

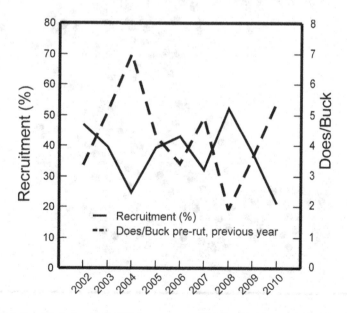

FIGURE 34.13 Relationship between recruitment rate and doe:buck ratio previous fall.

34.4.5 LOCATIONS OF HARVESTED DEER

Locations of deer harvested on the DDA and brought to checking stations were recorded annually. Locations aggregated in 2002–2012 (Figure 34.14) mimicked harvest locations on a yearly basis and demonstrated the inequality of harvest: many more deer were harvested per unit area for ANF portions of the DDA and far fewer deer were harvested per unit area on private properties within the DDA (Figure 34.15) to reflect change in Figure 34.14.

This disparity in harvest among the DDA ownerships was reflected in disparity of deer density following implementation of practices designed to reduce deer density (Figure 34.16). Initial deer density was similar among ownerships (excepting CP, which was higher) and declined similarly with advent of DMAP and Unit 2F permits. However, when DMAP and Unit 2F permits declined after 2006 and hunters increasingly used the permits to harvest antlerless deer on ANF portions of the DDA, density crept back up on CP, BWA, and FIA lands while remaining at goal on ANF lands

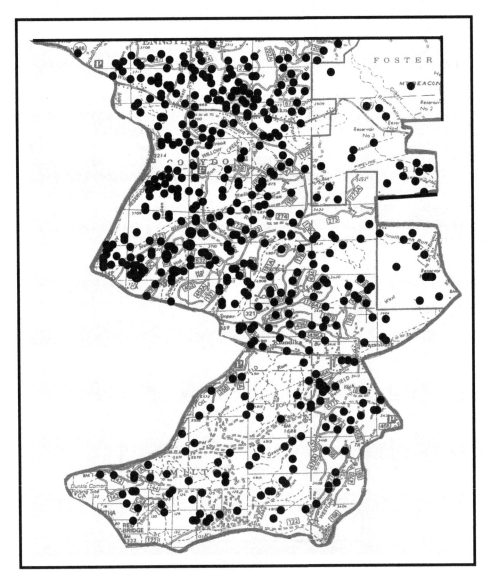

FIGURE 34.14 Loci of deer harvested on/adjacent to DDA 2002–2012.

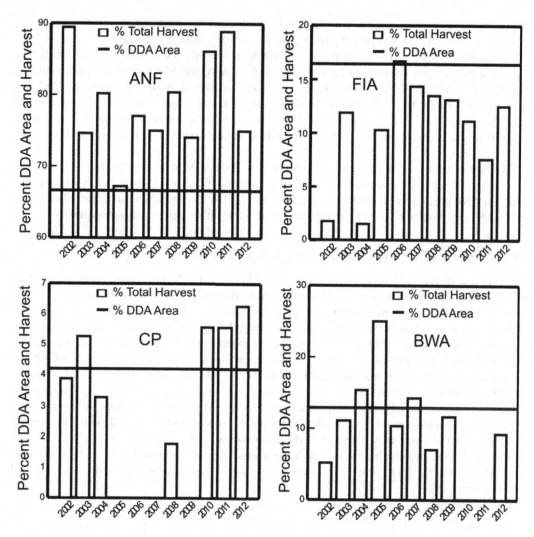

FIGURE 34.15 Comparison of proportion of harvest and DDA lands by ownership.

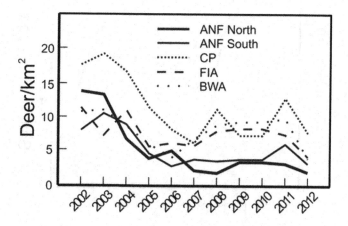

FIGURE 34.16 Comparison of deer density among DDA lands by ownership.

(Figure 34.16). The drop in deer density 2012 on FIA, CP, and BWA lands reflected the large increase in UNIT 2F permits in 2010–2011 (up from 22,148 in 2010 to 34,000 in 2011).

As an adaptive management strategy, the KQDC leadership team attempted to redistribute harvest such that more deer would be harvested on CP, BWA, and FIA lands by creating separate DMAP units in 2012. That portion of the ANF north of state highway 59 was designated as a separate DMAP unit (Unit 1986), and CP and BWA lands were combined to form a separate DMAP unit (Unit 1996). FIA lands were surrounded by ANF lands on the southern portion of the DDA, but the difficulty of creating a separate DMAP unit for FIA lands resulted in retention of the original unit (Unit 135).

The higher density of harvest loci on ANF portions of the DDA reflect differences in access and familiarity: the highest concentrations of harvested deer are close to well-identified and established roads on ANF portions of the DDA and much lower on FIA, CP, and BWA lands. Results over the following years were to be used to determine how to adjust this adaptive management step.

34.4.6 Hunter Education, Participation, and Satisfaction

One-day deer density and impact workshops were held on the DDA every spring from the project's inception until 2010 when insufficient demand for the workshop precluded its continuance. Workshop attendance exceeded 20 participants every year and hunters made up a majority of the audience. Hunters indicated in postworkshop evaluations that they better understood the relationship between deer density and impact and were more supportive of management based on balancing deer density with sustainable management of other forest resources. Hunters made up the majority of persons conducting the annual deer roadside counts on the DDA and continued to bring harvested deer to checking stations during hunting season.

To obtain an additional source of feedback, hunters bringing deer to checking stations in 2007–2011 (data for 2012 incomplete) were asked to fill out surveys on a scale of 1–10, with 1 being most dissatisfied and 10 being most satisfied. Satisfaction scores ranged between 5.8 and 7.5 during ups and downs in deer density, indicating that hunters generally were satisfied with the hunt and deer herd (Figure 34.17). Hunter satisfaction scores increased when calculated deer density increased and vice versa.

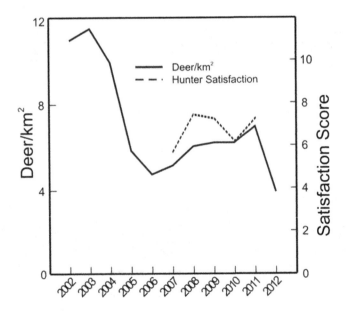

FIGURE 34.17 Hunter satisfaction scores 2007–2012.

Similarly, hunter satisfaction scores between younger hunters (<50 years old) and older hunters (>49 years old) were higher when they saw more deer hunting and lower when they saw fewer deer. Satisfaction scores were low in 2010 when deer density was actually higher than the preceding 6 years, but opening day conditions were difficult, with rain and fog all day and hunters seeing few deer. Almost all hunters said there were too few deer, but more than 90% said they would continue to hunt the DDA.

34.4.7 SELF-SELECTION OF ALPHA AND BETA HUNTERS

When deer density plummeted after 2004, many (assumed omega) hunters came to checking stations to vent their frustration at seeing very few or no deer—checking station volunteers received many complaints from angry (and unsuccessful) hunters who were sometimes verbally abusive. As time went on, and deer density remained low and at goal, fewer and fewer angry hunters showed up at checking stations to express their dissatisfaction. Hunters brought fewer deer to checking stations, and those with trophy deer said it was worth seeing fewer deer to bag a trophy. The average age of hunters bringing deer to checking stations declined consistently. Satisfaction scores of hunters bringing deer to checking stations remained just under 7 (6.9 on a scale of 10 being most satisfied and 1 being least satisfied) in 2007–2011, by which time deer density had been at goal for 7 years. Younger hunters (<50 years old and presumably more knowledgeable regarding deer management) averaged higher satisfaction scores than older hunters (>49 years old and presumably more inclined to be informed on deer management by their peers, including omega hunters).

We hypothesize that after deer density declined and remained low, many omega hunters stopped hunting on the DDA. The returning hunters were more heavily weighted to alpha and beta types, who were more successful in harvesting deer, and even though fewer deer were harvested, deer density was so low that the smaller number of assumedly more successful hunters was able to harvest enough deer to offset recruitment and keep deer density at goal.

34.4.8 ECONOMICS

Property owners on the DDA did not include cost–benefit calculations into their evaluations of the effectiveness of the KQDC program. Rather, they noted that they were able to discontinue fencing to protect regeneration sites from deer browsing and also noted the return of advance regeneration seedlings and increased sightings of formerly rare herbaceous plants such as cucumberroot, trillium, and wild lily-of-the-valley.

However, approximate costs and benefits of the KQDC program can be calculated. Annual monitoring costs, involving salaries of employees collecting density and impact, roadside, and checking station data, and fuel expenses traveling to and from roadside counts, data collection sites and checking stations averaged $0.33/ha ($0.80/acre). Benefits include elimination of annual costs of building, maintaining, and taking down fencing to protect regeneration sites from deer browsing, and costs of herbicide application to reduce prevalence of interfering plants (ferns and undesirable woody vegetation) caused by deer browsing, which promoted growth of interfering species in the absence of desired species (seedlings, shrubs, wild flowers) deer removed by browsing. Elimination of annual fencing costs across the DDA would produce a savings of ~$9.50/ha (Stout et al. 2013), for a cost:benefit ratio of approximately 1:30. Elimination of need and cost of herbicide application (estimated at ~$80/ha treated, generally in excess of 500 ha/year) would increase the ratio further. With numbers like these, it is easy to justify the costs of annual monitoring of deer impact and deer and habitat health.

EDITOR COMMENTS

The KQDC was successful in reducing deer density to the target level and maintaining it, with subsequent reductions in deer impact and improvements in deer and habitat health. The demonstration project used all the components of deer–forest management for sustainable natural resource use as espoused in this book and framed by adaptive management. It is impossible to separate out the individual importance of the adaptive management components of monitoring; goal setting and checking; communication; incorporation of values, cultures, traditions, and science; hunter access; and working with natural resource agencies and other stakeholders, but it can be said that the combined whole worked to solve an important biological, social, and economic problem in Pennsylvania that can serve as a model for deer management. Perhaps the most important message is that while the KQDC program worked, it must be continued in the future and considered an ongoing and perpetual cost/component of forest management.

David S. deCalesta, editor

REFERENCES

deCalesta, D. S. 2012. Collaboration among scientists, managers, landowners, and hunters: the Kinzua Quality Deer Cooperative. In J. P. Sands, S. J. DeMaso, M. T. Schnupp et al. eds. *Wildlife Science: connecting research with management.* Boca Raton FL: CRC Press.

deCalesta, D. S. 2013. Reliability and precision of pellet-group counts for estimating landscape level deer density. *Human-Wildlife Interactions* 7:60–68.

deCalesta, D. S. and S. L. Stout. 1997. Relative deer density and sustainability: a conceptual framework for integrating deer management with ecosystem management. *Wildlife Society Bulletin* 25:252–258.

Kosak, J. 1995. *The Pennsylvania Game Commission 1895–1995: 100 years of wildlife conservation.* Harrisburg PA: The Pennsylvania Game Commission.

Pierson, T. G. and D. S. deCalesta. 2015. Methodology for estimating deer impact on forest resources. *Human-Wildlife Interactions* 9:67–77.

Royo, A. A., S. L. Stout, D. S. deCalesta et al. 2010. Restoring forest herb communities through landscape-level deer herd reductions: Is recovery limited by legacy effects? *Biological Conservation* 143:2425–2434.

Stout, S. L., A. A. Royo, D. S. deCalesta et al. 2013. The Kinzua Quality Deer Cooperative: can adaptive management and local stakeholder engagement sustain reduced impact of ungulate browsers in forest systems? *Boreal Environment Research* 18:50–64.

35 Ward Pound Ridge Reservation Park
Convincing the Decision Makers

Daniel Aitchison

CONTENTS

> Information is a source of learning. When it is organized, processed, and available to the right
> people in a format for decision making, it's a benefit rather than a burden.
>
> **William Pollard (2011)**

35.1 INTRODUCTION AND HISTORY

Ward Pound Ridge Reservation (WPRR) is a forested public park, owned and operated by the Westchester County Department of Parks, Recreation, and Conservation. WPRR was acquired and created by the Westchester County Parks Commission in 1925 for camping and outdoor activities, similar to New York State Park areas in the Catskills and the Adirondacks, and was to be held as a great forest preserve (Herr and Kohl 2013, Pessoni 1995).

WPRR is located in the towns of Pound Ridge and Lewisboro, New York, in the northeastern corner of Westchester County, approximately 4 km west of the Connecticut border and 43 km northeast of New York City. WPRR is mostly bordered by town roads. Opposite the roads or directly abutting WPRR are forested private properties and estates, homes, horse farms, fallow fields, and swampland. WPRR is composed of 1508 ha of forested land and 111 ha of nonforested land, with 271 ha as riparian zone (Hubbard 2008). The "nonforested" land is mostly made up of the maintained meadows and fields that surround the main interior roads of WPRR. The landscape is composed of rolling hills, rocky ridges with rugged talus slopes, ravines, and the valleys and floodplains of small rivers, a result of the region's glacial history. WPPR water resources are part of the New York City watershed. The Cross River bisects the park, flowing east to west, and empties into the Cross River Reservoir. Feeding into the Cross River is the Waccabuc River, which enters the park in the northeastern corner, and a number of small streams originating from springs, ponds, swamps, marshes, and vernal pools. Due to the park's history, topography, hydrology, and varied soil compositions, there is a broad variety of habitats: sandy glacial moraines, stands of mixed deciduous hardwoods, spruce and pine groves, hemlock ravines, fields, meadows and edge habitat, expansive swamps and marshes, stony ridges, and steep cliffs. A forest management plan identified 26 different forest cover types based on tree species age and composition (Hubbard 2008). Predominant tree species include white pine (*Pinus strobus*), yellow birch (*Betula alleghaniensis*), black birch (*Betula lenta*), yellow poplar (*Liriodendron tulipifera*), eastern hemlock (*Tsuga canadensis*), white ash (*Fraxinus americana*), sugar maple (*Acer saccharum*), red maple (*Acer rubrum*), black cherry (*Prunus serotina*), red spruce (*Picea rubens*), Norway spruce (*Picea abies*), red oak (*Quercus rubra*), black oak (*Quercus velutina*), white oak (*Quercus alba*), and chestnut oak (*Quercus prinus*).

WPRR has a rich history of human occupation and use. The first hunter-gatherer Native peoples used stone shelters on the property about 10,000 years ago during the Paleolithic era. Later Native American groups used the forested land within and around WPRR for hunting, fishing, and farming small plots of land (Herr and Kohl 2013). Colonial farmers began working the land in the late 1600s to early 1700s, tilling the land for farming and pastures. Because of the steep hills and rocky landscape, many areas were not suitable for agriculture or livestock and remained forested, with bark and wood being harvested for tanning operations and the charcoal industry (Herr and Kohl 2013).

By the time Westchester County Parks Commission started land acquisition to create WPRR in the 1920s, some of the local agricultural and forest-based endeavors had been abandoned and once-cleared landscapes were reverting to forest. From 1933 to 1940, under Roosevelt's Emergency Conservation Work Act, a Civilian Conservation Corps (CCC) camp was established within WPRR. The CCC assisted with construction of roads and structures within the park as well as undertaking a massive reforestation effort to remediate the extensive damage caused to the forest by Dutch elm disease and the chestnut blight. After the diseased trees were cut and burned or used for construction of camping shelters, over 500,000 white and red pines (*Pinus resinosa*), hemlocks, and Norway and white spruces (*Picea glauca*) were planted to reforest areas of the park.

When WPRR was first established, deer hunting was not permitted. However, in 1925 "temporary rules" did allow for the "lawful taking of upland game ... by citizens of this county under individual permit ... hunting of foxes and other vermin may be done by special order to be obtained." Eventually, the laws of Westchester stated that: "No person, except under a lawful permit from the commissioner, shall molest, kill, wound, trap, hunt, take, chase, shoot or throw missiles at, remove, or have in his possession any animal, reptile, bird, bird's nest, or squirrel's nest, or remove the young of any such animal or the eggs or young of any fish, reptile or bird."

The white-tailed deer (*Odocoileus virginianus*) was extirpated from Westchester County and much of the surrounding area by the late 1800s. This was mostly due to overhunting and loss of habitat from agricultural and timbering practices in the region. The last deer was said to have been shot near Sing Sing in 1861 (Severinghaus and Brown 1956). However, a remnant population existing in the Berkshires of Massachusetts grew in numbers and eventually repopulated areas devoid of deer. To encourage and protect the growth of these new populations, the New York Conservation Commission prohibited deer hunting in Westchester and surrounding counties in the early 1900s and did not permit hunting of deer again until the 1940s. By then, the high human population density in Westchester County and complaints of deer damage resulted in the state allowing restricted deer hunting to bow hunting only for safety reasons. To enhance herd growth, the State originally issued permits for bucks only until 1952, after which either sex could be harvested. As declining farmlands (and forage resources for deer) were replaced by dense, young forestland and suburban development, deer foraging shifted to the increasing abundance of ornamental plantings and maintained lawns. Deer hunting failed to stem the increasing deer density and subsequent impact on suburban landscaping

FIGURE 35.1 Impact of high deer density on landscaping on private property, Westchester County. (Photo by Dan Aitchison.)

and forested understories. Large predators had been eliminated from the region hundreds of years before. Improving deer habitat, abundant food resources, and refugia from hunting provided by the growing suburban landscape, coupled with limited predation on deer, resulted in an explosion of the deer population.

As deer populations grew, Westchester County residents became concerned with increasing landscape damage (Figure 35.1), deer/vehicle collisions, and the perception that deer played a role in the spread of Lyme disease. At the request of the Westchester County Board of Legislators, a White-Tailed Deer Study Committee was commissioned in 1988 and reported on in 1991.

The report noted that "management of the deer situation in Westchester is totally the province of the New York State Department of Environmental Conservation (DEC). The County can recommend, the County can assist, and the County can support the State's management programs, but the County cannot act directly." It also stated that, "It should remain County policy that the extensive Westchester County park system be considered as nature preserves, and that no hunting be permitted within them. By providing such havens as a matter of policy, the County and City guarantee that white-tailed deer will not be extirpated again." The report did recommend that, "several different methods and policies by which the size of the deer herd might be reduced ... the most effective way of achieving a large scale reduction would be through hunting, both by recreational bow hunters and by highly proficient marksmen with firearms." But by passing the responsibility for deer management onto the State, the Westchester County Department of Parks, Recreation, and Conservation resolved to do nothing about deer populations in County Parks and County-owned properties.

While the White-Tailed Deer Report centered on human-related issues, naturalists and curators in the Westchester County Park system were expressing concern over the devastation to the forest understory (Figure 35.2). They predicted future problems as the forest understory disappeared and invasive plant species started spreading in parks. Besides the visible and tame herds of deer wandering the open spaces in WPRR, a major winter deer kill in the early 1990s provided stark evidence of the overpopulated deer herd and dwindling food resources.

The senior curator of County Parks wrote of deer in an unpublished article: "Their effect on park biodiversity has been devastating. Fields that once abounded in wildflowers are now barren except for the grasses, thistles, and milkweeds that deer find unpalatable. The herbaceous layer and shrub growth of Westchester's forests is rapidly disappearing. Ground-nesting birds that depend on undergrowth for cover, such as Ovenbirds, Worm-eating Warblers, and Black-and-white Warblers are declining. Ruffed Grouse, which were once abundant in the county, are now virtually extirpated. Tree seedlings and saplings, necessary for forest regeneration, are almost completely absent from many of our park's forests ... That deer are overabundant seems obvious to naturalists and park managers ... Already we're seeing many of the deer-sensitive species disappearing."

FIGURE 35.2   Depleted forest understory on WPRR prior to deer management. (Photo by Dan Aitchison.)

On November 18, 2005, Westchester County's *Conservation Café* hosted a "Conversations on Conservation" event titled "White-Tails in Westchester County." The forum presented a panel of experts for managing the situation. The conference organizers wrote a letter to Westchester County urging a comprehensive review of deer impacts on Westchester's forests and biodiversity.

In February 2006, Westchester County appointed a Citizen Task Force "to examine the problem of deer overpopulation" The primary mission of this task force was "to examine the impacts of white-tailed deer on forests, and to make recommendations for improved deer management for forest regeneration and native biodiversity." The task force included officials from Westchester County Government, representatives from the New York City Department of Environmental Protection (DEP) and New York State Department of Environmental Conservation (DEC), Department of State, Humane Society of the United States, local municipal governments, local environmental and community organizations, and private citizens.

After 2 years of reviewing deer management options, visits, communications, meetings, presenters, scientific papers, and applicable case studies, the task force published an 80+ page report concluding:

- "The lack of forest regeneration, the severe impact on biodiversity, the threat to water quality, and other detrimental ecological impacts, call for immediate action. The Task Force sets forth urgent recommendations in four areas: Deer management and monitoring, public education, legislation and funding, and the establishment of a public-private partnership for an adaptive deer management program."
- "A Spring 2008 deer pellet group count and browse impact survey … led by deer consultants from Pennsylvania and involving 36 state and local foresters in Ward Pound Ridge Reservation … indicated an average density of 24.6 deer/km^2, and no forest regeneration in 91.5% of the plots studied. The report concluded that, 'The deer density was the highest observed by either presenter, anywhere, on continuously forested areas throughout New York, Pennsylvania, Maryland and Vermont, in over 10 years of such work.' The report in part concluded: 'The deer herd within WPRR must be brought down to ecologically viable levels; on the WPRR this density is in the 2–4 deer/km^2 range … The ecological cost of not achieving deer herd reduction is collapse of the ecosystem.'"
- "We recommend specifically that the County law be amended to allow hunting in a minimum of three County Parks, including Ward Pound Ridge Reservation, Muscoot Farm, Lasdon Arboretum, Mountain Lakes Park, and Blue Mountain Park."

Separately from the task force report, the Watershed Agricultural Council prepared a Watershed Forest Management Plan after studying and mapping the landscape of the park. The report listed control and management of the deer population as a top concern and one of the first actions to be undertaken before attempting other forestry measures.

35.2 PILOT ADAPTIVE DEER MANAGEMENT PROGRAM

Following the recommendations of the Citizens' Task Force, Westchester County Parks commenced meetings in January 2009. An advisory committee composed of the Executive Director of Mianus River Gorge; a retired DEC biologist; the Forest Coordinator for the DEP; the Executive Director of Watershed Protection and Partnership Council, New York State Department of State; Westchester County law enforcement personnel; and Westchester County Parks Board and Westchester County Parks staff met over an 8-month period to determine the best design and implementation of a deer management program in Westchester County Parks. Coordinators for existing regional management programs from Mianus River Gorge Preserve, Town of Pound Ridge, Rockefeller State Park, and the town of Ridgefield Connecticut were consulted, and a pilot volunteer-based bow hunting program was developed as the most efficient, least intrusive, safest, and most cost effective. Primary emphasis was on safety of the surrounding public and volunteer hunters. Existing Westchester County laws permitted hunting.

Initially, the idea of using county law enforcement personnel to shoot deer over bait with rifles at WPRR was entertained, but later dismissed due to concerns over noise, safety, logistical planning, carcasses that would have to be handled and transported, and anticipated public disapproval. WPRR was slated to be the first park hunted, but county officials determined to first test and refine bow hunting in 2009 in a pilot program on smaller individual parcels (Muscoot Farm, Lasdon Park, and the Arboretum) not included within WPRR boundaries—areas little used by the general public during the regular deer hunting season. Fifty bow hunters and 15 alternates were chosen by lottery and screened for bow hunting proficiency, safety awareness, and deer management philosophy.

For admittance into the pilot program, hunters were required to be Westchester County residents, have a current New York State hunting license, and submit an application. Applications were made available through the Westchester County website. A lottery was drawn among applicants and those selected were required to pass a proficiency test requiring them to hit a 23-cm-diameter target at 23 m with three of three arrows. Applicants passing the test were required to attend a mandatory orientation covering the ecology of County Parks' forests; necessity of managing deer for forest regeneration; hunter safety; and the rules, regulations, and requirements of the program. At the orientation, applicant IDs and state hunting licenses were checked, and waivers of liability were signed and submitted. Packets with maps of the properties, copies of hunting rules and regulations, parking placards to identify vehicles, and reflective tree stickers to identify tree stands were given to the hunters. A unique "hunter number" was assigned to each hunter. After the orientation, participants were provided with Westchester County "hunter IDs" and paid an administrative fee of $25.

To hunt the properties, hunters were required to complete a series of tasks with each hunt. Upon arriving at a site, they parked in designated and posted parking areas. Some parking areas were in areas not accessible by public vehicles and had gates with combination locks. After parking, each hunter was required to "call in" to a voicemail box set aside for the deer management program. The hunter would record the date, their name, hunter number, and park that they were hunting for the day. At the end of the day, each hunter was required to "call out," providing the same information as the morning, but were also asked to make note of any harvested or unrecovered deer. The voicemail was checked throughout the day by staff. The call in/out system allowed managers to know how many hunters were in what parks at what times to relate hunter presence in parks on a real-time basis; to confirm Pilot Adaptive Deer Management Program (PADMP) hunter presence in an area; to investigate incidences of poaching by outside hunters reported by PADMP hunters; to know that hunters had left the parks safely at nightfall; and to cross-reference hunter entries of parks, harvests, and days and times hunted.

Kiosks at the parking areas contained maps of the properties with marked deer management units (DMUs) and safety zones. DMUs were established based on terrain, accessibility, distance from private property, and definable landmarks. Each DMU was assigned a code with letters and numbers specific to each park and were used by staff to specify individual hunting units for identifying hunter use and for collecting harvest data. DMUs facilitated communication between staff and hunters through referencing of DMUs. We established safety zones prohibiting hunting after the New York State mandate for 150-m buffers surrounding inhabited dwellings, and we placed them around areas of high public usage such as main site complexes, picnic areas, parking lots, and playgrounds. Hunters were not allowed to hunt within or shoot into safety zone areas. Safety zones were specified on maps and posted with signs in the woods.

Each kiosk contained push pins, each labeled with the numbers assigned to individual hunters. When entering the woods in the morning, hunters placed their pins in the location where they would be hunting on the map. Availability of hunting locations was on a first-come, first-served basis. Initially each DMU was labeled with a maximum number of hunters that it could accommodate, but we established that hunters self-regulated the locations themselves and the limits were removed. The push pin and map system allowed hunters to: (1) know other hunters' locations and to avoid transitioning through or hunting the same areas, (2) use known movements and locations of other hunters as a hunting strategy to predict the movements of pressured deer, (3) distinguish over- or underused areas, (4) note successful areas of harvest and deer sightings, and (5) resolve conflicts with each other through transparency.

The system also allowed WPRR managers to easily locate and respond to hunters in the case of an emergency, to respond to reports of poachers, and to assist with retrieval and tracking of deer.

Each kiosk contained paper "hunter logs," which hunters were to fill out upon the completion of a hunt. Required information included hunter name, date, DMU location, start and end time in stand, sightings of deer by sex, adult/fawn, antler points, harvests, woundings, and any unique wildlife sightings. These data were cross-referenced with call in/out reports and with written harvest records and entered into spreadsheets for analysis.

For safety reasons, hunters could hunt only from elevated tree stands. Only climbing stands were permitted, and they had to be removed by the end of each day. The elevated stand position meant that a miss or pass through would embed in the ground. It also offered better visibility of the surrounding landscape and targets. These stands provided hunters good visibility of the surrounding landscape and targets, and the flexibility to hunt in different areas of the properties. This flexibility enabled hunters to adjust to seasonal conditions and adapt hunting strategy. The first-come first-served structure of the program also meant that hunters were not guaranteed the same location to hunt each day. In 2010, WPRR managers purchased and emplaced four ladder stands for use in areas that were not conducive to the use of climbing stands and to entice hunters to hike farther into the properties. However, by the 2015 season, these stands were removed because deer harvests did not increase and the stands were hazardous for older hunters; caused disputes among hunters; required a significant amount of work and time to transport, set up, and maintain; and served as a visual reminder to the public that hunting was taking place.

While hunters usually tracked, retrieved, dressed, and transported their own harvested deer, WPRR managers assisted hunters in finding deer they had wounded but were unable to find. It was important that harvested deer be recovered to avoid unwanted public relations associated with unrecovered carcasses, and to promote an ethical standard among program hunters. The WPRR provided deer carts to hunters for retrieval of harvested deer. Managers used county-owned vehicles to drive into areas prohibiting vehicular access so they could assist hunters in retrieval of harvested deer from less accessible parts of the park. Managers would assist in tracking wounded deer to recover animals. The managers' assistance and intimate knowledge of the park often led to the recovery of harvested deer when hunters' efforts fell short. The managers used Deer Search, a volunteer-based dog tracking service certified by the DEC, to help hunters find deer they had wounded but could not find.

Transport of deer was the responsibility of the hunter, but managers would sometimes assist with venison donation. Managers had a good working relationship with a local deer butcher who accepted deer for the Hunters for the Hungry program, a national hunger relief organization. Managers often transported harvested deer not wanted by hunters to this butcher. Deer not wanted by hunters and not acceptable for human consumption were donated to the local Wolf Conservation Center. The multiple options for hunters to donate harvested deer once they "filled their freezer" encouraged them to harvest additional deer, lending to higher harvests for the program.

The PADMP applied for and received Deer Management Assistance Permits (DMAPs) from the DEC. DMAPs are for antlerless deer only and were given to the PADMP managers to administer on-site to hunters. The availability of the extra antlerless permits increased harvest of antlerless deer, especially as there was no time lag between hunters applying for and obtaining the permits.

Deer were harvested and brought to check stations where managers recorded harvest data and confirmed hunter-reported call-in harvests. Managers weighed deer at check stations, removed jawbones and aged deer, noted sex of harvested deer, and recorded beam diameters and number of antler points for bucks. Data collection mirrored methods used by the DEC for compatibility with DEC-collected data. These data focused on deer quality and provided managers with a visible trend of declining age structure and improving deer condition while the herd was being reduced. Harvest sex ratios indicated that hunters were harvesting more antlerless than antlered deer, attesting to their dedication and alignment with the goal of the program to reduce deer density and impact.

To select for and retain dedicated, successful hunters to harvest antlerless deer, we initially required participants to hunt a minimum of 5 days annually in the program and adhere to an "earn-a-buck" rule stipulating that they harvest an antlerless deer before harvesting an antlered deer.

In 2009, 45 deer were harvested without incident, and public response was mostly favorable. The pilot program was deemed successful and expanded to other parks the following season. Planning to open the entire WPRR to deer hunting in 2010 began.

In 2010, the Westchester County Adaptive Deer Management Program (ADMP) added the properties of WPRR and Sal J. Prezioso Mountain Lakes Park to the pilot program properties. One hundred and eight deer were harvested from WPRR the first year, representing an approximate 25% reduction in the 2008 estimated herd density.

35.3 MONITORING

As recommended by the task force, monitoring was and continues to be an integral part of the ADMP. Initial emphasis was on strengthening the structure and methods of the hunting program for reducing deer density as quickly and efficiently as possible. And, WPRR biologists realized it would be years before significant and positive vegetative responses would be measurable in the forest understory. Therefore, although monitoring vegetative response of the forest was a priority, initial monitoring centered on deer density and harvest. After the program matured, vegetation monitoring became the main focus and gauge of the effectiveness of the ADMP. Ultimately, changes in browse impact and regeneration will be used to direct management of deer density, including bow hunting and/or use of additional methods for population/impact reduction.

Monitoring of deer density was begun at WPRR in 2008 using deer pellet group counts (deCalesta 2013) on 1.28-m-radius plots at 30-m intervals; browsing impact (Pierson and deCalesta 2015) was monitored on every other plot on a 2.59-km^2 block on the western side of WPRR. Density and impact data were collected on a yearly basis, situation permitting. In 2014, three more 2.59-km^2 blocks were added to include the northern, southern, and eastern portions of the park and provide a more representative evaluation of the effects of management in WPRR.

Before the 2013 hunting season, through the direction of and with the assistance of Mianus River Gorge, trail cameras were installed in WPRR to estimate deer populations following Jacobson et al.'s (1997) methods. The ADMP continues to investigate new options for forest monitoring to combine with those currently in use. While adding new methods for evaluating deer impact now will not be able to use a prehunted WPRR as a baseline, ADMP managers feel that the slow progress of the bow hunting program and potential to institute new deer management methods as a supplement to the program within the next few years make use of emerging methodology worth the effort.

Initially we recorded field-dressed weight, age (estimated by tooth wear and replacement methodology), sex, and number of points per antler and antler beam diameter. We abandoned weight measurement (as a surrogate for health) in 2011 when we decided that improving herd health was not a focus of the program. The age of harvested deer declined over time; such reductions in a hunted population indicate that harvest intensity is working to reduce deer density as fewer deer survive hunting to reach older ages.

We use age structure of harvested deer to corroborate estimates of number of deer harvested and deer density reduction. We use sex ratio of harvested deer to indicate that hunters have bought into the overall mission of the program (reduce deer density and impact by increasing antlerless harvest) and that the majority of harvest continues to be antlerless deer. Favorable public response to this information indicates that the public does perceive that ADMP is geared toward reduction of deer density and impact rather than to trophy hunting. An additional value of collecting harvest data at check stations is that it increases staff interaction with hunters in the field. Encounters of ADMP staffers with nonhunting park users occurred occasionally at check stations. The users usually took an interest in check station activity, especially estimation of deer age, and staff used such opportunities to educate the users about the ADMP program.

35.4 STAKEHOLDER EDUCATION/INVOLVEMENT

Educating/involving stakeholders was associated with their acceptance of the ADMP. Major stakeholder groups were volunteer hunters, Westchester County commissioners, Westchester County Parks administrators, New York State Department of Environmental Protection administrators, homeowners with property abutting WPRR, animal protection groups, and the general public. The Senior Curator of Wildlife educated stakeholder groups with organized presentations and speeches, including speaking at municipal board meetings, and through telephone, e-mail, and personal conversations. While animal advocacy groups do not accept hunting to manage deer populations and associated impacts, the information the ADMP and others provided has been associated with members of those groups agreeing that deer do have a negative impact on the ecosystem. For recreational users of WPRR and residents bordering the property, personal conversations outlining the checks, balances, and safety measures of the program; offers to provide on-site tours to exhibit different aspects of the program; and follow-up calls and conversations by staff have seemed to reduce their concerns about the bow hunting program.

The personnel from the AMDP described the program and presented deer density and impact data to hunters at mandatory orientation meetings, through e-mail updates, and through personal conversations with hunters to help them understand their role as deer managers. Furthermore, our informed hunters described the mission of the ADMP, results, and their role in the program to members of the public. This interaction has helped gain public acceptance of hunting on WPRR (formerly taboo in Westchester County).

Westchester County administrators did feel the need to hold public meetings to gather public input prior to instituting the ADMP on WPRR. The ADMP program was incorporated easily into the Westchester County Parks system because of the recommendations of the 2008 task force, the support of the County Executive's office, and the fact that the WPRR was part of an existing governmental agency. Other parks within the system have had varying degrees of success in starting deer management programs: in 2010, the village of Croton-on-Hudson was criticized for assembling a deer management program without seeking input from stakeholder groups, and animal advocacy groups used this criticism to squash the program. In 2015, the City of Rye held public meetings to initiate conversations on deer management in the city and found an overwhelming amount of support from residents whilst seeking input from groups opposed to hunting to find a palatable, practical, and productive solution. In 2014, due to public opinion, Hastings-on-Hudson shied away from a plan to net and euthanize deer and is instead experimenting with use of contraceptives delivered by dart gun.

35.5   PARTNERSHIPS

Partnering with other organizations and agencies provided the ADMP with support for its programs, assistance, guidance, sharing of materials and data, and further research opportunities. While the DMPs in Westchester County are unique to their own landscapes and stakeholder interests, DMP administrators worked together to learn from each other for pursuing individual goals. Shared research has refined methods and eventually will result in comparable data from different sites in the county. Experimentation with different methods of hunter management has been shared among DMPs. Partnerships with other landholders who allow DMP access, such as the watersheds of the New York City Department of Environmental Protection or Aquarion, have led to the reduction of deer refugia surrounding DMPs. Granting access between DMPs enhanced the missions of all programs. For example, in 2011, WPRR allowed the Town of Pound Ridge DMP to hunt 152 acres in the southern portion of WPRR. ADMP hunters had had trouble accessing this part of the property, but the Town program used an established hunting parcel that abutted it for improved hunter access.

35.6   RECRUITING AND RETAINING QUALITY HUNTERS

One of the greatest challenges to the WPRR bow hunting program has been attracting, screening, and retaining quality hunters. While hunters supported the mission of the program, they were still unpaid volunteers with their own set of values; for example, some refuse to harvest yearling or fawn deer. The difference in values between hunters and managers was a challenge in achieving the desired harvest.

35.7   PROGRAM COSTS

It is difficult to quantify the costs for the ADMP, as no budget line has ever been established. Material costs are minimal because little is needed to run the program. There were some initial purchases made for use in all ADMP parks, such as deer carts, ladder stands, and a handful of small tools. Since that time, required equipment and materials often were purchased for use throughout the year for additional wildlife management programs and projects in County Parks. Some costs such as lumber have been absorbed by WPRR as already purchased and on hand. Tools are borrowed from WPRR maintenance or the nature center. The Senior Curator (wildlife) is a full-time employee of the County, and an hourly assistant was hired specifically for the deer season. Now, however, with the increase in wildlife-related projects in Westchester County Parks, the hourly assistant's duties are not specific to the ADMP, so the position has become a year-round position. The nonprofit Friends of Westchester County Parks purchased some equipment for the ADMP such as eight trail cameras. The yearly fee of $40 charged to each hunter is absorbed into the County's "general fund," and is not earmarked specifically for the ADMP or other wildlife-related projects.

35.8   ADJUSTING THE ADAPTIVE DEER MANAGEMENT PROGRAM

We used monitoring, log-in/log-out reporting, and other hunter data to make adjustments to the program. The effort required to harvest a deer increased over time, likely a reflection of reduced deer density. Initially, we required hunters to hunt on 5 separate days to qualify for hunting in the subsequent year. This requirement was increased to 10 days, and by 2013, we increased the requirement for continued participation in the program from 10 day-hunts to 15, with the exception that harvest of a deer superseded the requirement for a fixed number of day-hunts. As before, the increase in time requirement served to retain hunters who were committed to the program. A flaw with the days-hunted requirements was that there was no minimal amount of time defining a hunt-day. Taking advantage of this grey area, hunters who were not harvesting deer but still wanted to hunt the properties the following season began investing in "bogus hunts." After logging in, some hunters would spend less than an hour hunting. While it is possible to harvest a deer in this short time period, it is unlikely and disrupts others' hunting opportunities through hunters traveling through the woods during prime hunting hours.

To counter the problem of bogus hunts, the 2015 time requirements were changed from 15 day-hunts to 45 hours in-stand to assure that hunters were actively hunting. The increase served to retain hunters who were willing to invest time in the program and weed out those who were not. Additionally, the altered requirement for hunting in successive years on WPRR was in line with those on other, nonpark areas with similar management programs.

To better quantify this requirement, the Mianus River Gorge's (a sister park in the Westchester system of county parks) "Tunney Rule" (see Chapter 36) was adopted, wherein the average of hours required to harvest an antlerless deer in year X is multiplied by 1.2 to produce number of hours required to hunt in year $X + 1$.

Because we realized that some hunters could harvest a deer without hunting the required number of hours and that hunters who harvested a single deer still contributed to overall harvest, hunters who harvested one or more antlerless deer were exempt from the time requirement to hunt in the year after they harvested a deer. The alternative requirement of "or harvest of a deer" was added because the ADMP had not retained hunters in the past who did not meet hunt time/day requirements, but who had harvested one or more deer. Managers felt that hunters who harvested deer and contributed to overall harvest numbers should not be removed for not hunting the prescribed number of hunt-days.

The requirement that hunters harvest an antlerless deer as a precondition to being allowed to harvest an antlered deer was dropped, likely resulting in at least an antlered deer being harvested by individual hunters. Managers felt comfortable with this last change, especially after a DEC mandate (applying to the ADMP program) that only antlerless deer could be harvested the first 2 weeks of deer season and limited the number of buck tags allowed per hunter. Also, data indicated that most ADMP hunters did not harvest antlered deer.

Through time, ADMP administrators changed the process for accepting new hunters to meet the needs associated with increased number of volunteer hunters, increased size of included property, and associated paperwork. By 2014, the ADMP was managing a total of five other parks besides WPRR, plus some portions of DEP state property. Managers' time devoted to the program was diluted, as they had taken on additional roles in County Parks managing other wildlife species and their impacts as well as conducting research. By 2015, the program proficiency-tested hunters before accepting applications and drew applicants from those who passed the test. It accepted hunters from outside of Westchester County. The ADMP no longer required past program hunters to proficiency test with each new season, accepting the original proficiency result. Along with increased applications, legal waivers, and current DEC hunting licenses, it became necessary to submit copies of a DEP Access Permit for properties the ADMP program was allowed to include in its program. Hunters were still required to obtain Westchester County IDs, and the cost of the annual administrative fee was raised to $40.

Initially, there were no antler restrictions on deer hunting within the WPRR. Because hunters could amass a maximum of eight buck tags for the WPRR through exchange and transfer of tags with other hunters, and because our initial interest was in harvesting as many deer as possible, many young bucks were harvested in the initial years of the program, altering the population's male age structure on WPRR. As of 2013, NYSDEC restricted allocation of DMP permits on WPRR to antlerless deer only and allowed hunters only two buck tags. Because of the reduction in buck tags, hunters now select for more mature bucks with more antler points and harvest a higher proportion of antlerless deer.

35.9 RESULTS

The 2012 deer density estimate for the block in the western portion of WPRR was 17.5 deer/km^2, down from 24.6 in 2008, a 29% reduction. By 2014, deer density had declined further to 7.6 deer/km^2, representing a 69% reduction over 6 years. The "no regeneration" and "no impact" indices of deer impact reflected high impact in 2008: 91.5% plots had no regeneration and ~2.3% plots had no deer impact on the western plot. By 2014, those percentages had changed to ~87.69% plots with no regeneration and ~5.38% plots without impact, and the understory was much improved (Figure 35.3). There was a direct correlation between browse impact and the estimated deer density among the zones: the higher the estimated population, the higher the observed impact, except for in the northern zone.

These results are similar to those of the successful Kinzua Quality Deer Cooperative case history: large and rapid declines in deer density resulting in reaching target density using public hunting, paired with lagged but positive improvement in vegetative response. Because the understory was so heavily impacted on WPRR, and because there were no openings created (e.g., timber harvest or natural events resulting in opening of the forest canopy) to stimulate ground-level forage production, vegetative response/improvement will be limited and protracted. Hunting effort to harvest a deer also continues to increase. Hours per harvest (all deer) on the WPRR increased over the years: 15.6 hours in 2010, 20.2 hours in 2011, 16.0 hours in 2012, 19.8 hours in 2013, and 70.4 hours in 2014.

Deer were harvested by a minority of hunters. Data from 2013 were typical: Of 106 hunters registered for the program, 84% actually hunted. Of those, 64% harvested no deer, 15% harvested one deer, 12% harvested two deer, and 9% harvested three or more deer—only 30% of all hunters registered for the program actually harvested deer. Hunting effort and harvest declined on WPRR in 2014 because: (1) the number of bow hunters in Westchester County and surrounding areas declined over time; (2) some bow hunters hunted more than one DMP, reducing the time they hunted on individual DMPs and other (private) properties; and (3) the ADMP provides hunter access to multiple County properties without assigning hunters to specific properties, so it is difficult to maintain hunting pressure on any one park. With the reduction in deer density over time and the vastness of the landscape, hunters found WPRR increasingly difficult to hunt and too large to scout for new areas within the park. Desiring less formidable and more productive hunting opportunities, some began to hunt in other Westchester County parks, resulting in a reduced deer harvest on WPRR.

Reduced numbers of deer sighted by volunteer hunters over the years reinforced an assumed reduction in deer density resulting from hunting. However, the counterintuitive trend of hunters harvesting more deer while seeing fewer emerged in 2012, suggesting that successful hunters were

FIGURE 35.3 Regeneration in 2014 after reduction in deer density. (Photo by Dan Aitchison.)

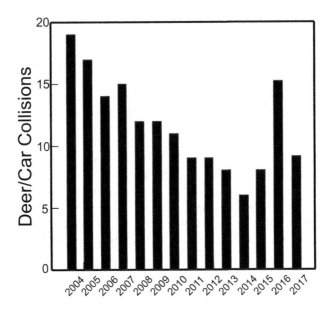

FIGURE 35.4 Change in deer/vehicle collisions associated with reduction in deer density. (Credit: Data provided by the Pound Ridge Police Department.)

becoming more familiar with hunting on WPRR and more efficient over time. The declining age of the deer harvested over time supported the expected reduction in age structure of the deer herd as an intended consequence of management. Composition of annual harvests stabilized at about 70%–80% antlerless deer: the ADMP hunters adjusted their selection of deer harvested to meet management goals. Additionally, 10% of the hunters harvested the majority of deer.

Damage to landscaping of affected residents declined and the number of deer/vehicle collisions in the town surrounding WPRR decreased as deer density declined in response to the town archery-based deer management program beginning in 2007 and the county's expansion of the ADMP to WPRR in 2010 (Figure 35.4).

35.10   STAFF/HUNTER INTERACTION

Staff interaction and communication with hunters contributed to the success of the ADMP in WPRR. Such "face time" made hunters feel their voices were being heard and that they were valued by the ADMP. A friendly relationship developed between ADMP managers and ADMP hunters: communication between them is continual and open. While hunters never indicated they felt they were being watched over, the on-the-ground presence of staff served to remind hunters they should monitor and check their behavior while on-site and in the public eye.

Hunters took pride in their participation in deer management and their role in the reduction of herd size and impact, which invests them as an asset to the overall mission of the ADMP.

The successful deer management on the ADMP has been extended to other landscapes by hunters applying the tools and experiences they learned in the ADMP to other DMPs. After his hunting experience on the WPRR ADMP, the manager for the Somers Anglefly DMP enlisted the top two hunters of the ADMP to manage a similar program on the Teatown Lake DMP; another top ADMP hunter structured a program for the Pound Ridge Land Trust.

WPRR staffers extended the scope of the ADMP success with deer management to other venues within Westchester County, consulting with administrators from other municipalities and organizations within the County. The ADMP collaboration with New York City Department of Environmental Protection (NYCDEP) resulted in ADMP hunters managing deer on NYCDEP properties adjacent to WPRR. The ADMP was allowed to issue replacement DMP tags to hunters, especially helpful in 2014 when the State decreased its number of visits to issue tags and withdrew from the local check station. The WPRR ADMP staff extended application of its ADMP program in 2015 by holding a deer symposium for municipalities, residents, landowners, and other stakeholders wherein they addressed the many facets and unique challenges associated with deer management in suburban communities. The symposium facilitated movement for a regional approach to deer management among municipalities; response from attendees was overwhelmingly positive.

35.11 ALIGNMENT OF GOALS AND VALUES, AND RETENTION OF HUNTERS

For a volunteer-based bow hunting program, it is important that the goals of the managers and the hunter/volunteers be similar, or at least co-acknowledged. In the early years of the ADMP, managers desired a reduction in deer populations and hunters desired new, unhunted, open landscapes and a high success rate. At that time both parties' goals were compatible and worked. However, as deer density declined and deer became wary during hunting season, hunters had to increase time and effort to harvest or even to see deer and began to lose interest in hunting the WPRR and other County ADMP properties. ADMP managers continued to promote harvest of antlerless deer to further reduce deer density. In response, hunters claimed they had helped achieve ADMP goals in reducing deer density and impact, and requested that ADMP managers reduce or at least temporarily suspend deer harvest on the ADMP properties. The hunters also asked that the ADMP provide them with new properties to hunt. Some hunters left the ADMP to hunt on private properties or other DMP properties outside the ADMP group of properties. Two of the top hunters from the WPRR ADMP left to hunt on and manage hunters at Teatown Lake Reservation for pay.

If the ADMP is to continue with volunteer archers, it will need to devise a system to reward those who are willing to invest time into the increasingly difficult hunting environment of WPRR and other parks. If this cannot be achieved, it is likely that hunters will continue to leave the program for other opportunities, and the bow hunting program may have to be supplemented with methods such as culling deer by sharpshooters or replaced by them altogether.

35.12 RETAINING THE ADMP VIA STAKEHOLDER INFLUENCE

Faced with budget shortfalls in 2012–2013, the Westchester County Commissioners voted to eliminate funding for the ADMP. However, complaints over the budget cuts from landowners adjacent to WPRR and hunters involved in the program carried sufficient weight that the commissioners reconsidered the cuts and restored the program in its entirety. Restoration of funding was critical to long-term success of the ADMP: research by McCullough (1979) on the George Reserve Deer Herd detailed how rapidly deer can repopulate to previous levels when predatory pressure is eliminated, as would occur on WPRR if the ADMP program were defunded.

The ADMP developed a successful deer management program in a public park system without limiting nonhunting public access; monitored effectiveness of the program; adapted it yearly to improve its effectiveness; educated the public; and collaborated with other DMPs, organizations, and agencies to reduce deer density and browse impact. However, to enhance forest regeneration, deer density needs to be reduced further and kept at goal levels by annual hunting using educated volunteer hunters.

35.13 A FINAL WORD

While it is important to achieve and maintain target deer density as an essential component of restoring forest regeneration, opening of the canopy through girdling of individual trees or timber harvest, planting of tree seedlings in areas devastated by storm damage and deer browsing, and removal of invasive plant species are additional management steps WPRR managers will need to incorporate into the ADMP to further promote regeneration and diversity of the forest overstory and understory.

EDITORS' NOTE

Recovery of ground vegetation, including seedling regeneration and the herbaceous layer, would be hastened/improved if WPRR were to allow timber harvest, or at least used silvicultural practices such as thinning or shelterwood to open the overstory and increase light reaching the forest floor to stimulate understory development. Timber harvest would produce income that could be earmarked for improvement in habitat and diversity on WPRR, remove fuel build-up in the understory and reduce fire hazard, and improve habitat for the wildlife community, especially species benefitting from early succession habitat. The detailed case history demonstrates how inclusion and education of stakeholder groups can provide political pressure on administrators to initiate, maintain, and support adaptive management of deer within forested landscapes.

David S. deCalesta, editor

REFERENCES

deCalesta, D. S. 2013. Reliability and precision of pellet-group counts for estimating landscape level deer density. *Human-Wildlife Interactions* 7:60–68.
Herr, B. and M. Kohl. 2013. *Images of America, Ward Pound Ridge Reservation*. Charleston NY: Arcadia.
Hubbard, A. 2008. Watershed Forest Management Plan, Ward Pound Ridge Reservation. Watershed Agricultural Council's Forestry Program internal document.
Jacobson, H. A., J. C. Kroll, R. W. Browning et al. 1997. Infra-red triggered cameras for censusing white-tailed deer. *Wildlife Society Bulletin* 25:547–556.
McCullough, D. R. 1979. *The George Reserve deer herd*. Ann Arbor MI. The University of Michigan Press.
Pessoni, P. A. 1995. *Historical notes on the Ward Pound Ridge Reservation 1640–1940*. Stamford CT: Landmark Document Services.
Pierson, T. G. and D. S. deCalesta. 2015. Methodology for estimating deer impact on forest resources. *Human-Wildlife Interactions* 9:67–77.
Pollard, C. W. (2011). *The soul of the firm*. Wheaton Illinois. The ServiceMaster Foundation.
Severinghaus, C.W. and C. P. Brown. 1956. History of the White-tailed Deer in New York. *New York Fish and Game Journal* 32:129–167.

36 The Mianus River Gorge Preserve

Incremental Adaptive Management

Mark Weckel

CONTENTS

> Unwavering incremental change can create remarkable and monumental results.
>
> **Ryan Lilly**

36.1 INTRODUCTION

The 309-ha Mianus River Gorge Preserve (MRGP), located in northeastern Westchester County NY, was created in 1953 as the Nature Conservancy's (TNC) first land project. The linear, fragmented preserve lies at a suburban–exurban interface approximately 75 km northeast of New York City and is surrounded by large estates (>1.6 ha) and protected reservoir land (Figure 36.1). The preserve is composed of approximately 200 ha of 70–100 year-old second-growth northern hardwood forest encompassing a 100-ha core of old-growth hemlock (*Tsuga canadensis*) forest, both of which protect wetlands and tributaries of the Mianus River and provide drinking water for 130,000 residents of New York and Connecticut. Today, the preserve is managed by the Mianus River Gorge Inc., a separate not-for-profit organization whose mission is to preserve, protect, and promote appreciation of the natural heritage of the Mianus River watershed through land acquisition and conservation, scientific research, and public education throughout the region. The MRGP is currently staffed with

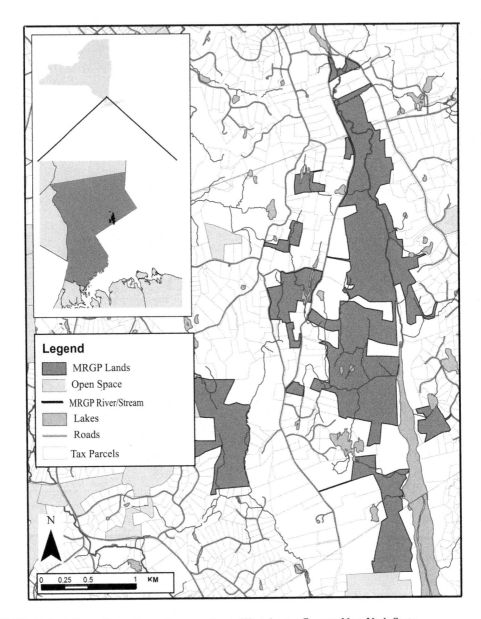

FIGURE 36.1 Mianus River Gorge Preserve. Inset: Westchester County, New York State.

three full-time biologists and a supporting part-time staff of four year-round employees. A Board of Trustees, composed primarily of stakeholders from the five towns of the Mianus watershed, oversees management of the preserve. I was one of the architects of the deer management plan and oversaw the program from 2004 through 2014.

From the preserve's inception until the late 1990s, MRGP staff had maintained a hands-off, preservationist attitude toward land management (Duncan and Duncan 2001). Reflecting the prevailing philosophy of the day (Landres 2010), managers sought to protect biodiversity through land acquisition followed by prohibition of most forms of natural resource use, including hunting and timber harvest. Nor did staff engage in proactive management activities (e.g., exotic plant removal, habitat management, canopy thinning). As has become a familiar story for many suburban nature preserves, the combination of a high proportion of edge habitat, few natural predators, and prohibition of hunting resulted in ideal conditions for the rapid growth of the local white-tailed deer (*Odocoileus virginianus*) herd. By the early 2000s, deer density estimates from unmanaged, forested areas adjacent to/within 7 km of the preserve averaged 23.4 deer/km^2 (Brash et al. 2004; CTWF 2008; Daniels et al. 2009)—exceeding levels known to suppress woody regeneration (Rooney 2001). On the MRGP, a browse line had become evident and many wildflower species were either no longer present or browsed before flowering (R. Christie, Executive Director, MRGP, personal communication).

Because it was recognized that the process—and costs—of developing a deer management program start well in advance of actual management actions, MRGP administrators began a multiyear process in 2001, identifying obstacles, collecting data, and building critical support though meetings with state and town officials, local hunters, MRGP neighbors, and the Board of Trustees.

As a prelude to deer management, MRGP staff initiated studies of woody and herbaceous vegetation to investigate the impact of the local deer herd on the preserve's plant community. Beginning in 2001, the first of what are now four deer exclosures, ranging in size from 0.15 to 0.85 ha, were constructed to quantify the impact of deer herbivory on herbaceous species, to protect remaining herbaceous populations and local seed banks, and to serve as a site for piloting restoration strategies. Within 1 year, the abundance of red trillium (*Trillium erectum*) and true Solomon seal (*Polygonatum biflorum*) in the first exclosure increased by 929% and 2400%, respectively (R. Christie, Executive Director, MRGP, unpublished data). By the fourth year, several new species, including round-lobed hepatica (*Hepatica americana*), rattlesnake plantain (*Goodyera pubescens*), and shinleaf pyrola (*Pyrola elliptica*), not recorded prior to the exclosure's construction, were documented within the exclosure, suggesting that viable seed banks may remain for species no longer represented aboveground.

In 2004, MRGP staff began exploratory communications with adjoining homeowners, hunters, and other stakeholders; collection of baseline forest health data; development of a written plan; and obtaining Board support for deer management. A survey of the woody vegetation was conducted to make historical comparisons with Bard's (1967) study in 1965 of the structure and composition of the Mianus old-growth hemlock forest. Bard (1967) documented healthy regeneration of several hardwood species and predicted the continued dominance of eastern hemlock. The 2004 survey found no hemlock saplings (0.3–0.9 m tall). There were broad changes in hardwood diversity, with a loss of 12 species in the sapling age class and a total decline in sapling abundance from 0.160 to 0.024 stems/m^2 (Weckel et al. 2006). Only American beech (*Fagus grandifolia*), a deer-resistant species, was well represented in the understory, reflecting a pattern described for other deer-impacted hemlock forests (Whitney 1984). At this point, MRGP biologists began investigating deer management options with the goal of promoting advanced woody regeneration and maximum floral diversity.

The biologists wondered whether deer density could be reduced effectively within an area as small as the MRGP in the face of a sustained and unmanaged high density regional deer population, and, if so, what the most suitable method would be. They considered but rejected immunocontraception (Miller and Killian 2000). While immunocontraception has been shown to be effective on island populations of white-tailed deer (e.g., Rutberg et al. 2013), this strategy was tabled because of the prohibitive costs (White Buffalo 2004), concerns over its efficacy (Gionfriddo et al. 2011)—especially in open populations, and the need for immediate herd reduction rather than slowed population growth characteristic of control via immunocontraception.

Based on the work of Mathews and Porter (1993) in the Adirondack State Park to the north, localized, small-scale lethal herd reduction appeared promising. Fidelity of female white-tailed deer family groups to small (7–20 km^2) localized home ranges raised the potential for removing most if not all members of these family groups, creating a hole of lowered deer density for several years that would not be immediately filled by immigrating deer (Porter et al. 1991; McNulty et al. 1997; Oyer and Porter 2004). Although questions existed (and still exist) over the efficacy of such localized deer removal (Blanchong et al. 2013; Miller et al. 2010; Williams et al. 2013; Weckel and Rockwell 2013), MRGP biologists thought it held the best potential for immediate action in the absence of a regional strategy for managing deer in Westchester County, NY.

Deer population reduction through hunting has been the method most often used on public forestlands and seemed the best candidate for deer herd reduction on the MRGP. Use of contracted sharpshooters was not feasible because Westchester County and local municipalities prohibited the discharge of rifles. In addition, at the time, sharpshooting had never been used in Region 3S, the New York State (NYS) Department of Environmental Conservation (DEC) management unit encompassing the MRGP. The NYS DEC did allow the use of shotguns outside of the regular hunting season under a special nuisance permit on some larger private residences and orchards in northern Westchester. However, while neighbors and the Board of Trustees supported deer reduction, there was a general concern, and some opposition to, the use of firearms.

Because bow hunting is legal in Westchester County, can be performed by volunteer hunters, is relatively discreet, and can be used to safely harvest deer in close proximity to residences, MRGP biologists selected controlled archery hunting as best alternative to hunting with firearms for reducing the deer herd. They were aware that bow hunting is less efficient than other lethal methods (Hansen and Beringer 1997; Krueger et al. 2002) and that it was unclear to what level bow hunters could reduce the deer herd. Therefore, they created the MRGP Deer Management Program (DMP) as an experimental initiative requiring continual review and adaptation, including revisiting alternative strategies if and when management goals were not met.

As suburban forests are fractured, so too is support for lethal deer management. While MGRP staffers were aware of increased interest for lethal management among the Westchester County public, they also recognized the potential for opposition. For nonprofit organizations that rely on public support and donations, the prospect of alienating contributors and conservation partners is of paramount concern. In addition, staff knew they needed to generate internal and external support for deer management, not only from a legal and financial perspective, but also because private residences composed a large portion of the landscape where the deer were to be managed: MRGP has over 200 resident humans in homes directly abutting the preserve. These neighbors are financial supporters and their properties are potential management areas. MRGP staff knew that finding and maintaining neighbor support would be crucial to the success of the deer management program, and that it would take time and had to be transparent and accountable.

Staff recognized a number of challenges unique to the MRGP in implementing bow hunting. The preserve is long and narrow, fragmented, and not well roaded; it is surrounded by many large, wooded home sites; and there was neither a hunting culture nor an established corps of bow hunters familiar with the immediate area. As with many forestlands with deer overabundance issues, there was a diverse set of stakeholders, each with differing views on deer and forest management, including the use of lethal means for herd reduction.

36.2 THE PROGRAM

In response to these challenges, and the uncertainty of the chosen approach, MRGP staff incrementally implemented its deer management program, starting small and low-key and gradually increasing the scope and scale of what they did. Based on private conversations, they knew that there was support for deer herd reduction from landowners adjacent to the preserve. However, in 2003, there was backlash to a controlled archery hunt conducted at the nearby Greenwich Audubon preserve in Fairfield, Connecticut, <10 km south of the MRGP (K. Dixon, Executive Director, Greenwich Audubon, personal communication).

36.2.1 PEOPLE MANAGEMENT

MRGP staff knew they had to establish and maintain communications and a working relationship with hunters who would be effecting deer herd reduction through bow hunting. In 2004, MRGP staff began a discrete pilot program on a fraction (2.3 km²) of the MRGP and delayed a planned public education campaign. Hunting was permitted only during December when the preserve was closed to the public. To further minimize interaction with the public, hunters were not permitted to park on public roads. Staff focused outreach efforts on stakeholders immediately adjacent to areas being hunted. These actions allowed staff to concentrate on the logistics of the hunt without the distraction of negative publicity. However, such lack of transparency is probably not possible for most public entities, nor is it good long-term strategy. To have a more effective program and to meet its education mandate, staff became more vocal about the DMP in following years. In year two, hunting was permitted while trails were open to hiking and signs were posted in the woods to warn hikers from deviating from the trail. By year four, MRGP staff was regularly discussing the rationale for its DMP and program logistics at public events. Hunters had a more visible presence in MRGP woods and on roads.

Engaging homeowners and increasing public outreach became increasingly important to deal with unexpected challenges. In 2008, staff discovered that non-DMP hunters were posing as MRGP hunters to gain access to neighbors' properties included in the DMP. In addition, staff realized that access to private land would be an important strategy for harvesting deer whose home ranges included MRGP lands and adjacent residential properties. Through letters and door-to-door visits, staff initiated conversations on deer management with interested neighbors to resolve problems with trespassing hunters and to add lands surrounding private homes to the DMP area. By 2011, MRGP hunters had access to 4.1 km^2, with private residences accounting for 33% of the total management area.

Managing hunters is similar to organizing any volunteer labor pool. To be successful, hunters had to be well trained, have sufficient support from staff, had to receive proper oversight, and had to feel appreciated by the organization they were working with. These components were addressed though regular meetings, working with hunters in the field (e.g., helping to retrieve harvested deer, encouraging them to scout new sites), maintaining regular phone and e-mail conversations, and by sponsoring (occasional—not every year) hunter appreciation dinners. In addition to establishing rapport with hunters, staff spent time afield resolving disputes over hunted areas, responding to reports of poaching, discouraging off-trail hiking by the general public, and increasing boundary patrols during deer season. Most importantly, to make the deer program more inclusive and responsive to hunters' concerns, staff created a Deer Council in 2007, which was a body of 2–3 persons elected to represent hunters' interests and serve as a conduit for two-way communication between MRGP managers and hunters. During this time, the Deer Council developed recommendations for regulating number of hunters, hunting rules and regulations, enforcing DMP policy, and designating minimal number of antler points for identifying legally harvestable deer.

Hunter access has been a continuing issue with relatively few access points: many parts of the MRGP are only accessible by foot paths and more remote sections requiring 20–30 minute walks to get to hunting spots, with few opportunities for motorized vehicles. As a result, there was a patchy distribution of hunter effort. In 2008, after consultation with the Deer Council, staff experimented with organizing mandatory group hunts (deer drives) to direct hunting effort into interior areas where hunters rarely went and which may have served as deer refugia. While successful in increasing hunter effort, deer drives proved too difficult to organize and frustrated many hunters. Specifically, hunters complained that micromanaging specific hunts took away from their independence and deprived them of a primary characteristic of bow hunting: solitude. As a result, many hunters referred to the deer drives as "work."

To replace the unpopular deer drives while addressing the need for maintaining hunting effort, staff worked with the Deer Council in 2009 to develop a "Tunney Rule." The rule required that hunters hunt a minimum number of hours in a year to be eligible to hunt in subsequent years. The minimum number of hours hunted was defined as 125% of the number of hours it took to harvest an antlerless deer averaged over subsequent years of the program. As a hypothetical, if from 2004 to 2010, harvesting an antlerless deer required on average of 100 hours hunting, then an unsuccessful hunter would need to have spent 125 hours hunting in the 2011 season to be eligible to hunt in 2012. However, harvesting an antlerless deer superseded the hours of hunting commitment: a hunter could meet the Tunney Rule by harvesting an antlerless deer regardless of hours spent hunting. The Tunney Rule was well received by hunters because it rewarded effort and success, was fair and transparent, and provided hunters with a measure of independence in how they met their obligation to the MRGP. For MRGP managers, the Tunney Rule was successful, as it provided a mechanism to maintain hunting effort as the herd size decreased (and as harvesting a deer required more time hunting). Tabulating harvests and hours did not require additional effort on the part of hunters or MRGP staff: hunters were required to call or e-mail ahead of hunts, as well as complete daily hunting logs that included time in the stand, wildlife observed, and information on harvested deer (see the Ward Pound Ridge Reservation case history for more details).

36.2.2 Monitoring

Monitoring deer density and impact was representative of the entire MRGP. In 2004, we established 22 permanent plots to monitor changes in the woody and herbaceous vegetation using a modification of the protocol Bard (1967) employed so we could compare current with historical data. At each plot, overstory trees (1800 m^2 sampling area) were inventoried at 10–15 year intervals, while understory woody species (119 m^2 strip transect) and the herbaceous layer (13 1−m^2 quadrants) were monitored every 3–5 years. The four deer exclosures built in 2001 to protect seed banks and to experiment with various restoration methods (e.g., species reintroduction, canopy thinning, exotic plant control) were checked monthly to repair damage caused by wind and storm events. Starting in 2009, staff used digital trail cameras to track annual changes in deer abundance (Weckel et al. 2011a). To obtain feedback from hunters, including attitudes about the program, we combined informal and regular conversations with hunters with the results of a formal survey of Westchester hunters (Weckel et al. 2011b).

36.2.3 Program Costs

Support of applied ecological research is part of the mission of the MRGP: costs were met through the general operation budget and external grants. Program costs were of two kinds: (1) monitoring deer and forest resources and (2) developing and maintaining effective interpersonal relationships with and between hunters, private landowners, and other stakeholders. Dollar expenditures for monitoring (materials and staff time) were easier to calculate; monetizing human capital (the primary resource) expended in dealing with stakeholders, including landowners and hunters, was more subjective. With funds provided by a grant, staff paid approximately $10,000 for 20 trail cams in 2009. These initial cameras have been in use for 5 years and are only now showing signs of irreparable wear and tear. Deer exclosure materials (2.3-m-tall heavy-duty polypropylene fence, 3 m, 2 cm diameter rebar, cedar poles, and tension wire) totaled approximately $7500. The smallest exclosure (∼0.15 ha) was completed with a team of three individuals over 2 days. Inventories of the herbaceous and juvenile woody vegetation are conducted every 3–5 years by a single observer over 5–7 days. The overstory will be inventoried every 15 to 20 years by three observers over approximately 9 days per inventory year.

Human capital costs were incurred in interacting with adjacent homeowners via direct mail campaigns and door-to-visits, in arranging and continuing meetings between hunters and private landowners, and in everyday interactions with concerned stakeholders. In 2006, when the MRGP made hunting access to adjacent home sites a priority, the program coordinator spent a full week on the recruitment process. For the first few years of the MRGP DMP (2004–2006), the coordinator spent approximately 50% of his time working with stakeholders from October through January. As the program matured, so did the relationships with hunters and landowners, and the coordinator's time spent interacting with stakeholders declined to less than 15% October–January.

36.3 RESULTS

36.3.1 Herd Reduction and Vegetation Response

Hunters removed 288 deer at a female-to-male ratio of approximately 2.3–1 between 2004 and 2013. As of 2010, the MRGP deer herd averaged 17.2 deer km^{-2} (Weckel and Rockwell 2013).

The MRGP does not have estimates of prehunt deer density specific to the management area; however, estimates from unmanaged, forested areas adjacent to the MRGP averaged 23.4 deer/km² (SE = 0.76; see Weckel and Rockwell 2013).

Between 2004 and 2010, there was a 500% increase in density of seedlings <0.3 m tall and having >1 set of postcotyledon leaves; but this height class represents the *potential* for recruitment and survival into the next stand. Response of saplings 0.3–0.9 m tall, representative of successful advanced woody regeneration, was more modest. Excluding American beech, the density of native tree saplings increased from 0.14 to 1.18 stems km⁻² (unpublished data), an eightfold increase. Saplings of American beech, a browse-resistant species, increased dramatically (~142%) following hunting. Most telling, there was a 48% decline in total stem density in the height/diameter class (>0.9 m tall and <2.54 cm dbh) representative of seedlings vigorous enough to survive to become adult trees, sans destructive deer browsing. With the MRGP herd much more abundant than the 5.8 deer/km² cited as a threshold deer density for negative impact on advanced forest regeneration (Tilghman 1989), these results are not surprising. Staff continues to conduct both annual camera trap monitoring and regular vegetation surveys.

36.3.2 THE TUNNEY RULE AND HUNTER RETENTION

The Tunney Rule was associated with retention of a central, dedicated corps of repeat hunters, likely because it was not overly complicated, was been well received by hunters, and did not micromanage the hunting experience. It set a clear requirement for continued participation in the MRGP DMP. Distribution of hunter effort remains patchy, however, and was not directly addressed by the Tunney Rule. MRGP staff continues explore new options for incentivizing hunting in more remote, difficult-to-access regions of its preserve.

It remains to be seen how increased hunting effort required to harvest a deer will affect hunter participation. It is likely that the minimum deer density attainable in the MRGP DMP will largely be determined by hunter motivation. Interestingly, a core group of MRGP hunters expressed continued commitment to the MRGP DMP even after the crushing 2012 season when hours per antlerless harvest reached nearly 90 (Figure 36.2). Their reason for staying: the preserve itself. The forests are relatively old, the topography craggy, hunter access is restricted, and dogs are kept out. In sum, the MRGP forest feels wild and hunters feel part of something that is theirs.

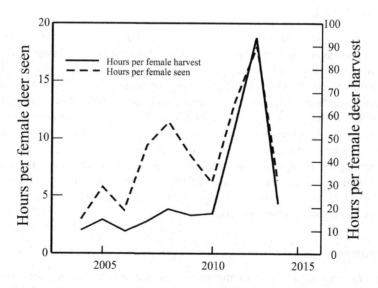

FIGURE 36.2 Hours per female deer seen vs. hours per female deer harvested.

36.3.3 IMPORTANCE OF ADJACENT LANDSCAPE

Despite our success in obtaining hunter access to residential areas, managing deer and hunters on private home sites was a continual challenge. Suburban lawns and ornamentals on properties immediately adjacent to MRGP likely provided a reliable, and in many cases preferred and highly nutritious, food resource (Gorham and Porter 2011), keeping even highly abundant herds healthy. Such herds can exhibit relatively high maximum sustained yield/reproduction while severely depleting forest understories (see Chapter 15) of food and cover. Swihart et al. (1995) reported that deer foraged more intensively closer to houses where available plant species richness was greater during winter in the nearby towns of Bethel and Newtown in Connecticut. Such off-preserve feeding becomes increasingly problematic during years with a poor acorn mast crop, when deer spend even less time on-preserve during hunting season foraging for acorns, a fall staple. In 2011 and 2012, for example, deer spent less time in the MRGP forests and were observed foraging on or near suburban lawns, resulting in a dramatic increase in hours-per-harvest and lower harvests (Figure 36.2). In this manner, private properties not only provide an additional food resource, but also serve as deer refugia from hunting.

36.3.4 MANAGEMENT PARADOX

While the ultimate goal of the MRGP DMP was (and continues to be) enhancing forest biodiversity, the immediate priority was to design a controlled hunt that would motivate hunters toward maximizing and sustaining harvests, thereby reducing deer density and impact. Therein lay the paradox of deer management: as hunters drive deer density down and experience diminishing returns for their efforts, they eventually will reach a threshold deer density where they are incapable of, or unwilling to, reduce the herd any further. Bow hunting is especially sensitive to this threshold due to the inherent challenges of the bow compared to other weapons (e.g., shorter killing radius, greater effort per harvest). On the MRGP, hours per female (yearlings and older) harvest and hours per female sighting have increased over time (Figure 36.2), making hunting on the MRGP increasingly challenging. For hunters, the program may be less rewarding than it once was. The manager's enduring challenge is to design a DMP that sustains harvests by motivating hunters to continue hunting and harvesting deer despite decreasing prospects for harvest success.

36.4 LESSONS LEARNED

36.4.1 PUBLIC OUTREACH

As mentioned earlier, public outreach is paramount, especially when recruiting private lands for inclusion in the deer management program. Participating homeowners can bring new and productive hunting lands into a program such as the MRGP. This can be a morale booster for hunters, as well as a way to eliminate refugia. Once a property has been included, it must be treated as a distinct management unit regardless of its size. While many properties come with no strings attached, many homeowners will have their own set of requirements, such as restricting hunting hours, designating specific parking spots, or asking hunters to call ahead. While fostering good relations between hunters and homeowners is necessary, responsibility for hunter management rests with the manger. It is here that managing multiple private properties can become time consuming and challenging regarding establishing and maintaining good relations among land manager, hunter, and private landowners. Regular communication at the beginning, middle, and end of each season is important in retaining private landowners and hunters.

36.4.2 MONITORING

Managers are best served by a mix of monitoring protocols that capture both short- and long-term dynamics. Wherever possible, managers should adopt monitoring protocols used by local land owners or state agencies to facilitate regional collaboration. The MRGP staff relied on trail cam surveys (see Jacobson et al. 1997; Weckel et al. 2011b) to track changes in deer abundance, long-term vegetation plots to monitor changes in the woody and herbaceous vegetation, and deer exclosures to protect seed banks and to experiment with various restoration methods (e.g., species reintroduction, canopy thinning, exotic control).

Some assert that limited resources should be spent primarily monitoring changes in vegetation. However, restricting monitoring to vegetation surveys limits the ability to demonstrate progress on political time scales, especially if recovery of forest understory lags dramatic deer herd reduction by years. Managers must be able to correlate reduction in deer abundance with forest recovery, especially where there is opposition to lethal herd reduction; hence the need to monitor deer density. Site-specific factors that can negatively impact regeneration other than deer browsing (e.g., closed overstory canopy, soil disturbance) should at least be recorded anecdotally for reference when/if changes in deer density do not seem related to changes in understory vegetation.

36.4.3 START-UP CONSIDERATIONS

The process—and costs—of developing a deer reduction program start well before the first deer is harvested. At the MRGP, a concentrated effort to develop a written plan and acquire Board support began in advance of obtaining permission to hold a deer hunt: exploratory conversations and collection of baseline forest health data began as early as 2001. Land managers should be prepared for a multiyear process of identifying obstacles; collecting and analyzing data; and building critical support though meetings with local governing boards, local hunters, neighboring homeowners, and state natural resource agency personnel.

36.4.4 THE MIANUS RIVER GORGE PRESERVE DEER MANAGEMENT PROGRAM: SUCCESS OR FAILURE?

As noted above, we believe deer density declined from 23.4 to 17.23 by 2010 on the MRGP DMP. From this perspective, the MRGP bow hunting program has been, at best, only modestly successful. Nevertheless, the MRGP DMP, as the first controlled hunt on preserve land in Westchester County, provided a test case for local deer management in Westchester County and has been successful in moving the discussion of deer overpopulation forward from nearly two decades of debate into action. Since the MRGP was opened to bow hunting in 2004, two dozen additional DMPs within Westchester County have been created. Many of these programs are less than a decade old and result from a recent modeling study (Weckel and Rockwell 2013) suggesting that the degree to which bow-only hunts reduce deer populations may not be evident until the second decade of management. It remains to be seen to what final extent the MRGP DMP and other programs can reduce local deer herds.

In Westchester County and other similar suburban counties, land managers need to take advantage of gains made in generating public support of bow hunting when exploring other potentially equally contentious management strategies (e.g., irregular hunts by sharpshooters armed with high-powered rifles to achieve 1-year herd reduction sufficient to get deer density to goal—see KQDC case history). This requires being transparent about the progress, success, and limitations of localized bow-only controlled hunts and setting a clear vision for future adaptive management efforts. In this vein and to end on a cautionary note, while fine-scale DMPs may help in starting to alleviate overbrowning pressure on specific properties, such localized strategies should not be confused with a regional approach to deer overabundance. Since 2004 when the MRGP began its program, over 10,000 acres of land have been opened to bow hunting in the Westchester County; however, the county deer take has been flat (Figure 36.3). The decision of nature preserves and county parks to reverse their hunting bans provided additional opportunities for hunters, and in doing so diluted hunting effort over the expanded hunting landscape—the limited number of cooperating bow hunters was stretched thin by

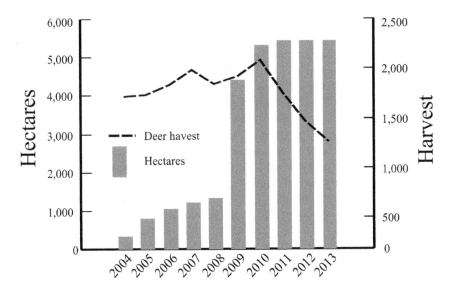

FIGURE 36.3   Hours per observation and per harvest for female deer (≥1.5 yrs) at the Mianus River Gorge Preserve; 2004 to 2013. New York State Department of Conservation harvest statistics for Westchester County, NY and acreage of land managed by independent deer management programs in Westchester County, NY, 2004 to 2013. http://www.dec.ny.gov/outdoor/42232.html.

providing them with competing properties for hunting. Developing a regional, coordinated solution to high deer densities needs to proceed in concert with the actions of land managers of smaller suburban nature preserves and regional wildlife management agencies.

EDITOR COMMENTS

This case history attests to the value of protracted, two-way communications between stakeholders and managers—it took a long time to get the adjacent landowner stakeholders to agree to participate in a program involving removing deer by hunting, albeit bow hunting, but they did, to its success. Involvement of various stakeholders in an integrated managing committee was an additional useful practice. The case history also demonstrates that a deer control program can be employed over a fragmented DFMA and that an incremental approach to adaptive management (making small steps over time rather than attempting a large number of activities without a plan for assessing which were effective and which were not) can be effective. The innovative rule devised to winnow out unproductive hunters and emphasizing alpha hunters is a good model to follow. This case history and another (Chapter 35) illustrate the difficulty of controlling deer density by archery only hunting—one wonders if using sharpshooters with silenced rifles might be a next logical step.

David S. deCalesta, editor

REFERENCES

Bard, G.E. 1967. The woody vegetation of the mature forest of the Mianus River Gorge Preserve. *Bulletin of the Torrey Botanical Club* 94:336–344.

Blanchong, J.A., A.B. Sorin, and K.T. Scribner. 2013. Genetic diversity and population structure in urban white-tailed deer. *Journal of Wildlife Management* 77:855–862.

Brash A.R., E.V.P Brower, L. Henry, and D. Savageau. 2004. *Report on Managing Greenwich's Deer Population*. Town of Greenwich Conservation Commission. Greenwich CT.

CTWF (Citizen's Task Force on White-Tailed Deer and Forest Regeneration). 2008. Final report. Westchester NY.

Daniels, T.J. R.C. Falco, E.E. McHugh et al. 2009. Acaricidal treatment of white-tailed deer to control *Ixodes scapularis* (Acari: Ixodidae) in a New York Lyme disease-endemic community. *Vector Borne and Zoonotic Diseases* 9:381–387.

Duncan, J.S. and N. Duncan. 2001. The aestheticization of the politics of landscape preservation. *Annals of the Association of American Geographers* 91:111–115.

Gionfriddo, J.P., A.J. Denicola, L.A. Miller et al. 2011. Efficacy of GnRH immunocontraception of wild white-tailed deer in New Jersey. *Wildlife Society Bulletin* 35:142–148.

Gorham, D.A. and W.F. Porter. 2011. Examining the potential of community design to limit human conflict with white-tailed deer. *Wildlife Society Bulletin* 35:201–208.

Hansen L. and J. Beringer. 1997. Managed hunts to control white-tailed deer populations on urban public areas in Missouri. *Wildlife Society Bulletin* 25:484–487.

Jacobson, H.A., J.C. Kroll, R.W. Browning et al. 1997. Infrared-triggered cameras for censusing white-tailed deer. *Wildlife Society Bulletin* 25:547–556.

Krueger, W.J., J.B. McAninich, and D.E. Samuel. 2002. Retrieval and loss rates of white-tailed deer by Minnesota bowhunters. In R.J. Warren, ed. *Proceedings of the First National Bowhunting Conference*. Archery Manufactures and Merchants Organization. Comfey MN.

Landres, P. 2010. Let it be: A hands-off approach to preserving wildness in protected areas. In D.N. Cole, and L. Yung, eds. *Beyond Naturalness: Rethinking Park and Wilderness Stewardship in an Era of Rapid Change*. Island Press. Washington DC.

Mathews N.E. and W.F. Porter. 1993. Effect of social-structure on genetic-structure of free-ranging white-tailed deer in the Adirondack Mountains. *Journal of Mammalogy* 74:33–43.

McNulty, S.A., W.F. Porter, N.E. Mathews et al. 1997. Localized management for reducing white-tailed deer populations. *Wildlife Society Bulletin* 25:265–271.

Miller, B.F., T.A. Campbell, B.R. Laseter et al. 2010. Test of localized management for reducing deer browsing in forest regeneration areas. *Journal of Wildlife Management* 74:370–378.

Miller, L.A. and G.J. Killian. 2000. Seven years of white-tailed deer immunocontraception research at Penn State University: A comparison of two vaccines. *Proceedings of Wildlife Damage Management Conference* 9:60–69.

Oyer, A.M. and W.F. Porter. 2004. Localized management of the white-tailed deer in the central Adirondack Mountains, New York. *Journal of Wildlife Management* 68:257–265.

Porter W.F., N.E. Mathews, H.B. Underwood, R.W. Sage, and D.F. Behrend. 1991. Social organization in deer: Implications for localized management. *Environmental Management* 15:809–814.

Rooney, T.P. 2001. Deer impacts on forest ecosystems: A North American perspective. *Forestry* 74:201–208.

Rutberg, A.T., R.E. Naugle, and F. Verret. 2013. Single-treatment porcine zona pellucida immunocontraception associated with reduction of a population of white-tailed deer (*Odocoileus virginianus*). *Journal of Zoo and Wildlife Medicine* 44:S75–S83.

Swihart, R.K., P.M. Piccone, A.J. DeNicola, and L. Cornicelli. 1995. Ecology of urban and suburban white-tailed deer. In *Urban Deer: A Management Resource?* J.B. McAninch, ed. *55th Midwest Fish and Wildlife Conference*, St. Louis MO.

Tilghman, N.G. 1989. Impacts of white-tailed deer on forest regeneration in northwestern Pennsylvania. *Journal of Wildlife Management* 53:524–532.

Weckel, M. and R. Rockwell. 2013. Can controlled bow hunts reduce overabundant deer populations in suburban ecosystems? *Ecological Modelling* 250:143–154.

Weckel, M., F. Secret, and R. Rockwell. 2011a. A modification of Jacobson et al.'s (1997) individual branch-antlered male method for censusing white-tailed deer. *Wildlife Society Bulletin* 35:445–451.

Weckel, M., J.M. Tirpak, C. Nagy et al. 2006. Structural and compositional change in an old-growth eastern hemlock *Tsuga canadensis* forest, 1965–2004. *Forest Ecology and Management* 231:114–118.

Weckel, M., A. Wincorn, and R. Rockwell. 2011b. The sustainability of controlled archery programs: The motivation and satisfaction of suburban hunters. *Wildlife Society Bulletin* 35:330–337.

White Buffalo 2004. Experimental control of a suburban population of white-tailed deer using immunocontraception. Princeton Township NJ. *Interim Summary Report*.

Whitney, G.G. 1984. Fifty years of change in the arboreal vegetation of Heart's Content, an old-growth hemlock–white pine–northern hardwood stand. *Ecology* 65:403–408.

Williams, S.C., A.J. DeNicola., T. Almendinger et al. 2013. Evaluation of organized hunting as a management technique for overabundant white-tailed deer in suburban landscapes. *Wildlife Society Bulletin* 37:137–145.

37 Forests of Hemlock Farms
Educate, Educate, Educate

Marian Keegan

CONTENTS

37.1 INTRODUCTION

Hemlock Farms (HF) is a private, gated residential community that was organized in 1963 as a 501(c)4 social welfare corporation under the title Hemlock Farms Community Association (HFCA). Governance and operations are regulated by a Declaration of Covenants, Deed Restrictions and Easements; Articles of Incorporation; By-Laws; Codes; and an Environmental Mission Statement. A nine-member board of directors, hereinafter referred to as The Board, is the governing body, which is advised by a number of committees. A HF manager is contracted by The Board, and department directors and staff report to the manager. Value is maintained with Board-approved budgets, and financial stability is assured with standard auditing and accounting procedures.

Located on the Pocono Plateau of northeastern Pennsylvania in Pike County, HF is a 1860-ha forested recreational residential community surrounded by the Delaware State Forest and hunting lands (Figure 37.1).

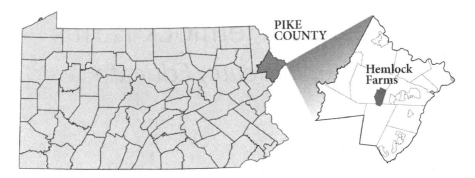

FIGURE 37.1   Location of Hemlock Farms within Pike County, PA.

Oak-heath forest is the typical ecological community in HF, dominated by chestnut oak (*Quercus prinus*), with associated species of red, white, and black oaks (*Q. rubra, Q. alba, Q. velutina*); red maple (*Acer rubrum*); white pine (*Pinus strobus*); pitch pine (*P. rigida*); black birch (*Betula lenta*); grey birch (*B. populifolia*); sassafras (*Sassafras albidum*); witch hazel (*Hamamelis sp.*); and eastern dogwood (*Cornus florida*). Shrubs include mountain laurel (*Kalmia latifolia*), blueberry, and huckleberry (*Vaccinium sp.*). HF is situated in the headwaters of two major tributaries to the Delaware River and within a high-elevation glaciated landscape. A diversity of wetland communities (vernal pools, shrub swamps, forested swamps dominated by hemlock [*Tsuga canadensis*], and former peat bogs) is an important ecological feature on the landscape. There are approximately 3200 wooded developed home sites, 800 wooded undeveloped lots, and common areas within the community. Common areas host forests and meadows: 114 ha of 12 wetlands and impounded water bodies including two 40-ha lakes and 12 lesser and variously sized lakes and ponds; beaches, boat launches, pools, sports fields and courts, playgrounds, dog run, hiking trail; 187 km of paved roads; and facility buildings, churches, and a private 103-hectare 18-hole golf/country club.

During the 1800s, market hunting nearly eliminated white-tailed deer (*Odocoileus virginianus*) from Pennsylvania and other eastern states. In response, the Pennsylvania Game Commission (PGC) was created in 1895 to protect and restore populations of deer and other wildlife (Kosak 1995). Prior to and during this period, natural predators of deer (mountain lion [*Felis concolor*] and gray wolf [*Canis lupus*]) had been extirpated, extensive and repeated logging created vast amounts of deer forage, and restrictions on deer hunting led to a deer population explosion (see Chapter 14, "National and Regional Perspectives on Deer Hunting"). Residential development of HF and prohibition of deer hunting dramatically increased forest edge habitat—perfect for deer—and deer numbers, leading to a superabundant deer population. The high deer density reduced or eliminated sensitive plant and animal populations, and fostered the spread of undesirable exotic plants, such as Japanese barberry (*Berberis thunbergii*).

In 1927, George Brewster purchased the land that is now HF and developed the forests, meadows, and swamps into a working farm and hunting grounds: lakes and ponds for fishing; orchards, open fields, and stocked game for a wild game preserve; bridle trials for horseback riding; and a sheep pasture, steer barn, cow barn, turkey house, and buckwheat field for farming.

As stated in the Environmental Mission Statement, the primary goal of HF is to maintain the forested rural environment that attracted members to HF. Reaching that goal requires educating the residents and staff on ecological realities, including requirements for enhancing and preserving native plants, trees, animals, and habitats unique to HF, and management provides staffing and educational opportunities. Educational efforts include identifying and implementing actions to mitigate or prevent human activities that damage or degrade the natural setting and situations where environmental conditions may negatively impact members. One such environmental condition was an overabundant deer herd that damaged the ecosystem and members' properties. After several decades of living with an overabundant deer herd, HF members became increasingly aware of the high deer density and resultant impacts. This situation did not escape the attention of HF community leaders, who recognized the damage by deer to members' properties, meadow and forest vegetation, and ecological services (e.g., quality water resources and recreational opportunities) provided by the natural systems in HF. Community leaders were determined to develop a management system to minimize and mitigate negative deer impacts.

37.2 RECOGNIZING IMPACTS OF OVERABUNDANT DEER

Realization that too many deer were causing too much damage to community resources grew over time among community leaders and members. People recognized the impacts of the overabundant deer herd: the distinct deer-caused browse line that reduced the privacy of wooded lots, the barren forest floor devoid of beautiful and diverse wildflowers and shrubs, the decimated appearance of unhealthy deer, and the notable reduction in diversity and abundance of songbirds and small mammals. Natural assets are integral to residents' love of nature and life in the woods, and asset losses were felt by the residents. They bore the associated economic and human safety impacts resulting from vehicle collisions with deer, removal of deer carcasses, ineffective efforts to protect native and ornamental plantings from browsing deer, risk to maintenance of structure and composition of forest vegetation, tick-borne diseases vectored by deer, deer feces on beaches and ballfields, a lack of healthy wetlands and forests that purify their drinking water, and flood damage when heavy rains fell that would have been less if forests and meadows were intact ecological systems functioning to absorb and infiltrate the rains. HF residents had been concerned about these issues for many years but were conflicted by their love of deer as creatures symbolizing Nature's beauty. Clashes between members grew over how to manage the deer issue. The Board committed to addressing the issue and developing and implementing a deer management program.

37.3 INITIAL STEPS TO MITIGATE IMPACTS OF OVERABUNDANT DEER

In 1997, a controlled public archery hunt was approved by The Board to reduce deer density. HF developed a management process that included identifying and selecting skilled and eligible bow hunters and directing them to large, contiguous forested area within HF during deer archery season in Pennsylvania. However, only nine deer were removed by the hunters, which was far fewer than the number needed to reduce the herd to a sustainable population. HF members who opposed the action protested the hunt and filed a court action, resulting in a legal agreement requiring a majority vote of approval at the annual membership meeting for future "deer hunts."

To be more inclusive and facilitate input and agreement among HF residents for future deer management, The Board approved the formation of the Environmental Committee in 1998. Studies were commissioned from experts in deer biology, ecology, and conflict resolution to quantify and describe impacts of the overabundant deer herd and seek solutions acceptable to HF residents. A deer necropsy study confirmed that the "deer behavior and physical condition appears to have been seriously altered due to unnaturally high deer density resulting in an unhealthy depauperate forest and a lack of adequate nutrition to maintain 'healthy deer'" (Natural Resource Consultants 1998). Two fenced deer exclosures were erected in 1999 along a hiking trail close to the HF Mail Room facility and ballfield (high human traffic areas) to demonstrate that vegetation can germinate and grow under a forest canopy and in an open meadow if protected from excessive deer browsing. Dr. Ann Rhoades, University of Pennsylvania senior botanist and author of *The Plants of Pennsylvania*, inventoried vegetation in the deer exclosures, inspected forest conditions in HF, and noted that the impact of white-tailed deer on the plants and forest: "Hemlock Farms is certainly one of the most severely impacted sites I've seen anywhere" (Rhoades 2000).

A contracted Forest Stewardship Plan documented widespread damage of HF forests by deer (Hyde and Hall 2000). A 2001 communitywide Deer Questionnaire revealed that "being close to nature" was a major reason people live in HF, but that conflicts with deer were prevalent. In 2003, a field tour of fenced areas on nearby state forestlands was presented by Pennsylvania Bureau of

Forestry foresters and attended by HF residents and board members to gain understanding about the impacts of deer on forest regeneration.

37.3.1 ESTIMATING THE NUMBER OF DEER

Concerned HF members wanted to know "How many?" How many deer were in HF, how many deer represented a sustainable population, and how many deer had to be removed for how many years? HF commissioned a deer survey (Natural Resource Consultants, 1998), and counts were taken by community volunteers from 1999 to 2004. Results showed a highly variable range: 120–741 deer within HF. Furthermore, the Natural Resource Consultants estimated deer density ranging from 15 to 46 deer/km². Three aerial surveys using forward-looking infrared surveys (FLIRS) were taken May 2004, May 2005, and November 2005; counted 264, 141, and 149 deer, respectively; and noted that most deer were browsing along the roads, at the edges of developed properties, and on the golf course (Bernatas 2004, 2005). The variability of the results from various counting methods provoked thought and argument about setting a number of deer for removal.

37.3.2 COMMUNICATING THE IMPACTS

Simultaneously, HF management developed and delivered educational material to residents, committees, and The Board. In 1999, a code prohibiting feeding deer and bear was recommended by the Environmental Committee, approved by The Board, and enforced beginning in 2001. Commissioned studies were made available to HF residents. Relevant deer impact issues were emphasized in HF's monthly newspaper and articles, e.g., citing needs of birds and other animals for diversity and abundance of trees, shrubs, and other plants for forage and cover, and results from the Deer Questionnaire revealing among other things that "being close to nature" is a key resident value.

37.3.3 DEVELOPING A DEER MANAGEMENT PLAN

In 2002, Dr. Gary Alt and Dr. Marrett Grund from the PGC presented deer management information to the Environmental Committee and related the information to the deer situation in HF. Shortly thereafter, the United States Department of Agriculture Animal and Plant Health Inspection Service (USDA APHIS) Wildlife Services (hereinafter referred to as WS) issued an environmental assessment of white-tailed deer management in Pennsylvania (USDA 2003).

Particularly influential was this quote by Dr. Gary Alt, internationally recognized expert on deer and bear biology for the PGC: "I have spent much time in Hemlock Farms over the past 30 years studying the bear and deer, and have seen the enormous impacts of too many deer. It is critical that the deer herd be controlled and maintained at about 8–10 deer per square mile (3–4 deer/km²) to allow the vegetation in your community to regenerate. Tough decisions HF makes today will greatly affect the health and sustainability of your forests for wildlife, and people to enjoy, in the decades to come."

37.3.3.1 A Framework

By 2002, a framework of goals, assumptions, limiting realities, and evaluation criteria for deer management on HF had been developed to direct an educational effort for obtaining consensus among HF residents for a deer management program.

Proposed goals relating to deer management in HF included: (1) maintain and restore the forest throughout as much of HF as possible, (2) substantially reduce deer impacts to landscape plants, (3) substantially reduce problems and costs of deer carcass removal and disposal, (4) substantially reduce risks of vehicle–deer collisions, (5) minimize risks of contracting tick-borne illnesses, and (6) substantially reduce nuisance of deer feces at recreational facilities.

Certain assumptions and realities limited the types of possible options. Assumptions: (1) a population of healthy deer in HF was desired by residents; (2) acknowledgement that efforts to manage the deer population, once initiated, must be pursued for at least 15 years or risk failing; (3) efforts to manage the deer population, once initiated, will require accurate, annual estimates of the deer population; (4) management actions must be in accord with state and community laws and policies; and (5) lethal population reduction would require support from a majority of residents, as indicated by community vote in the annual elections, and be approved by the PGC.

Criteria to evaluate deer management options were established: (1) viable deer management options should be safe, effective, humane, and economical; and (2) the options should be flexible and allow for adapting to changing circumstances. In addition, it was acknowledged that the deer herd and hunting policies on the Delaware State Forest and other hunting grounds surrounding HF influenced management outcomes on HF.

A spreadsheet was developed for scoring management options based on criteria of availability, safety, humaneness, effectiveness, specificity, initial cost, long-term cost, and reversibility. Potential for each management option was scored based on perceived chances of success for each criterion, where a score of 1 = not likely to succeed, 2 = moderately likely to succeed, and 3 = most likely to succeed.

This framework of goals, assumptions, criteria, and scoring were distilled into a working document to facilitate informing and educating residents, and eventual choice of management option.

37.3.3.2   Options not Viable

Based on this scoring, management options rejected as not viable—with reasons—were: (1) introduction of historical deer predators (e.g., mountain lions and gray wolves)—failed all criteria; (2) trapping and transferring deer elsewhere—stringently regulated by the PGC because of concerns about diseases such as chronic wasting disease, requires veterinary evaluation of every animal, difficulty of finding recipients to take deer, high stress to deer, and prohibitive cost; and (3) poisoning deer—not allowed by state, dangers to humans and other species.

37.3.3.3   Viable Options

Management options rated as potentially viable—with identified concerns—were: (1) no action—fails to address deer impacts on forest and human conflicts; (2) selective fencing/protecting planted tree seedlings with individual protectors/roadway reflectors (for frightening deer)—limited effectiveness; (3) a boundary fence/driving deer out of fenced area/lethal removal of deer broaching fence—high capital cost, possible impacts on bear and deer movements, genetic pool in the fenced deer population; (4) immunocontraception—currently only effective with small "captive" deer populations within isolated landscapes such as islands or fenced landscapes, not likely to be approved by PGC; (5) public hunting requiring special "deer depredation permits" to allow hunters to harvest multiple antlerless deer/limited to "Green Belt" areas, on HF golf course, and/or on abutting private property/volunteer hunters—requiring prior PGC approval, human safety concerns, difficulty in managing and doubtful effectiveness; (6) contracting sharpshooters to hunt from designated bait/feeding stations—potentially effective and relatively economical, human safety concerns; and (7) trap and euthanize deer—safe and effective, meat and hide donated to agencies for distribution to people in need, costly ($300+/animal).

37.3.3.4   Optimal Options

Three options were selected as optimal for HF: (1) trap-and-kill; (2) sharpshooters; (3) boundary fence/trap-and-euthanize, sharpshooters, "deer drives," or immunocontraception to reduce the deer population in the fence.

37.3.3.5   Setting a Sustainable Deer Density Goal

Based on multiple informed sources, and with the assistance of the WS (USDA 2016), HF management developed a deer management plan. A goal of 4 deer/km² of available deer habitat (13 km²), which is a count of roughly 50 deer in HF, was established as a sustainable deer density that the habitat in HF could support. Available habitat was the total area within the community (18.7 km²) minus the area occupied by roads, lakes, and ponds (5.7 km²)

37.3.4   Enrolling Stakeholders in the Deer Management Plan

37.3.4.1   Education, Education, Education

In 2004, the HF Environmental Committee recommended that The Board develop a communitywide educational effort explaining how the goals for deer management were developed and the assessment of viable options. The plan was to seek a vote of HF residents at the 2005 annual membership meeting on a selected deer management option. Thus began a year and a half of seminars on viable options, deer ecology, forest restoration, and expert panels. A series of "Deer Fact Sheets" was widely disseminated to members, and articles in the *Hemlock News* further educated HF residents on the importance of adhering to the "Do Not Feed Deer and Bear." Education efforts included publishing the HFCA code, explaining the ecological and public health impacts of too many deer, reporting results of commissioned studies, and describing how recommended deer management options would be implemented with associated costs.

Coincidentally, in January 2005, Audubon Pennsylvania and the Pennsylvania Habitat Alliance convened a 1-day statewide forum to compile and examine pertinent research, enlist expertise, weigh the issues, and produce a vision of what ecosystem-based deer management might entail in large forested areas of eastern United States, using Pennsylvania as an example (Latham et al. 2005). Changes to deer management at the state level, regulated by the PGC, were occurring that could help or hinder development of deer management goals in HF. HF community leaders were active participants at the forum, where they voiced urgency in helping communities such as HF to resolve the conflicts and mitigate damage resulting from too many deer.

The consequential nature of actions being studied by HF management was not lost on The Board. In early spring 2005, The Board selected the sharpshooter option for managing deer. Government wildlife biologists who were expert sharpshooters would cull the deer at night, dress the carcasses, and deliver the meat for processing and donation to local food banks. The decision for accepting and implementing this option was to be placed before HF residents for a vote at the July 2005 annual membership meeting. To assist in the educational effort, a public relations firm was hired in early spring of 2005 to intensify and focus the educational effort and ensure the information was presented to HF scientifically, authentically, and with clarity. An informative brochure and poster were designed, printed, and presented to HF.

37.3.4.2   The Ballot

The ballot proposition presented to members was to: "Authorize the Board of Directors to use lethal means of deer population control, if needed, to maintain our goals that are to restore our forest and healthy deer. This authorization shall expire on July 8, 2006, or on the date of the 2006 Annual Membership Meeting, whichever occurs first." A detailed rationale accompanied the proposition. The proposition passed with a 58% majority. The proposition was voted on annually, and educational efforts continued each year ahead of the vote. Every year, HF members voted to continue with the sharpshooter option, voting for continuance with a majority vote as high as 72%.

37.4   IMPLEMENTING THE DEER MANAGEMENT PLAN

37.4.1   Deer Density Survey

The initial aerial surveys of deer density commissioned by HF (aerial FLIRS) were the starting point for estimating the number of deer to remove during the first deer cull. Every year thereafter, distance sampling methodology (Willy Wenner, wildlife biologist, USDA APHIS Wildlife Services, personal communication) was employed to collect information for estimating deer density along an 80-km survey route passing through representative portions of HF.

Survey teams consisting of a driver, data recorder, and two observers recorded distance deer were from the closest point on the survey route by time, gender, and age class. Data were collected from a survey vehicle moving 6–9 km/hour. Surveys began around dusk and generally concluded around midnight; observations were collected using hand-held FLIR units, spotlights, binoculars, and range finders. Distances of deer observed from the vehicle were entered into a database. Distance sampling calculations were employed to calculate deer density for determining the number of deer to be removed each year from HF to reach the deer density goal.

The initial survey route was limited to one of three townships (which included 90% of HF) because the PGC required that a municipality apply for the deer control permit for removals by sharpshooters under contract with HF. A few years later, the PGC allowed private communities to apply for the permit directly. As a result of this change in PGC policy, the survey route was adjusted to include all of HF (Figure 37.2). Survey data are entered into a database that calculates average observation distances, area surveyed (square km), deer density estimates (deer/km²), and buck to doe ratios.

It is important to note that individual surveys are snapshots of deer density on the nights of data collection. There are many factors affecting deer observations that must be considered when interpreting data. Weather conditions, seasonal movements, and hunting pressure can increase or decrease deer observations on any one night. Thus, density data represent trends more than actual population estimates. Multiple estimates taken prior to beginning deer removal operations are averaged to determine the number of deer to remove each year. Estimates are useful for comparing deer density trends over time.

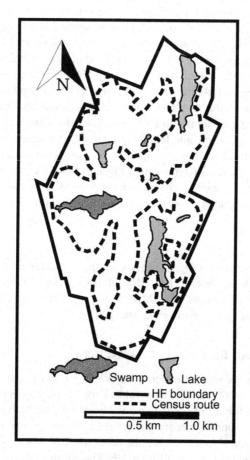

FIGURE 37.2   Deer census route on HF.

37.4.2 Culling Deer

Reducing deer density to goal is accomplished each year with a combination of public hunting and removals by contracted sharpshooters. As a condition of the deer control permit that allows sharpshooters to remove deer, the PGC requires that an active Deer Management Assistance Program (DMAP) for hunters be implemented to the fullest extent possible in the project area. HF management developed a process and purchases hunting insurance from Quality Deer Management Association (QDMA) for controlled hunting on a designated 32-hectare huntable area in HF. HF applies to the PGC for 10 DMAP coupons and opens a sign-up period for members who are licensed hunters. DMAP hunters complete HF registration and permission forms and submit the coupons to the PGC for extra doe licenses. The DMAP hunters hunt deer during deer seasons on the designated huntable area, report their takes to HF and PGC, and may keep and consume deer taken through public hunting. Another condition of the deer control permit is a strictly enforced "No Feeding" policy for deer in HF. HF residents are informed that any interference with public hunting or the government contractors is a violation of state law and carries fines.

Deer density surveys are conducted prior to deer removal by the sharpshooters and after the statewide public hunting deer season closes. Surveys provide the quantitative basis for the PGC to issue a specified number of tags to HF management for sharpshooting operations. Deer culling by sharpshooters is based upon a protocol collaboratively established among the HFCA, the PGC, and WS. Sharpshooter removal operations were initially scheduled from October through March. After the deer density goal was achieved at the end of removal operations in March 2008, culling by sharpshooters was limited to January through March to provide optimum hunting experiences for public hunting on HF. Bait sites for culling are placed to draw deer out of dense cover and into common areas to position them for safe shooting during sharpshooter operations. Sharpshooting is conducted with 0.223- and 0.243-caliber rifles equipped with silencers. FLIR units and night vision goggles are used to locate and observe deer in complete darkness and enhance the ability to ensure safe removal operations by detecting civilians, rocks, and homes located on the property. Nighttime shots are taken using spotlights and filtered lights. Deer selection is conducted by targeting antlerless deer first. Large-antlered deer are not selected for removal. Remaining deer are removed on a first opportunity basis, provided safe shots can be taken. Adult does are targeted when more than one deer is present. Data (sex, age, time, and location) from culled deer are recorded. Ages of deer are determined by evaluating tooth wear and eruption. Ages are recorded in half years from October through December and whole years from January through March. Each deer taken is tagged. Field-dressed deer carcasses are delivered directly to a meat processor the morning following each night of sharpshooting operations, and HF donates the venison to local food banks.

37.5 RESULTS

A successful deer management program relies on safe, humane, effective, and economical implementation. Bringing sharpshooters into HF at night to shoot deer and instituting public hunting in HF had to be safe. After 12 seasons, HF DMAP hunters and contracted sharpshooters culled deer professionally without a safety or controversial incident.

37.5.1 Deer Surveys and Removals by Contractors

Annually, preculling surveys were conducted by WS biologists to estimate deer density on HF and provide a target number of deer for removal to drive density to goal. Deer densities dropped significantly after culling (Table 37.1), with two notable exceptions, both of which were attributed to movement of deer into HF—one movement was to seek forage because a failure of acorn production forced them out of their typical home ranges; the other movement was to seek a safe haven during hunting season after two seasons of no culling in HF. The number of deer removed annually by WS

TABLE 37.1

History of Deer Surveys and Removals during Deer Culling in HF

Date	Deer/km²	Number Deer Culled
10/26/2005–01/19/2006		176
02/06/2006–03/29/2006		190
10/16/2006	37.0	
10/23/2006–12/14/2006		134
01/03/2007	15.8	
01/22/2007–03/28/2007		67
04/24/2007	6.6	
10/02/2007	15.8	
10/15/2007–12/18/2007		61
01/16/2008	11.2	
02/05/2008–03/13/2008		59
04/08/2008	6.9	
10/08/2008	11.6	
10/15/2008	10.8	
10/20/2008–03/19/2009		102
03/25/2009	3.1	
09/21/2009	4.8	
09/22/2009	7.8	
12/14/2009	7.6	
12/15/2009–12/18/2009		28
12/13/2010	6.4	
12/14/2010	5.2	
01/18/2011–03/8/2011		25
12/13/2011	17.1	
12/14/2011	19.1	
01/09/2012–01/13/2012		77
01/17/2012	8.5	
01/19/2012	4.6	
01/23/2012–01/24/2012		10
12/11/2012	2.2	
12/14/2012	2.0	
12/17/2013	1.2	
12/19/2013	5.1	
12/15/2014	14.2	
12/17/2014	12.7	
01/20/2015–01/23/2015		34
03/23/2015	18.1	
04/01/2015–04/09/2015		56
04/21/2015	9.6	
12/15/2015	3.6	
12/17/2015	9.2	
01/05/2016	4.0	
02/02/2016–03/04/2016		22
12/12/2016	3.9	
12/14/2016	4.2	

sharpshooters reflected the attempt to tie deer removals to annual estimates of preculling density to reach the sustainable goal.

WS sharpshooters removed 1054 deer in 2005–2016. The goal of 4 deer/km^2 was achieved after four seasons of culling (the 2009 season). As expected, annual recruitment of fawns and young bucks increased the deer herd in HF by about 30% or 20–25 deer for the next 2 years (2010 and 2011). During these 2 years of deer culling, two deer density surveys were conducted after the regular firearm deer hunting season and just before deer removal operations to determine the number of deer to be removed by sharpshooters. Precull deer surveys revealed an increase that would have been expected from normal fawn and young buck recruitment into the HF herd.

In 2011, a regional-low acorn mast production occurred in the eastern United States (including within the landscape surrounding HF) and caused a disruption in the regional deer herd for several years. During the winter of 2011–12, many deer migrated into HF seeking food on the ballfields, along the roadsides, and on the golf courses. An increase in number of deer was counted during the December 2011 surveys, resulting in an average density of 18 deer/km^2 (approximately 235 deer within HF within the area of deer habitat). A preliminary goal was to cull 185 deer; however, due to the significant increase of deer over the previous year, WS recommended 1 week of removals followed by two surveys to reassess deer density. Over 4 nights, WS removed 77 deer and then performed midremoval surveys, estimating an average density of 4.3 deer/km^2 on HF. Ten more deer were removed and operations ceased for the season. When normal mast crop production returned in 2012, precull surveys in December of 2012 and 2013 revealed that deer densities were under goal density, and zero deer were culled. The reason for the abnormally low deer density was assumed to be that nonresident migrant deer had returned to their normal home range.

In December 2014, an unexpectedly high number of deer was observed. After two seasons of significantly reduced hunting pressure in HF, e.g., no deer culls, WS assumed that deer on the adjacent lands sensed that HF was a "safe haven" from hunters and sought "refuge" in HF. Two precull surveys were performed, averaging 13.4 deer/km^2. Thirty-four deer were removed, and another survey resulted in 18 deer/km^2. Fifty-six more deer were removed and a survey revealed 9.6 deer/km^2. Even though this density was above goal (4 deer/km^2), further removals were stopped for the season because of the highly variable surveys and unexpected deer behavior.

In December of 2015, three precull surveys resulted in highly variable densities averaging 5.6 deer/km^2. Twenty-two deer were removed over four nights during the winter; then culling ceased for the season. The following season produced similar estimates of 3.9 and 4.2 deer/km^2. These densities were close to goal density and no culling occurred (USDA 2016 and 2017).

37.5.2 Deer Removals by Deer Management Assistance Program Hunters

HF DMAP hunters harvest is a small portion of the total deer cull (Table 37.2). However, public hunting opportunities provide a way for residents to contribute to stewardship of the HF deer herd.

Once residents ceased feeding deer, deer behavior normalized. Deer are rarely seen grazing on roadsides or ballfields and golf course at all hours of the day and night. They remain in the forests and meadows, establishing regular feeding routes more suited to their crepuscular nature. Members notice that the deer appear to be larger and do not exhibit signs/symptoms of injury or disease.

TABLE 37.2

Number of Deer Removed Annually by HF DMAP Hunters, 2005–2016 (HFCA Internal Records)

2005	2006	2007	2008	2009	2010	2011	2012	2013	2014	2015	2016
1	2	4	1	0	2	4	1	2	1	1	2

37.5.3 REDUCTIONS IN VEHICLE–DEER COLLISIONS AND CARCASS REMOVALS

The number of collisions between vehicles and deer dramatically decreased once culling began (Figure 37.3). HF residents benefit with lower insurance and vehicle repair bills. HFCA insurance and vehicle repair costs also decreased, which saves on member's dues. Travel by the 10,000 residents and guests on the 120 km of roads in HF is much safer, which is a significant assurance in this gated community.

Prior to deer culling, the number of deer found dead in the community was very high. Carcasses were found along HF roads from collisions with vehicles, or members complained of dead deer on their private property. Managing disposal of carcasses was too time consuming for the HFCA Public Works Department, and a contractor was hired for this job. The number of carcasses decreased notably after beginning deer culls (Figure 37.4). Costs of removing dead deer declined correspondingly to the point where costs were removed as a budget item in 2011.

37.5.4 DONATIONS TO FOOD BANKS

Significant goodwill was generated by total donation of 14.2 tons of lean venison to local food banks between 2006 and 2017 by HF through Hunters Sharing the Harvest, a Pennsylvania-based charity that facilitates the donations (Table 37.3).

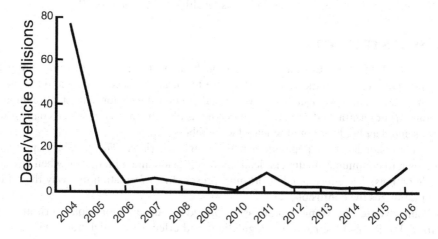

FIGURE 37.3 Number of collisions between vehicles and deer reported to HFCA.

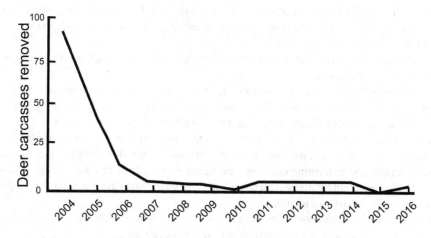

FIGURE 37.4 Number of deer carcasses removed from HF.

TABLE 37.3
Kilograms of Venison Donated to Local Food Banks through Hunters Sharing the Harvest

2006	2007	2008	2009	2010	2011	2012	2013	2014	2015	2016	2017
4911	2517	1658	1411	374	431	1255	0	0	1320	323	0

37.5.5 RECOVERY OF FOREST REGENERATION

Research-quality data on forest regeneration were not collected. However, recovery of understory vegetation in HF is well noted. For example, a highly visible field that previously was a sheep pasture has regenerated from a closely cropped field of weeds to a diverse cover of herbs, shrubs, and deciduous and coniferous trees, a portion of which was established and certified as a Monarch Butterfly Habitat. Photographs and members' testimony of the recovering and restored forests were sufficient to produce a majority "Yes" vote of approval for the continued use of lethal means to manage the deer population as needed.

After almost two decades of examining options, educating members, and implementing an integrated deer management plan, management reduced the deer herd in HF to sustainable levels, where it has remained. As a result, deer and forest health were restored in HF.

37.6 LESSONS LEARNED

- Education. Education. Education. Education laid the foundation for community trust and provided a benchmark that members use to measure success. Education is ongoing because new members constantly move into HF and must be informed for the annual vote to continue or terminate deer culling. Education is also the means by which changes to the forests and deer herd are communicated to members.
- Surveys of abundant deer populations may be highly variable. The variability presents an unknown to community members, leading to confusion and a hindrance when advancing viable options. The educational effort must present the information in a way that informs and minimizes the confusion.
- In Pennsylvania, the Game Commission provides various regulations permitting deer culling. HF learned the rules and regulations and effectively collaborates with state and federal agencies to manage its deer herd.
- Deer surveys are most critical at the time of deer removal operations—at the beginning to determine numbers for removal, sometimes in the middle when the deer population has been unpredictable, and sometimes after removal operations to verify that the sustainable goal has been reached.
- Safe, humane, effective, and cost-controlled methods are available to manage deer populations in communities.
- As with all natural resources, monitoring the resource is paramount to successfully managing the resource. Carefully chosen and forward-thinking "photo points" that represent the diversity of habitats and sites within HF would have provided a tool to visually document the progression of forest recovery and inform members on the progress.
- Ecological changes affect deer behavior. A consistent method to monitor deer population density trends and determine deer removal numbers is critical to establish at the onset of operations, and adaptive management policies and procedures are necessary to respond to changes so that goals can be met.
- Natural resource professionals are needed to assist, facilitate, collaborate, innovate, control, adapt, and oversee deer management programs in communities.

EDITOR COMMENTS

This case history, like most others, made use of an effective communications effort to educate stakeholders and administrators on the deer–forest situation, review potential management actions, and convince administrators to embark on a program to reduce deer density by hunting/culling. And, like most other case histories, the manager enlisted the help of established professionals to assess the deer management situation and offer qualified advice on course(s) of action to consider. What was unusual about this case history was the ability of the manager to successfully promote the idea of using sharpshooters and resident hunters to reduce deer density in a residential area using rifles rather than bow and arrow—certainly grist for the other case history managers to consider in enhancing their reliance on bowhunters to reduce deer density.

David S. deCalesta, editor

REFERENCES

Bernatas, S. Vision Air Research. May 9, 2004; May 13, 2005; November 23, 2005. *White-tailed deer survey for Hemlock Farms*. Lord's Valley PA: Studies Commissioned by HFCA.

Hyde, P. J. and B. Hall. 2000. *Forest Stewardship Plan for the property of Hemlock Farms Community Association*. The Forest Management Center. Program administered for private managers by the Pennsylvania Department of Conservation and Natural Resources Bureau of Forestry.

Kosak, J. 1995. *The Pennsylvania Game Commission, 1895–1995: 100 years of wildlife conservation*. Harrisburg PA: Pennsylvania Game Commission.

Latham, R. E., J. Beyea, M. Benner et al. 2005. *Managing white-tailed deer in forest habitat from an ecosystem perspective: Pennsylvania case study*. Harrisburg PA: Report by the Deer Management Forum for Audubon Pennsylvania and Pennsylvania Habitat Alliance.

Natural Resource Consultants, Inc. 1998. *Deer necropsy report for Hemlock Farms*. Hawley PA: Study commissioned by HFCA.

Rhoades, A. 2000. Deer exclosure plant inventory. Study commissioned by HFCA.

USDA APHIS Wildlife Services. 2003. *Environmental assessment and decision and Finding of No Significant Impact (FONSI) for white-tailed deer damage management in Pennsylvania*. United States Department of Agriculture Animal Plant Inspection Service.

USDA APHIS Wildlife Services. 2005 through 2017. *Activities summary report white-tailed deer management program, Hemlock Farms Community Association, Lords Valley, Pike County PA*. Annual reports of deer management activities. United States Department of Agriculture Animal Plant Inspection Service.

USDA APHIS WILDLIFE SERVICES 2016 (Updated). *White-tailed deer management plan; Hemlock Farms Community Association, Lords Valley, Pike County, Pennsylvania*. United States Department of Agriculture Animal and Plant Health Inspection Service.

38 Hamilton Small Woodlot Case History

Managing Deer and Forests via Comprehensive Silviculture Aided by Grants

Sue Hamilton and Jeff Hamilton

CONTENTS

> Persistence and determination are always rewarded.
>
> **Christine Rice**

38.1 INTRODUCTION

Growing up hiking and camping in the Pine Barrens of New Jersey instilled a love and appreciation for forests—my husband and I purchased 90 ha of northern hardwood/Allegheny hardwood forestland in Sullivan County and moved our family there to live, work, and nurture respect for the natural world in our four children. A primary forest road and secondary trails provide good access to most of the property for timber management and recreational use. Pole Bridge Creek runs through our property, which is 85% forested. An 8-ha lake with a meandering shoreline provides opportunities for recreation and wildlife viewing. Its open water, shoreline, and wetlands habitats attract beaver, otter, mink, heron, bald eagles, osprey, water fowl, deer, and bear.

Our property is located less than 800 m from the small town of Laporte, Pennsylvania, surrounded by "big woods" (large tracts of northern hardwood forest including hemlock [*Tsuga canadensis*]). Our intent has been to manage the property for recreation (hiking, cross-country skiing, running, swimming, observing wildlife, and fishing) and for sustainable management of other forest resources, including timber. However, the forest did not look healthy to our inexpert eyes when we purchased it. Accordingly, we contacted Extension forestry specialists from Pennsylvania State University and local consulting foresters for evaluation, advice, and recommendations. With their help, we created a stewardship plan, setting goals and specifying activities for managing our forest resources. At their recommendation, we attended the Penn State Forest Stewardship (PFS) program, which provided us 40 hours of forestry education in weekend sessions in return for our promise to use this education to help others in our community.

38.2   ASSESSMENT

Our stewardship plan identified five timber stands on the property, each with different tree and understory species and of differing age and vertical vegetative structure. A federal cost share program, which paid for part of the plan, required a timber stand appraisal, which surprised us by identifying over $100,000 of sawtimber value, primarily black cherry *(Prunus serotina)*. We also learned that timber from the stands generally had been overharvested, leaving few good seed trees of desirable species (black cherry, red and sugar maples [*Acer rubra and A. saccharum*], and hemlock) in the overstory and with relatively little stocking of replacement seedlings. Over the previous decades, our forest had also suffered the higher elevation decline of ash (*Fraxinus americana*) and sugar maple, beech bark disease, outbreaks of elm spanworm (*Ennomos subsignarius*), forest tent caterpillar (*Malacosoma disstria*), and anthracnose fungus. Deer browsing had eliminated most seedlings of desirable trees, leaving understories dominated by undesirable tree species resistant to deer browsing (American beech [*Fagus grandifolia*], birches [*Betula sp.*], striped maple [*Acer pensylvanicum*]) and ferns. The stand that was dominated by hemlock was threatened by a possible/ probable hemlock woolly adelgid (*Adelges tsugae*) infestation, and/or hemlock elongate scale disease (*Fiorinia externa*), which might lead to high hemlock mortality over the next decade.

38.3   MANAGEMENT RECOMMENDATIONS

Deer move through the property regularly despite poor forage conditions. The forestry specialists and consultants stated that deer were and would continue to be a major cause of management failures and recommended reducing deer abundance and impact through hunting over the entire property. They emphasized harvest of antlerless deer to reduce reproduction. If deer harvest could not reduce the local deer herd sufficiently to reduce their impact, the foresters recommended deer exclusion fences to protect seedlings in the three stands receiving vegetation management.

Following the foresters' recommendations, we established goals (provide healthy and diverse forest regeneration for the environment, wildlife, and timber production) and drew up a management plan to achieve the goals. Treatments focused on removing undesirable vegetation, fencing, planting, and hunting to reduce deer density and impact on regeneration. The foresters detailed treating understory with herbicides to kill beech and birch brush and fern in one stand, thinning to remove undesirable trees from the overstory in another, and underplanting conifers and other species in groups in established small openings in a third. They suggested a wait-and-see attitude for the remaining two stands to determine whether existing seedlings would develop to regenerate one of the stands and whether the incipient woolly adelgid infestation in the hemlock stand would require or pre-empt hemlock management in the other.

38.4   REDUCING DEER DENSITY

Hunting was a new adventure for us. We had no prior experience but learned that hunting is an important tool for deer management and a source of meat. We took hunter education training, got hunting licenses, and hunt our property with rifle and crossbow. We augment our hunting effort by inviting friends to hunt, but we are not comfortable opening the property to public hunting or to persons we do not know. Our close proximity to town, where deer are protected, resulting in high deer density in the immediate landscape, and finding the time to hunt have been our biggest challenges. Our hunting consists of small drives with friends and hunting from tree stands and ground blinds. We harvest approximately one deer per year, which does not result in appreciable reduction in deer impact. We are trying to find ways to make our private hunting more effective. We have not had much communication with neighbors regarding hunting. Neighboring lands are an amalgam of hunt clubs, vacation home owners, leased forest land, and a state forest.

38.5   DEER MONITORING

We have not conducted scientific monitoring of the deer density on the property. Simple observation of animals and their impact on the forest, communication with neighbors' sightings, and use of a game camera make clear that there are too many deer visiting the property to allow healthy forest regeneration without reducing deer density or building fences to protect vegetation.

38.6   MANAGING THE HEMLOCK FOREST

The forest around the lake and adjacent to the creek is dominated by hemlock. Since the 2002 survey, they have developed an elongate hemlock scale (*Fiorinia externa*) infestation, causing much hemlock mortality. We have not performed a salvage hemlock harvest for aesthetic reasons, as well as the wildlife benefits and some seedling protection provided by dead and downed hemlock trees. Following our forester's recommendations, we continue to underplant these areas with other conifer species, as they are important for the protection of waterways and as wildlife habitat. We planted 1650 red spruce (*Picea rubens*), 300 Norway spruce (*Picea abies*), 750 white spruce (*Picea glauca*), 250 black spruce (*Picea mariana*), 10 Douglas-fir (*Pseudotsuga menziesii*), 50 red pine (*Pinus resinosa*), 55 white pine (*Pinus strobus*), and 400 Meyer spruce (*Picea meyeri*) seedlings in this area. Other seedlings and bushes planted around the lake for diversity and wildlife benefits include 50 swamp white oak (*Quercus bicolor*), 10 black gum (*Nyssa sylvatica*), 10 silky dogwood (*Cornus amomum*), 100 elderberry (*Sambucus nigra*), 10 winterberry holly (*Ilex verticillata*), 25 dwarf willow (*Salix herbacea*), and 100 buttonbush (*Cephalanthus occidentalis*).

Generally speaking, we do not protect the conifer seedlings with tree shelters, except some close to the lake to prevent beaver (*Castor canadensis*) from felling and eating the seedlings when they reach heights of 1 m or greater. Building, transporting, and placing so many of these shelters throughout a forest would be a monumental task, as we do most of the work ourselves. The deer may browse a bit on the spruce seedlings, and when the trees reach a height $\geq$60 cm, we find some antler-rubbed by deer.

In addition to impact from deer and beaver, many factors affect survival rates of seedlings. Matching the site conditions to a seedling's growth requirements is important, and all require adequate amounts of water. Seedling plugs cost approximately 60% more than bare root seedlings, but we find their survival rate approximately double that of bare root seedlings. Shipped seedlings often arrive in poor condition and mortality rates often are greater than those of seedlings we can purchase at nurseries close to home.

38.7   PLANTING AND PROTECTING SHRUBS AND
TREES FROM DEER BROWSING

We wanted to improve wildlife habitat by planting recommended tree and shrub species for wildlife. We knew we would not be able to reduce deer density by hunting sufficiently to ensure survival and development of planted shrubs and trees, so we began protecting plantings with a variety of exclusion devices. During our forestry training (PFS), we learned about, but more importantly, observed the value of deer exclusion fencing. Many folks were using plastic tree tubes to protect seedlings and

FIGURE 38.1   Individual tree protector around a white oak seedling. (Photo by Sue Hamilton.)

enclosing small areas up to 8 m in diameter with 1.2- or 1.5-m high fencing. Deer are hesitant to jump into small enclosed areas of this size.

We protect hardwood seedlings and shrubs from deer with fences or cages. We have planted many seedlings and experimented with different types of tree shelters. We tried plastic tree tubes and 45-cm-tall plastic tubes topped with 90-cm black plastic netting, but they became playthings for local bear and did not protect the seedlings. We tried several types of plastic mesh or lengths of wire fencing to enclose seedlings. The only shelters that have worked for us are of 1.2–1.5-m-high welded wire fencing, formed into 45–60-cm-diameter cylinders and held in place by two sturdy stakes (Figure 38.1) to protect individual planted trees. We have used these tree shelters in forest openings (up to 0.2 ha) and to interplant seedlings in the forest. This type of tree protection is very expensive in terms of cost (currently approximately $7/shelter) and labor. Nevertheless, the shelters have been a successful management tool and are reusable.

We have tried using several types of material to enclose small areas 7.5 m in diameter. We recommend using 30-m rolls of welded wire fencing 1.2 m high and 10 heavy-duty 1.5–2-m steel T-posts for each fence (Figure 38.2). If the fence is not sturdy, deer and/or bear will break through or heavy snow will collapse the fence. We tie strips of marking tape every 3 m or so along the top of the fence in order to make them more visible. The cost is approximately $125 for each enclosure, but they are manageable for a landowner. They can be used in smaller forest openings, they are reusable, and deer did not enter them.

In our single largest project, we employed a variety of silvicultural practices to regenerate a 12-ha forest stand. The area held much of our current timber value in black cherry as well as potential for producing subsequent timber harvests and wildlife benefits. Beneath the understocked but valuable canopy, we had extensive beech and birch brush and hay-scented fern (*Dennstaedtia punctilobula*) with little species diversity. Beech bark disease (*Cryptococcus fagisuga* scale insect and *Nectria coccinea* fungus in concert) would prevent the beech from ever becoming useful timber, and black birch (*Betula lenta*) is taking over as a dominant species with little prospect of future value. If we

FIGURE 38.2 Small area fencing to protect seedlings. (Photo by Sue Hamilton.)

had harvested the timber from this stand in 2002, the forest would never have returned to productive or diverse growth.

The first step was an herbicide application to eradicate hay-scented fern and the undesirable beech and birch brush understory. Then, knowing we would be unable to negate browsing by an overabundant deer herd, we enclosed the regeneration site with a 2.4-m-tall deer exclusion fence (Figure 38.3).

The herbicide application successfully removed the fern, but by the second growing season, it was clear it had failed to remove the beech and birch brush understory, which was just a few feet higher

FIGURE 38.3 Fencing surrounding regeneration site. (Photo by Sue Hamilton.)

than the reach of the herbicide spray. Because the tops escaped the spray, the brush trees recovered. We needed to try something else. We tried hand-clearing and herbicide, treating the small stumps in a 1/2-ha test plot within the fence. Several spot treatments to eliminate undesirable seedlings followed. This was effective but very labor intensive. Despite a limited overstory to provide seed, red maple and cherry seedlings grew. We planted with 50 red oak, 50 sugar maple, 25 tulip poplar (*Liriodendron tulipifera*), 25 serviceberry (*Amelanchier* sp.), 50 winterberry holly, and 25 Fraser fir (*Abies fraseri*), providing much improved species diversity and wildlife habitat.

Next we tried a "hack-and-squirt" treatment to the large beech and black birch trees in a second small test plot as well as a basal spray treatment to the large beech and black birch in a third test plot. Both of these treatments killed the trees and brush, but, like the method above, were not cost effective for use on large areas due to the amount of labor required.

The regrowth of desirable tree species in these cleared areas was encouraging. After hand-clearing another 4 ha, we learned of a cost share program that enabled us to hire a contractor to machine-mow brush in the remaining 7 ha of the fenced area. Next, a logger felled and stump-treated the large beech and birch to provide sunlight, deter root sprouts, and eliminate their seed dispersal. He collected the logs to sell for firewood and paid $100 per truckload, our only forestry income of note. We removed several poor-quality black cherry trees with interfering crowns. Additionally, we planted 100 red spruce and 100 red oak seedlings to increase the number and diversity of seedlings that would form the next stand. During the late summer, 2 years following the mowing, we twice spot-treated undesirable seedlings that had sprouted inside the fence. Regeneration of the area has been successful, and the forest is ready for a harvest of the overstory. The developing trees will be high enough (above deer browsing height) to remove the fence in 1–2 more years.

In August 2009, a federal grant provided us funds for a mast tree release. This removed trees surrounding mast-producing trees (black cherry and beech) to increase production of cherries and beech nuts on 2 ha. We left tree tops as downed brush in hopes it would be enough to protect emerging seedlings. This technique worked in a few areas where the brush pile was dense and >2 m high and >4 m wide (Figure 38.4), but was not generally successful. We have since installed

FIGURE 38.4 Piling brush over regenerating seedlings to protect them from deer. (Photo by Sue Hamilton.)

tree shelters and small exclusion fences to protect the seedlings in this area. We have planted 260 red oak, 250 pitch pine (*Pinus rigida*), 100 witch hazel (*Hamamelis* sp.), 25 red pine, 50 shagbark hickory (*Carya ovata*), 25 arrowwood viburnum (*Viburnum dentatum*), 50 basswood (*Tilia* sp.), 50 chestnut oak (*Quercus prinus*), and 25 swamp white oak (*Quercus bicolor*) and direct-seeded 5 oz. of red spruce seed for additional mast, diversity, and wildlife benefit.

In 2014, we began a direct seeding trial in an attempt to increase plant diversity. In the spring, we planted red spruce seed to four 2-m² plots. Two received 50%–70% shade and two were completely shaded. In the fall, we direct-seeded red spruce seed to twenty-five 0.6-m² plots in the forest. We obtained seed from local nursery/seed stores.

We attempted to deter deer from two areas 15 m in diameter with an enclosing fence made from brush piled about 2 m high and about 3–4 m wide, expecting to replenish as it decomposes. We purchased and planted 2 g Sassafras seed, 5 g basswood seed, 3 g black gum seed, 0.2 g common serviceberry seed, 0.1 g red alder (*Alnus rubra*) seed, 50 g shagbark hickory seed, 0.1 g mountain alder seed (*Alnus incana*), and 1 g balsam fir seed. We also planted 30 acorns, variety 30 Chinese chestnut (*Castanea mollissima*), and seed from 20 cucumbertree (*Magnolia acuminata*) seed pods that we collected locally in the fall of 2014.

38.8   VEGETATION MONITORING

We routinely monitor the forest, check fences, tree shelters, and project sites for deer/bear/other disturbance as well as tree mortality, and we repair/replant/move/change as needed.

We take note of changes in the forest, observe the kinds of trees that do well and those that do not, noting location, sunlight, and other factors that might affect growth. We remove invasive plants when we find them, and our enclosed areas are constantly in need of clearing. Our records are imperfect but include documenting some treatments with photos. Initial observations on the fate of direct-seeded plants indicated that red spruce germinates in full sun and full shade, but survival rates were best under 50%–70% shade and worst under full shade. Some seeded species never germinated (balsam fir, basswood, and red alder), Chinese chestnut germinated but died after the first year, and only one seeded shagbark hickory survived. Seedlings resulting from planting seed from other species (cucumbertree, sassafras, black gum, and serviceberry) germinated and are surviving 2 years after planting (Figure 38.5).

38.9   PROJECT FUNDING AND ECONOMICS

Government grants and cost share programs are often available to forest landowners to fund forest and wildlife management practices. Assistance with our written forest management plan; installation of a deer exclusion fence; purchase, planting, and fencing seedlings; brush mowing; and herbicide treatments such as fogging, foliar spray, and stump treatments to eliminate undesirable and invasive species are forest management practices we have implemented using cost share funding. Financial assistance did not always cover the entire cost of treatments, but by doing our own work, we lowered costs substantially. Sometimes the obtained funds exceeded our project expenses when we did the work. Our local forest landowners' association, area service forester, Pennsylvania State Extension foresters, and our consulting forester were good sources for information and assistance regarding these programs.

38.10   FINANCING OPERATIONAL COSTS

Drafting our 2002 forest management plan cost $1325 plus $800 for a required timber appraisal. We received $825 in cost share fund, putting our out-of-pocket cost at $1300. Our 2004 twelve-ha understory removal and fence treatment included an herbicide application, fence installation, road access construction, and consulting forester fees totaling $14,463. Forest Land Enhancement Program

FIGURE 38.5   Serviceberry 2 years after seeding. (Photo by Sue Hamilton.)

(FLEP) cost share funds covered $4740 of this amount, leaving our out-of-pocket expense at $9723. The 2010 The Natural Resources Conservation Service (NRCS), Environmental Quality Incentives Program (EQUIP) provided a cost share of $750 per ha through their Forest Stand Improvement Program to implement our 2010 "Mast Tree Release" on 2 ha. The $1500 payment covered the total cost for a contractor to herbicide the smaller brush and cut larger brush trees to release the mast-producing trees from competition.

The 2011 brush mowing inside the 12-ha deer exclusion fence qualified for $7700 EQUIP funding for "Early Succession Habitat Management." We cleared brush and stump-treated approximately 4 ha of the area ourselves. A contractor removed the remaining brush by mowing at a cost of $4945. The remaining EQUIP funds essentially paid us for our labor.

The 2012 follow-up herbicide spot treatment to remove undesirable sprouts and invasive species inside the 12-ha deer exclusion fence and the 2-ha mast tree release area cost about $200 for the herbicide and sprayer plus our time (we performed the treatment ourselves). The EQUIP "Upland Habitat Management" cost share payment of $1530 more than covered the cost.

The 2012 EQUIP cost share payment was $5375 for "Tree/Shrub Establishment" in our forest. Since we did the work ourselves, the funds covered the purchase of 2281 seedling plugs 2–3 years old and bare root seedlings, as well as supplies for fences and tree shelters for the 228 hardwood seedlings. By doing the work ourselves, we were able to plant many more trees than the cost share plan requirement called for.

In addition to these funded projects, we spend hundreds of hours each year working in our forest, planting and protecting seedlings, removing undesirable vegetation, monitoring for progress and/

or failure, and making and maintaining roads and trails. Financially, this has saved us quite a bit. On the other hand, many of these projects are done with hand tools and require a great deal of time. They are small-scale and less costly compared to larger-scale projects that require heavy machinery and skilled labor, such as the 12-ha reforestation project in 2004. In addition to the out-of-pocket expenses mentioned above, we have invested approximately $70,000 in forestry expenses over a 14-year period for such things as seedlings, fencing, tools, herbicide, seed, hunting licenses and tags, forestry vehicles, training, and road construction.

38.11 SUMMARY: HAVE WE MET OUR INITIAL GOALS?

Over the past 15 years, we have learned a great deal regarding forestry techniques that work and those that do not work in this location. We used a variety of silvicultural practices along with fencing the deer out rather than substantially reducing the deer population. We have successfully initiated regeneration and improved health to a substantial part of our forest. We saw problems in the forest and learned that the forest was indeed in decline. We made a conscious effort to educate ourselves about forest and deer management and to network with other small woodlot owners facing the same challenges. We attended the Woodland Owners Conference at Penn Tech and joined the Bradford/Sullivan Forest Landowners Association. We served on that organization's Board of Directors and participated in the Pennsylvania Forest Stewards training, both organizations contributing to our knowledge of deer and forest management. We also volunteer to promote education regarding the importance of forest health at a local reforestation project level (Eagles Mere Youth Reforestation Project, High School Green Career Day, and Envirothon). We discovered that there are government programs such as Clean and Green, DMAP, and cost shares available to landowners to help with deer control and forest and wildlife habitat management. Organizations such as a local forest landowner group and the Pennsylvania State Extension offer forest landowners helpful information and assistance. We found working with our consulting forester very worthwhile.

Education gave us a better understanding of the problems we faced and provided ideas on how to address them. We learned that intervention would be time consuming and costly, but it made sense because forest stewardship was valuable for many reasons. A well-cared-for forest could produce a substantial income as well as provide good wildlife habitat, clean water, and aesthetic beauty.

Our forestry education included the very clear financial incentives to allow our standing timber to grow rather than harvesting it immediately. In addition to the growth in tree diameter and height, smaller trees grow to meet the 36-cm minimum diameter to be sawlogs, and larger trees improve in quality grade and value. Our initial timber value of over $100,000 could be expected to increase over 10% per year while left to grow. Additional benefits of proper forestry practices coupled with the timber sale delay would provide healthy and diverse forest regeneration as well as future income. Unfortunately, during this waiting period for the regeneration to take hold, the value of timber, particularly black cherry, dropped significantly. While it does appear that the predicted growth rates were somewhat optimistic, the collapse of the timber market was historic and due entirely to outside forces. Presently, we would be doing well to sell the timber that is ready for harvest for half of what it was valued in 2002, one-third in real dollars.

We enjoy working in the forest. It can be very rewarding but very frustrating at other times. It is labor and time intensive. Depending on the extent of the project, it can be expensive. Projects are ongoing and need regular maintenance. They are often unsuccessful. It takes a lifetime for a forest to grow. The forest is always changing and often unpredictable.

EDITOR COMMENTS

The Hamiltons made excellent use of outside educational opportunities and funding to finance their habitat improvements/deer fencing. Their reticence to allowing hunters they do not know to hunt their lands is not atypical of small woodlot owners. The small size of their holdings results in deer coming from adjacent properties to damage their understory vegetation. Like other small woodlot owners, they were able to take advantage of local expertise in forestry and wildlife to obtain professional advice and obtain some financial assistance from government programs. This property would provide a good evaluation of the Rose Petal hypothesis wherein harvest of most if not all deer using their property during hunting season might result in creation of a null-deer area that would persist for several years and not require annual removal of deer. Such removals might be effected by the landowners, family, and friends if pursued intensively. Fencing protected tree seedlings did nothing to protect seedlings, shrubs, and herbaceous vegetation outside fenced areas: protection of those plants required additional, single-plant exclusion devices.

David S. deCalesta, editor

39 The Brubaker Small Woodlot
Innovative Management of Deer Forage and Deer Harvest

Roy Brubaker

CONTENTS

We abuse land because we regard it as a commodity belonging to us. When we see land as a community to which we belong, we may begin to use it with love and respect.

Aldo Leopold (1949)

39.1 INTRODUCTION: OUR LANDSCAPE

We bought our property in 1998 when I was finishing my master's degree in forestry at Pennsylvania State University. Our land is in south-central Juniata County in the Ridge and Valley geographic region of Pennsylvania. The wooded portion consists of 8 ha of mixed red and white oak (*Quercus rubra* and *Q. alba*) and sugar maple (*Acer saccharum*) at the northern end and a 4-ha white oak woodlot at the southern end at the base of a small ridge. The 8-ha woodlot forms a large hollow at the foot of Shade Mountain and acts as a natural funnel from the agricultural fields on either side of us to deer bedding and escape cover on the mountain. Our land includes 20 ha of pasture (along with another 30 ha of pasture that are rented), and 2.5 ha in wetlands and poor pasture ground that is reverting to woods. The surrounding land is privately owned and ranges from large leased (for hunting) parcels of 400 ha acres or more to small woodlots like ours. While there are few parcels in the local landscape that are not hunted, most are hunted only lightly and the harvest is highly skewed toward antlered deer. We have often counted 60 or more deer feeding in neighboring fields within a few weeks following the close of hunting season, so deer pressure on our woods can be quite heavy.

39.2 OUR VALUES

Our management of our lands is framed by our values and experiences as small woodlot owners who enjoy hunting deer and tending the forest. We thought it critical to explore, develop, and articulate our values and beliefs before designing and implementing a management plan for our natural resources. We felt that if we could not identify what motivated us, we would have trouble understanding and affirming common values we share with neighbors and others whose motivations and preferences for managing natural resources appear to conflict with ours. Moreover, we were concerned that if we used motivations and impulses associated with these values and beliefs as guides, we might base management on incorrect information and not accomplish our goals.

Values driving our management philosophy relate to tangible forest outputs and personal well-being. Tangible values include: (1) managing all resources sustainably and ecologically; (2) growing and harvesting timber resources; (3) enhancing and harvesting game species; (4) enhancing diversity and abundance of nongame wildlife and plant resources; and (5) growing, harvesting, marketing, and consuming agricultural and farm products. We have a strong kinship with these resources and are keenly interested in the welfare and integration of all as we plan and manage for them. As Mennonites, we have additional spiritual and ethical attachments to our land and animals.

Personal values focus on "well-being" resulting from: (1) spending time in the woods, since that is where we feel most alive and centered, and (2) creating sufficient income from forest (and agricultural) resources to ensure that our family is well cared for and has options to pursue in life.

Regarding deer, we had to reconcile conflicting values we held of harvesting large antlered deer, harvesting antlerless deer for food, and balancing deer with resources on our property. We did not want excessive deer browsing to hinder our ability to produce sustained outputs of forest products, commercial and noncommercial, or other desired outputs. Because we shared a traditional deer hunting value (opportunity to harvest large antlered bucks), we were empathetic with hunters with a similar value.

39.3   THE MANAGEMENT SITUATION

The prior owner was harvesting logs from the 8-ha woodlot when we bought the property, cutting only the highest quality sawlogs. We made an offer on the property with the stipulation that the timber harvesting stop immediately. At that point, the basal area was around 16 m^2/ha, but only around 72% of that was in acceptable growing stock (trees of desirable species and quality to make a valuable sawlog.). Stocking density was extremely variable and ranged from sample plots where there were no sawtimber-sized trees of any potential value to areas with a basal area of 23 m^2/ha or more of acceptable growing stock.

The smaller woodlot was dominated by white oak trees. Most were around 60 years old and had grown up in an old pasture. This woodlot was well stocked at 100 feet^2/acre (23 m^2/ha) basal area. There was a well-developed midstory layer of musclewood (*Carpinus caroliniana*) and ironwood (*Ostrya virginiana*) in this stand, particularly along a small creek that runs through it, and some understory white pine (*Pinus strobis*) developing in one corner of the woodlot in spite of heavy deer pressure.

Conditions across both woodlots were generally at the lower margin of what foresters would consider fully stocked with timber of commercial value. Because the previous owner was an avid squirrel hunter, he harvested no hickory trees (*Carya* sp.), as he felt that would negatively affect his squirrel hunting—a value related to wildlife rather than timber.

More troubling than the condition of the larger woodlot overstory was the fact that there was no tree regeneration anywhere other than some suppressed black birch (*Betula lenta*) seedlings/saplings in the understory. There was a 1-ha patch of dense midstory striped maple (*Acer pensylvanicum*), a shade-resistant tree that robbed other more desirable tree species of sunlight and nutrients. Other areas contained well-developed midstory canopy layers of back gum (*Nyssa sylvatica*), birches, and red maple that cast heavy shade on the forest floor and suppressed regeneration of more desirable species. There was a distinct deer browse line throughout the property and very little ground-level vegetation. There were also a few areas of very dense, mature witch hazel (*Hamamelis virginiana*) and spicebush (*Lindera benzoin*).

Like many small woodlot owners, the previous owner regarded standing timber in his woodlot as something of a bank and marketed a few loads of sawlogs every now and then to supplement his farm income. Over time, this selective "removal of the best and leaving of the rest" resulted in poor stocking of desirable seedling species and severe interference caused by shading of undesirable tree species and ferns. Our challenge was to find cost-effective ways to remove some of the interfering low shade from the forest so we could begin regenerating and stimulating the oaks, maples, and other higher-value species that were becoming scarce in the overstory. And, we had to protect the burgeoning seedling regeneration of desired tree species from overbrowsing by deer. Thus, we had to deal with the impacts of past forest management practices and an overabundant deer herd. To date, most of our forest management efforts have been invested in addressing the status of the larger woodlot, as that was where the most pressing need for our intervention lay.

39.4 MANAGEMENT ACTIVITIES

39.4.1 MANAGING INTERFERING VEGETATION

As a forester, I had been taught there are "desirable" and "undesirable" tree species—usually based on their perceived current value as a timber crop, in spite of the fact that species wax and wane in value (e.g., black birch was highly desirable in the 1960s and now is considered an interfering species; red maple was considered an interfering species decades ago but now brings good prices at timber sales). We knew that areas of the woodlot dominated by dense layers of striped maple, an interfering species, had to be treated to permit seedlings from desirable tree species to regenerate and grow. The midstory of the woodlot was heavily stocked with birch, red maple, and gum in the 15–25 cm diameter class. These species cast considerable shade on the forest floor around the oaks, sugar maple, white pine, yellow poplar (*Liriodendron tulipifera*), cucumber wood tree (*Magnolia acuminata*), and hickory, suppressing seedling establishment of these desirable species. Our approach was to fell as much striped maple as possible and, time permitting, also fell the lowest-quality stems of the other undesirable species, with the restriction of waiting until late July to avoid disturbing nesting birds.

39.4.2 CREATING ALTERNATIVE FORAGE

To reduce deer use of our property, we fenced all our pastures with six-strand high-tensile wire. While the deer did go through and over our fences, we generally did not see the large groups of deer feeding inside the fenced areas that we saw feeding in our neighbors' fields. But deer funnel through our woods daily and frequently during the day—except during hunting season when it seems they only pass through at night.

We knew it would be impossible to keep deer from browsing on our forest vegetation unless we protected it with additional fencing we couldn't afford. We didn't think we could reduce deer abundance sufficiently with hunting, so we searched for alternative strategies.

The first few years we distributed the brush resulting from cutting interfering saplings and poles over the ground in areas, we were hoping to establish regeneration of desirable tree species. Concurrently, there was a 2-year acorn crop that benefitted from opening the overstory and piling of brush to restrict deer access: we got good germination and establishment of oak advance regeneration.

Increased light on the forest floor resulting from removal of interfering sapling and pole trees increased growth of a variety of species: seedlings from desirable and undesirable tree species and herbs, forbs, and shrubs. We observed that this "smorgasbord" of herbaceous and woody vegetation served to lessen the impact of deer browsing on desirable advance regeneration due to the great variety and amount of alternative forage species. And so began the concept of the "salad-bar" approach to reducing deer impact on desirable plants—a revelation that came to me while working in a restaurant with a salad bar.

I knew the chief concept in managing a salad bar—make sure people can get to the cheap, filling stuff (bread, chili, potatoes, iceberg lettuce, bread board, etc.) as easily as possible. Put these offerings in the largest bowls, right next to the plates, and make sure they look as fresh, plentiful, and appealing as possible. The expensive items (e.g., the shrimp and crab legs) were placed at the end of line where people had to reach to get them.

What we observed over the next several years in our woodlot was illuminating. Although understory striped maple, birch, and black gum trees were supposedly undesirable to deer, their stump sprouts appeared highly palatable to, and were heavily browsed by, deer. Red maple and black gum stump sprouts were so highly preferred by deer that even some of the oak and white pine seedlings that germinated in the newly found open patches actually got established. Black birch and striped maple, on the other hand, appeared as so much garnish; they made things look fresh and green, but the deer didn't eat them. And, it appeared that the "salad bar" had to be replenished every few years by subsequent cutting of saplings of undesirable tree species. However, it was becoming clear that the few spindly oak and pine seedlings that survived in these areas would be either covered in garnish or picked into oblivion by the browsing deer without more brush management.

By then we were experimenting with some pasture management concepts featuring introducing animal impact (both browsing and trampling) at different parts of a plant's life cycle to increase or decrease the plant's competitiveness. For example, grasses are adapted to compete under intense, short-term periodic disturbances. To give grasses a competitive edge over, say, multiflora rose (*Rosa multifora*), the trick is to postpone cattle grazing until the grass is just about to head out in mid-May, when the multiflora rose has just come into full leaf. Introduce cattle at high stocking densities, and in a short period they will graze or trample everything flat. At that point, remove the cattle: the short-term burst of high-intensity browsing will set back the grass and multiflora rose. Because grass can sprout immediately from the crown (growing point just above the roots) and will resume photosynthetic capacity within a few days of being grazed, and it will be thick and competitive for light, nutrients, and water, while the multiflora rose is only beginning to push out new leaves. If this cycle of directed grazing is repeated, it will enhance the competitive advantage grass has over multiflora under intense grazing disturbance and limit the amount of multiflora in the pasture without resorting to mowing or herbicide. We incorporated this concept with additional conceptual tools to tend the salad-bar business I had going in our woodlot, using my chainsaw and brush saw to begin the competitiveness program.

About this time, we became aware of the work of Fred Provenza and the Behavioral Education for Human, Animal, Vegetation, and Ecosystem Management (BEHAVE; http://www.behave.net) center at Utah State University. Provenza's research suggested that herbivore preferences for different plants are not nearly as "hardwired" as once thought. Rather, much of it is: (1) learned from observation and experimentation and (2) dependent on complex biochemical feedback loops based on overall nutrient intake. Provenza's thesis is that under high predation levels in landscapes with a wide diversity of forage species, deer will browse different plants in roughly equal proportions, and relative differences in plant preferences will be low in such contexts. When there is little predation pressure and/or low plant diversity, deer will develop increasingly pronounced preferences for some plants over others. Often, these preferences exacerbate or accelerate homogenization of landscape level plant diversity toward least-preferred species. Though we often think of this process as happening because deer abundance exceeds landscape level carrying capacity, Provenza's research suggests the trend toward reduced plant diversity dominated by least-preferred species can actually happen over time in the absence of predation pressure even at relatively low population densities.

We altered our salad-bar approach based on Provenza's work. Because we couldn't get deer to eat the "garnish" species (striped maple and birch), we cut these (along with introduced shrubs, e.g., multiflora rose) in spring after leaf out to drain their energy reserves and keep them from competing with salad-bar species and seedlings of desired tree species during the peak of the growing season. We sometimes returned to these areas the second spring with a very selective application of glyphosate to treat resurging garnish species. In midspring, we cut preferred salad-bar species (red maple, gum) during the winter when their energy reserves were below ground, encouraging a flush of alternative desirable forage species for the deer during the growing season. We also covered vulnerable seedlings with cut understory brush (slash). We hoped that the sounds, sights, and odors produced by our presence in the woods would create some "predator presence stress" when deer were using our woodlot and encourage them to feed in the neighbors' alfalfa fields instead of in the understory of our woodlots.

We continued to pile cut brush on oak, sugar maple, hickory, and pine seedlings, and on some stump sprout clusters of red maple and black gum so they could produce useful forage over time as the cut brush decayed and became less protective. After the cut brush decayed to the point where it no longer protected desirable seedlings and sprout clumps, we'd return and recoppice the red maple and black gum sprouts to get another flush of deer forage.

From time to time, we noticed that when we cut patches of mature hawthorn (*Crataegus* sp.) and black haw (*Viburnum prunifolium*) on forest edges, we created microsites wherein oak and hickory seedlings/saplings put on rapid growth along with the cut hawthorn and black haw stump sprouts. The cage-like branching structure of the hawthorns and black haw protected at least some leaves and twigs of the oak and hickory seedlings, even under heavy deer browsing pressure, to the point where, in a year or two, oak and hickory saplings grew above deer browsing height and were also protected from antler rubbing. These associations of hawthorn and black haw with woody seedling species affirmed the work of Vera (2000), who postulated that continuous grazing by large herbivores is a primary cause of forest successional pathways and that there are shrubs and other plants that facilitate establishment and protection of early and late successional forest trees under browsing pressure from ungulates.

In the process of turning our woodlots into a well-managed salad bar for the local deer herd, we can now predict where to head into my woods after a rain for a meal of oyster mushrooms, chanterelles, or honey mushrooms. The areas we cut in are always the best early-season archery stands, as the deer inevitably head past these areas to grab some salad on the way up to their bedding areas. And, after a year or so, we'll revisit these areas and fell a few more stems in the area to keep the salad bar fresh. We directionally fell the new tops onto the new cohort of seedlings on any open ground in the area, replenishing cover over 2- and 3-year-old seedlings in decaying tops from the last time we'd been in that area.

Our constant tending actions have yielded other benefits: we've been protecting ginseng (*Panax ginseng*) we seeded in, and have begun seeing pink lady slippers (*Cypripedium acaule*), Indian cucumber root (*Medeola virginiana*), and jack-in-the-pulpit (*Arisaema triphyllum*)—wild flowers we never saw in our woods until just a few years ago. Research shows these plants can be indicators of forest habitat recovery if they reach a foot or more of height and achieve reproductive status (Royo et al. 2010).

Does all this mean that our salad-bar approach to tending woods can be a restorative prescription for managing small forest areas under high deer pressure? Perhaps, if taken in conjunction with efforts to reduce deer density.

39.4.3 REDUCING DEER DENSITY

We calculated there were approximately 150 deer living within a 5-km radius of our woodlots, including 60 deer usually counted within neighboring agricultural fields. Although we could only guess at how many of these deer browsed our forest vegetation, we were sure we could not harvest enough of them by hunting our property alone to bring their density in line with environmentally acceptable levels of impact on our forest vegetation: many of them would not even be on our property on the days we hunted them.

We came to learn that to achieve a meaningful reduction in deer density, hunting had to be approached from the viewpoint that it was simply another management task requiring persistence and concerted effort, rather than a pleasurable part-time recreational experience. And, we needed additional firepower. What we hoped for was to reduce deer density using our deer harvest and that of invited hunters sufficiently to bring deer density in line with available forage such that they did not negatively impact regeneration of commercial tree species or cause loss of understory plant diversity, including wildflowers and shrubs.

To increase efficiency of deer hunting on our woodlots, we made sure that we personally had the maximum allowed number of permits for harvesting antlerless deer. We made sure that hunters we invited from the local community to hunt our woodlots also possessed the maximum number of antlerless permits. Many of these hunters had hunted out woodlots prior to our ownership and their experiences, memories, and hunting histories were deeper than ours: we anticipated that they could harvest deer from our woodlots at least as efficiently as we. Over time, we identified those invited hunters who held off harvesting a doe while waiting to bag an antlered deer and politely but firmly disinvited them to participate in future deer hunts, making it clear that the privilege of hunting on our lands was only for those who would harvest antlerless deer. We ended up with a dedicated core of meat hunters who, every year, including our contributions, would harvest about eight antlerless deer per year. This is a small portion of the estimated 150 deer within the local landscape, but we came to realize that this annual reduction of deer density, in league with the increased forage provided by the salad bar, seemingly brought the population of deer using our woodlots for browsing in rough balance with available forage such that we achieved successful regeneration of tree species of commercial value and resurgence of vulnerable understory forbs and shrubs, ending up with a much more diversified understory.

Our integration of providing deer with alternative forage and annually reducing deer abundance over 16 years resulted in enhanced forest diversity, forest resilience, and commodity value while allowing the deer, ourselves, and other hunters unfenced access to our woods.

ACKNOWLEDGMENTS

My hours in the woods have been, and forever will be, infused with the presence of my father, Roy L. Brubaker, and uncle Dale Brubaker. They were farmers committed to the stewardship of land, nature, and community. They taught me to pay close attention to plants, the knowledge of others, and the workings of my brain, and to claim my role as a deer predator.

EDITOR COMMENTS

This case history highlights the problems faced by small woodlot owners—their lack of control over landscape features that affect and compromise their deer–forest management actions. As with other small woodlot owners, the Brubakers were only partially successful with using public hunting to resolve their overabundant deer issues. And, like other small woodlot owners, they had to improvise to find workable solutions, in this case the use of coppicing to create an effective concept in deer damage management: manage alternative sources of forage created by silvicultural practices to protect regeneration by creating barriers to deer foraging and creating an alternative source of more readily available forage.

David S. deCalesta, editor

REFERENCES

Leopold, A. 1949. *A Sand County almanac and sketches here and there.* New York. Oxford University Press.
Royo, A. A., S. L. Stout, D. S. deCalesta et al. 2010. Restoring forest herb communities through landscape-level deer herd reductions: Is recovery limited by legacy effects? *Biological Conservation* 143:2425–2434.
Vera, F. 2000. *Grazing ecology and forest history.* New York. CABI Publishing.

40 Competing Interests
Forest Regeneration or Increased Deer Abundance?

Paul D. Curtis, Michael L. Ashdown, and Kevin Virkler

CONTENTS

It is as fatal as it is cowardly to blink facts because they are not to our taste.

John Tyndall (1879)

40.1 INTRODUCTION

The Adirondack Mountain ecoregion has a long history and tradition of deer hunting and forest management. Within the 2.5-million-ha Adirondack Park boundary (Figure 40.1), there is a mix of New York State Forest Preserve and private lands. A few large, private preserves were established in the late 1800s to promote hunting, fishing, and other outdoor recreation activities. One of these, the Adirondack League Club (ALC) near Old Forge, New York, was established in the 1890s for the benefit of its members. The area is primarily forested with northern hardwoods, and contains more than 13,000 ha. Guiding principles of the ALC are: (1) preservation and conservation of the Adirondack forest and the propagation and proper protection of fish and game in the Adirondack region; (2) establishment and promotion of an improved system of scientific forestry; and (3) maintenance of an ample preserve for the benefit of members for the purpose of hunting, fishing, rest, and recreation.

There has been a long history of fisheries, forest, and wildlife management on this wilderness preserve. Forest management has been part of the ALC since its inception as a means of generating income while preserving the outstanding recreational resources (Webster 1965). Revenue from sustainable timber harvest provides an important contribution to operating expenses. Concerns about conserving a portion of the western Adirondack region were well justified, as much of the

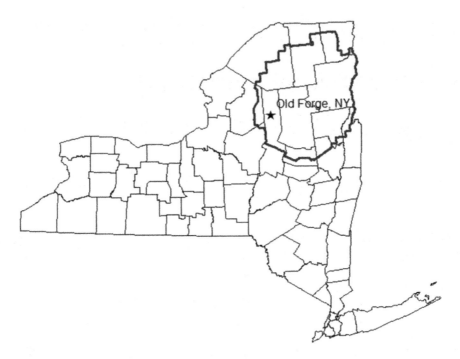

FIGURE 40.1 Map of New York State showing county boundaries, the boundary for the Adirondack Park, and the location of Old Forge.

surrounding lands were purchased by large lumber companies. While the Adirondack timber harvest peaked in 1905, new technologies and war-related shortages allowed the timber companies to harvest more remote areas for softwoods well into the 1940s, with the last timber drive on the South Branch of the Moose River in 1948 (McMartin 1994).

In the 1950s, as the use of hardwoods for pulp increased and prices rose, ALC and other forest owners began cutting the mature hardwoods that had been left standing during the previous timber harvest (Comstock and Webster 1990). In 1954, an assessment of the Gould Paper Company's holdings on the Moose River Plains revealed that approximately 90% of the hardwood stands had never been cut, and the subsequent harvest for the next decade was "as complete as anywhere," with the harvest of both spruce and hardwoods (McMartin 1994). McMartin (1994) described large areas of the Gould tracts as "filled with dense thickets of trees no bigger than an arm" which, if not exclusively beech brush, would have been ideal habitat for deer.

Deer thrive in early successional forests because of the readily available browse and the security that the dense cover provides for them. Deer also need escape and resting cover, so forest management activities should provide a mix of stand ages in both a spatial and temporal context (Hewitt 2011). Natural events and extensive timber harvest that create large openings in the forest will provide the conditions best suited for deer to thrive. Conversely, though, as the forest matures and seedlings grow out of reach of the deer, carrying capacity declines, and so will deer density. Much of the Adirondack Park is in the New York State Forest Preserve and contains mature forests with restrictions on timber harvests. For example, since 1964 when Gould Paper Company sold its holdings on the Moose River Plain to the State of New York (McMartin 1994) and large-scale logging practices diminished, the western Adirondacks forest has been aging (Alerich and Drake 1995). This U.S. Forest Service report illustrated that increased area of forest in the western Adirondacks region continued to shift into the older timber stand classes. This trend results in lower deer densities unless timber harvests create additional early successional habitat that favors higher deer abundance.

On November 25, 1950, a large blow-down created many thousands of ha of early successional habitat that favored deer population growth on and surrounding the ALC in the Moose River Plain.

This natural "disaster" leveled 25%–100% of the trees on 172,000 ha of the Adirondack Park, including portions of the ALC. One of the hardest-hit areas was the Moose River region, including some ALC lands (Brown 1985). The damage was so extensive that no conservation groups objected to New York issuing logging contracts for clean-up of downed timber on State Forest Preserve lands to prevent fires. The magnitude of this clean-up was marked by the fact that it took 5 years to complete and created a temporary glut in the timber market (McMartin 1994).

This combination of a long history of regional exploitation by the logging industry, vast areas affected by the blow-down of 1950, and conservative deer hunting laws provided conditions for an explosion in the deer population of the region. Record hunter take occurred at ALC in the late 1950s and throughout the 1960s and reflected the increased deer abundance (Figure 40.2). However, alarm over increasing winter deer kill prompted the ALC to contract with Cornell University to study the possible reasons for the deer mortality. Following a study by Neth (1956), ALC hunters began harvesting female deer because of forest overbrowsing.

Neth (1956) analyzed three factors that were directly related to deer population levels. First, in spring 1955 after a survey of traditional winter deer ranges at ALC, winter kill from the previous season was estimated at 6.3 deer/km^2, with 62% being fawns. Winter snow depth and duration have a major impact on northern deer herds (Hewitt 2011). Deer accumulate fat reserves during fall and use those resources when nutrition is limited. In a typical winter, deer may lose 20%–30% of their body mass (Mautz 1978). Parker et al. (1984) demonstrated the high energetic costs of deer moving through deep snow. Malnutrition, although probably always causing some deer deaths even during moderate Adirondack winters, may cause massive mortality when deep snow cover exceeds 100 days (Saunders 1988). Deer also voluntarily restrict food intake and activity during winter (Thompson et al. 1973), as the cost of searching for food may be greater than the energy obtained from low-quality winter browse.

Second, a survey of available and consumed browse conducted in two winter yards at ALC identified 66% and 78% of the available browse as low-palatability species (Neth 1956). For the high-palatability species, 80% or more of available browse was consumed, with 18%–32% of the small trees dead from over browsing.

And third, Neth (1956) analyzed field-dressed weights and antler-beam diameters for deer harvested on the northern portion of the club in 1955. In general, the mean antler beam diameter and dressed weights of the ALC bucks for all age classes was smaller when compared to deer of the same

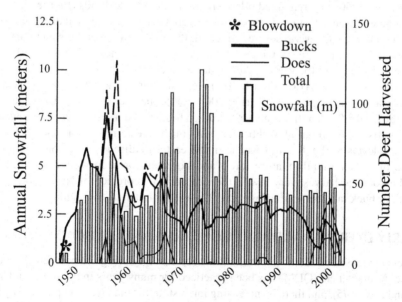

FIGURE 40.2 Snowfall totals for the Old Forge region with the annual totals of deer harvested at the Adirondack League Club near Old Forge, New York. (Courtesy of Northeast Regional Climate Center, Ithaca, NY.)

age class taken in the central or peripheral sections of the Adirondacks. The 3½-year-old and older bucks at ALC weighed less than those in the other regions by 4.5 kg or more. These comparisons suggested that there was more competition for forage at ALC than in the other nearby regions.

Another deer management study conducted by Debbie (1961) used the ALC's winter deer-feeding program as a means to obtain population estimates during the winters of 1960 and 1961. Similarly to Neth (1956), he examined deer winter kill and conducted browse surveys. One outcome of the winter feeding count was the low number of fawn observations (16.4% in 1960 and 14.1% in 1961). Acknowledging difficulties with age determined by the relative size of individual deer, this was assumed to be a low estimate. A deer winter-kill survey was carried out by walking transect lines and counting deer carcasses using the same protocol as Neth (1956). The winter mortality rate after several years of increased hunter harvest at ALC (including the record total take of 138 deer in 1959) was down to 2.7 dead deer per km^2 in 1960, and 1.1 per km^2 in 1961.

Similar to Neth's (1956) findings, browse surveys completed by Debbie (1961) showed high densities of spruce and beech in the sample plots and high percentages of available palatable tree seedlings being consumed. Debbie (1961) advised that while winter feeding of deer would reduce mortality rates, it would only be a temporary fix, and that deer populations needed to be kept within the carrying capacity of their range. He went on to state "where the forest was already heavily browsed … the deer population should be lowered below the carrying capacity of the range to allow the browse species to produce a density high enough that the deer will not consume the majority of it."

Some of these management recommendations were eventually implemented on ALC property, although it is unlikely that these studies were the catalyst. In recent years, the older-growth canopy has been opened up for tree regeneration with even-aged harvesting. In 2002, the New York State Department of Environmental Conservation (NYDEC) changed Environmental Conservation Law to prohibit the feeding of deer. This put an end to a long-time tradition at the ALC, but not before deterioration of forest regeneration had occurred. More than 40 years of chronic deer overbrowsing had negatively impacted tree species diversity and forest productivity.

Deer numbers and harvests have changed over time depending on prevailing forest habitat conditions on and near the property (Figure 40.2). Historically, deer harvests were low (20–40 annually, mostly bucks), and similar to the rest of the Adirondack region. Prior to the 1950s, only bucks were allowed to be taken under NYSDEC regulations. Deer numbers and harvests at ALC peaked in the late 1950s and remained relatively high through the 1960s (Figure 40.2). Does were harvested along with bucks from 1957 through 1970, contributing to the total deer harvest. However, with maturing habitat and several severe winters in the 1970s, deer numbers and harvest dropped. During 1970–1978, there were six winters with >30 m of total snowfall (Figure 40.2), impacting deer abundance throughout the western Adirondack region.

Recent deer harvests at ALC (Figure 40.2) were similar to the 1928–1939 period, when the average take was 11 bucks per year. During 2000–2005, the average deer harvest at ALC was 16 bucks per year from 8,900 ha (0.12 bucks per km^2). These deer harvest levels were lower than those in the other parts of the Town of Webb near ALC in the western Adirondacks, but were similar to deer harvest levels in the Town of Keene in the eastern Adirondack region (Figure 40.3). The combination of predominantly mature forest and potential for deep winter snowfall keeps deer density and associated harvest very low in the Adirondack ecoregion of New York. Typically, the annual index of buck take per km^2 is less than 0.4 deer for much of the region.

40.2   EARLY DEER MANAGEMENT CONCERNS AND ACTIONS

Some hunters expressed concerns about the low deer harvest during the past few decades at ALC. Quality Deer Management (QDM) has been practiced for many years in the southern United States (Brothers and Ray 1975), and there is increasing interest in this management philosophy in northern states, particularly Wisconsin, Pennsylvania, and New York. However, forest types, winter severity and snow depths, growing season length, and soil types are very different between the Adirondack

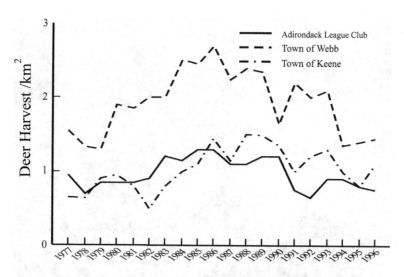

FIGURE 40.3 Total deer harvested per km² for the Adirondack League Club, Town of Webb and Town of Keene 1977 to 1996 (NYSDEC 1996 *New York State 20-Year Deer Book*, WE-173-G).

region and southern hardwood and pine forests. It is not clear if deer management techniques used elsewhere in the United States can be applied directly in the Adirondack region.

To evaluate this management approach, an outside consultant established a 2025-ha Quality Deer Management area combining several hunting territories near the center of the ALC preserve. Deer food plots were created, and weights and ages of harvested deer were recorded each season. ALC staff applied for DEC Deer Management Assistance Program (DMAP) permits to increase the harvest of female deer on the property. This program had two goals: (1) increase deer harvest to satisfy the interests of some hunters and (2) reduce the density of female deer to promote better forest regeneration in areas with recent timber harvests.

40.3 RESEARCH TRIALS

In 2004, Cornell University staff established a research project to evaluate the QDM program and forest regeneration at ALC. There is clear evidence in the literature that deer impact forest regeneration and the composition of hardwood seedlings (Tilghman 1989; Waller and Alverson 1997; Rawinski 2014). In the Adirondack region, deer show preferences for feeding on maple (*Acer* spp.) and yellow birch (*Betula alleghaniensis*) seedlings. Consequently, American beech (*Fagus grandifolia*), striped maple (*Acer spicatum*), and other less desirable woody species will compete for space and establish advanced regeneration when deer are present. In mature woodlands with low species diversity and several years of chronic deer overbrowsing, even low densities of deer (6–8 per km²) hindered establishment of desirable hardwoods (Horsley et al. 2003; Rawinski 2014) in northern Pennsylvania.

Replicated, standardized deer exclosures (7.6 m by 7.6 m; Figure 40.4) with paired control plots were established in four recent timber harvests (Combs Brook, Rock Pond, Oxbow, and Sylvan stands) during 2004 and 2005, the year following timber harvest in each stand.

A total of 48 test plots (24 fenced, 24 control) was established in the four recently-logged sites. Six paired plots, fenced and unfenced, were located randomly off a transect line in each stand to provide an assessment of the cut area. At the end of the initial two growing seasons, plant growth response was evaluated (counts of seedlings, saplings, and understory trees) on all test plots. Seedlings were separated into 3 height classes: 2–25 cm, 25.1–100 cm, and 100–200 cm or greater. All plots were again sampled 5 years after the timber harvest to evaluate seeding growth and deer impacts. In addition, soil pH and cation exchange capacity was tested at each of the four test stands.

FIGURE 40.4 Sampling vegetation in a deer exclosure at the Adirondack League Club near Old Forge, New York, during the year the plot was established in 2004.

40.4 RESULTS

Despite overall low deer densities in the Adirondacks, there was sufficient browsing pressure to inhibit forest regeneration outside of the fenced deer exclosures at ALC. There were few overstory sugar maples (*Acer saccharum*) in or near the four harvested stands, so the seed source, and subsequent counts of maple seedlings, were relatively low both 2 (Figure 40.5) and 5 years (Figure 40.6) after timber harvest. It is interesting to note there were low numbers of maple seedlings in the 3–25 cm height class, and counts were similar both inside and outside the deer exclosures. However, none of these maple stems grew into the greater height classes after 5 years, and it did not matter if the

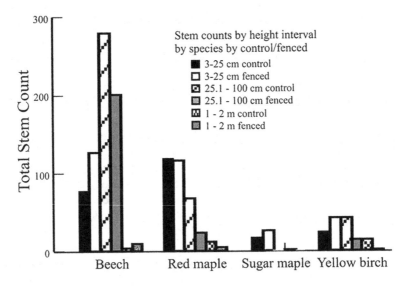

FIGURE 40.5 Number of hardwood stems in the fenced vs. control plots 2 years after the four stands were cut at the Adirondack League Club, near Old Forge, New York.

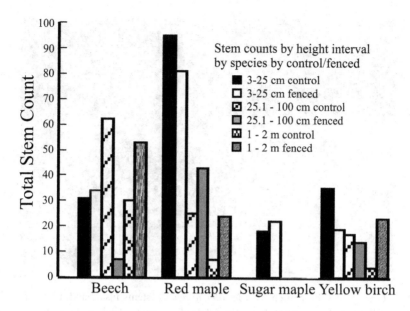

FIGURE 40.6 Number of hardwood stems in the fenced vs. control plots 5 years after the four stands were cut at the Adirondack League Club, near Old Forge, New York.

seedlings were fenced or not. Sugar maple regeneration is either impacted by the acid soils and/or is much slower growing than red maple, yellow birch, and beech seedlings.

Stem counts for yellow birch and American beech seedlings were similar, as 2 years after the stands were harvested, there were more seedlings in the 3–25 cm height class for paired control plots outside of the deer exclosures (Figure 40.5).

However, trends reversed in the 25.1–100 cm height class for both tree species. There were more birch and beech seedlings within the fenced exclosures than for control plots unprotected from deer foraging. There were also more red maple seedlings in the 25.1–100 cm height class within fenced deer exclosures vs. control plots. Seedlings below 25 cm in height were likely protected by snow throughout the winter in the Adirondacks. As seedlings grew taller, more stems were above snow level and were potentially subjected to winter deer browsing outside the fences. After 2 years, some yellow birch and red maple seedlings ($n = 26$) managed to grow into the 100.1–200 cm height class within the fenced exclosures, but few tall stems ($n = 6$) survived in control plots (Figure 40.5).

Five years after the fences were installed, there were still similar numbers of maple, birch, and beech seedlings in the 3–25 cm height class both inside and outside the deer exclosures (Figure 40.6). No sugar maple seedlings survived and grew into the higher height classes after 5 years both inside and outside of deer exclosures. Red maple stem counts were two to three times greater within fences in the two higher height classes than for control plots. Counts of tall beech stems were also higher within fences, but the difference was not as dramatic. Yellow birch had six times as many stems in the tallest height class (>1 m) within fences ($n = 23$) when compared with nearby control plots ($n = 4$, Figure 40.6). It is clear that deer exert browsing pressure on forest regeneration, even at relatively low Adirondack deer densities (Figure 40.7).

At the four experimental stands, soil pH ranged from 3.8 to 4.2, and calcium:aluminum (Ca:Al) ratios ranged from 0.33 to 1.62. Chronic acid soils complicate forest management in the western Adirondack region and add to the complexity of the system. However, they are not necessarily responsible for regeneration failures. Yellow birch, red maple, and American beech regenerated well when protected from deer, even on sites with very acid soils and low cation exchange capacity (Figures 40.5 and 40.6). Sugar maple regeneration and growth of mature trees do appear to be negatively impacted by the acidic soils in the western Adirondacks (New York State Energy and Research Development Authority [NYSERDA] 2013).

FIGURE 40.7 Growth of yellow birch (*Betula alleghaniensis*) stems inside and outside of a fenced deer exclosure 6 years after the site was harvested at the Adirondack League Club near Old Forge, New York.

40.5   IMPORTANT CONSIDERATIONS

40.5.1   TREE SEEDLING REGENERATION

Our findings are consistent with the conclusions of Sage et al. (2003), that there are "windows of opportunity" for successful hardwood regeneration in Adirondack forests. Light reaching the forest floor (resulting from silvicultural practices or disturbance), deer herbivory, and site-specific factors (competing vegetation and soils) are the primary determinants of successful hardwood regeneration. However, there are many other interacting factors (weather, predation, wind, fire, deer hunting pressure) that can significantly influence the outcome and whether regeneration is successfully established. These systems are complex, and forest managers should manipulate the three key variables that can be controlled (light, deer herbivory, and competing vegetation) to create the greatest opportunity for successful regeneration.

The first factor is appropriate silviculture to create sufficient light at ground level to stimulate tree growth. Shelterwood cuts can provide adequate light to promote regeneration while maintaining a seed source for desirable tree species. Sage et al. (2003) determined that leaving 7.1–14.3 m²/ha residual basal area in the overstory resulted in a well-stocked and diverse seedling response in Adirondack forests. However, the best seedling response was obtained only when herbicide treatments and deer exclosures were added to the experimental design because of the beech dominance in the understory. Failure to address deer herbivory and competing vegetation resulted in inconsistent regeneration regardless of the intensity of overstory removal (Sage et al. 2003).

40.5.2   DEER DENSITY

White-tailed deer (*Odocoileus virginianus*) are key regulators of forest regeneration based on their preference for, and avoidance of, different tree seedlings (Tilghman 1989; Waller and Alverson 1997; Sage et al. 2003; Rawinski 2014). Maintaining deer densities in balance with forest management objectives can be very difficult, especially if landowners want to maintain high deer harvests. Often, deer hunters want to see densities well above what is needed for successful tree regeneration or biodiversity goals. In mature woodlands with a history of deer overbrowsing, densities of 6–8 deer per km^2 hindered establishment of some desirable hardwoods (Horsley et al. 2003; Rawinski 2014). Given the shorter growing seasons and acidic soils in the western Adirondacks, even densities as low as 3–4 deer per km^2 inhibited successful regeneration of tree seedlings in our experiment. Telling hunters that they may need to cut those already low deer densities in half to see successful tree regeneration was resisted by some ALC club members.

In areas with chronic deer overbrowsing, deer-resistant beech brush or ferns may dominate the forest floor (Sage et al. 2003; Rawinski 2014). Dominance of beech in the understory of mature hardwood stands resulted in exclusion of other desirable hardwood species (Tierson 1967; Sage 1987). Spraying the beech understory with herbicide prior to cutting the main canopy prevented the herbicide from damaging overstory trees (Sage et al. 2003) and reduced beech competition with desirable hardwoods. However, this can be expensive, and specialized equipment is needed to treat large areas.

In highly seasonal winter environments, winter home ranges of deer are smaller than summer ranges (Tierson et al. 1985), and deer congregate in traditional wintering areas called deer yards (Rongstad and Tester 1969). Some hunters have suggested that the declining deer numbers at ALC may be due to the ban on winter deer feeding, which may have resulted in "losing the winter deer herd." Generally, deer show fidelity to their home ranges during both summer and winter (Hurst and Porter 2008). Female deer, especially, will return to their summer ranges each year, usually near where they were born. Recent studies in the Adirondacks documented shifts in traditional winter deer yards. Forage availability, coupled with winter weather conditions, appeared to have been the driving forces for these winter range shifts (Hurst and Porter 2008). Discontinued deer feeding at ALC could very well have caused the deer to shift their winter yard locations. In mature woods with chronic deer overbrowsing, deer may not find enough natural food to sustain them, and the deer move to other wintering areas. However, most deer will not shift to winter ranges until after the fall hunting season ends during November; thus, changes in winter deer yard use should not have impacted deer harvests at ALC.

An effort was implemented to decrease the deer herd size within the 2025-ha QDM area at ALC, and doe harvest was increased through the use of DEC DMAP tags during 1999–2005. However, hunter opinion about the taking of female deer was split in the ALC membership, and some hunters were vehemently opposed to taking any female deer. Total deer harvest at ALC was already much lower than desired, and far below "the good old days" in the late 1950s and 1960s. Hunters argued that taking female deer would further depress local deer numbers and harvests, and that acid rain, not low deer densities, negatively impacted forest regeneration. Despite these hunter claims, browsing damage to tree seedlings was observed in our four experimental stands (Figures 40.5 and 40.6), indicating the deer population still exceeded carrying capacity of the forest in some parts of the ALC.

40.5.3 Acid Deposition/Soil pH

Acid rain in the western Adirondacks complicates the forest management system and makes regeneration of sugar maple stands more difficult. Maple regeneration has been shown to be sensitive to soil pH and cation exchange capacity (NYSERDA 2013), and effects of acid rain may interfere with growth of maple seedlings and trees. Trees growing on soils with poor acid-base chemistry (low exchangeable calcium and base saturation) exhibited little to no sugar maple seedling regeneration, relatively poor canopy condition, and short- to long-term growth declines compared with study plots having better soil condition and lower levels of acidic deposition. Plots in the western Adirondacks that contained sugar maple seedlings had significantly higher ($p < 0.01$) soil base saturation and exchangeable calcium compared with plots that lacked sugar maple seedlings. Abundance of sugar maple seedlings was lowest on plots with upper B horizon soil base saturation less than 12%, and highest on plots with base saturation greater than 20% (NYSERDA 2013).

While establishing food plots throughout the QDM area at ALC, soil pH was identified as a concern. The pH levels of soils observed during this study in our experimental stands were considered extremely acidic and unsuitable for root zone establishment or optimal sugar maple growth. At our four experimental stands, soil pH ranged from 3.8 to 4.2, and Ca:Al ratios ranged from 0.33 to 1.62. High acidity interferes with exchangeable cation ratios (NYSERDA 2013). Park and Yanai (2009) found that failure of sugar maple regeneration on acidic soils in the Adirondack region was consistent with their finding that sugar maple seedlings were highly sensitive to nutrient availability. In experiments involving transplanted seedlings, only sugar maple showed significant increases in relative diameter growth of seedlings in plots amended with $CaCl_2$ vs. controls (Kobe et al. 2002). American beech, yellow birch, balsam fir (*Abies balsamea*), and red spruce (*Picea rubens*) showed no significant responses in diameter growth following $CaCl_2$ treatments. This sensitivity could ultimately contribute to the replacement of sugar maple by American beech in regions with low pH and cation exchange capacity.

Harvested areas in more remote parts of ALC have sometimes experienced good regeneration of sugar maple with surprisingly low soil pH levels. Prior regeneration cuts at ALC during 1996 and 1997 (Fox Pond and Higley Mountain) had similar low pH levels (average 3.7), but much better Ca:Al ratios (3.78). For example, Higley Mountain had good sugar maple regeneration following a 1997 harvest (ALC unpublished data), and this was likely influenced by favorable Ca:Al cation ratios within the B-horizon soils. Another stand (Gravel Bank) that was cut about the same time had higher soil pH (4.5) and a low Ca:Al ratio (0.39), but seedling regeneration failed at this location. It is clear that soil acidity is a complicating factor for maple regeneration in the western Adirondack region (NYSERDA 2013).

40.6 MANAGEMENT CONSIDERATION: ENHANCE REGENERATION WITH FENCING

The goal of our research was to enhance forest regeneration and sustainable forest management at ALC. At this point, management options are very limited. Fencing would seem to be the best remedy for protecting seedlings in regeneration areas. However, this is expensive, and costs for a 2.4-m-high, page-wire fencing can range between $5 and $6.60 per linear m, with annual maintenance expenses exceeding $1000 per exclosure (Smith 2005). Given the low economic value of stands at ALC (sugar maple and yellow birch are the trees with highest value), and long growing rotations, fencing is not economically feasible.

40.7 MANAGEMENT CONSIDERATION: CONTROL DEER DENSITY/INCREASE FORAGE

Sage et al. (2003) saw adequate hardwood regeneration in Adirondack forests if deer densities were below 5.2 deer per square km, and pockets of low deer density could be created by using DEC DMAP permits (tags allocated to landowners for taking antlerless deer during hunting season). Crimmins et al. (2010) suggested that the combination of deer population control and increasing the amount of timber harvest across the landscape could reduce herbivory to levels that may not impede growth and survival of forest vegetation. However, it will be difficult to increase the harvest of female deer, as several ALC members are opposed to doe hunting. This is understandable, as the ALC was founded to provide excellent outdoor recreation opportunities. With low Adirondack deer densities, and about 130 km^2 of forest preserve, hunters can be afield for days without seeing a legal buck. This situation is not unique to ALC, and creates a dilemma for deer and forest management in the western Adirondack region. But even if hunters supported lowering deer densities further with DMAP permits, would it be possible on a 13,000-ha preserve with limited road access? This is a difficult question with no clear answer. Hundreds of thousands of ha in the western Adirondacks have similar deer and forest management issues.

40.8 MANAGEMENT CONSIDERATION: INCREASE FORAGE BY TIMBER HARVEST

One approach that might be implemented to encourage successful regeneration is controlling the size of timber cuts and the proportion of the area being regenerated. In Appalachian northern hardwood forests, Miller et al. (2009) saw deer browsing rates decline as the amount of area in early successional habitat increased. They recommended having at least 14% of an area in well-distributed, even-age-managed forests, and this could substantially reduce deer impacts to tree regeneration. Given the low Adirondack deer densities, it may be possible to overwhelm family groups of local deer with seedling response if cut areas are large enough. Ten years after harvest, we saw good tree regeneration in two of our experimental stands (Sylvan and Oxbow). These cuts were larger than the Rock Pond stand, which still showed heavy deer browsing pressure after 10 years (Figure 40.8). Even with large harvest areas, deer herbivory may temporarily set back regeneration and add an additional 5–10 years to the typical harvest rotation. It appeared the Rock Pond stand was slowly being converted to softwoods (Figure 40.9) because of the intense deer herbivory during the past decade.

FIGURE 40.8 Tree regeneration inside (background) and outside (foreground) of a fenced deer exclosure 10 years after the Rock Pond site was harvested at the Adirondack League Club near Old Forge, New York.

FIGURE 40.9   Heavy deer herbivory over 10 years is slowly converting the Rock Pond stand from hardwoods to softwoods at the Adirondack League Club near Old Forge, New York.

It is clear that there is no simple solution or prescription for regenerating northern hardwoods in the western Adirondack region of New York. These forest systems are complex, and regenerating stands successfully depends on several interacting factors (Sage et al. 2003). Managers should do all that they can to reduce deer herbivory and competing vegetation prior to a timber harvest, then use even-aged management over a large enough area. If everything is done correctly, then the chances for successful regeneration are substantially increased. If the primary interacting components of the system (light, deer herbivory, competing vegetation) are not all addressed, then there is a good chance that hardwood forest regeneration will fail. Stands with adequate light may be converted to either beech brush or softwoods depending on stand history, advance regeneration, and deer browsing pressure.

40.9   CASE SUMMARY

ALC members were initially interested in increased deer harvests and scientifically based deer management. They hired a consultant to create a 2025-ha experimental quality deer–forest management area within their 13,000-ha forest preserve and followed prescriptions for establishing food plots and recording deer harvest data. However, members were reluctant to harvest additional female deer given the already low densities on club grounds. There has been a strong cultural tradition against harvesting does on ALC lands, similar to many other hunting clubs in the Adirondack ecoregion.

Even though data from the fenced deer exclosures clearly showed deer browsing impacts on forest regeneration, club members hired another consultant (D. deCalesta) to assess the forest management situation and make management recommendations. Based on his observations in several forest stands and regeneration areas, the primary recommendation was to harvest more female deer to bring herd density in balance with available regeneration. However, based on their perceptions and values, club members again disregarded this scientific assessment. The annual application for DMAP tags to take antlerless deer ceased, and only a few does are currently harvested each year during archery and muzzle-loader seasons on the 13,000-ha property. Deer densities remain low on the ALC lands, and forest habitats continue to be negatively impacted.

In this case, culture and values outweighed scientific evidence for the majority of club members who were deer hunters. There was a great deal of internal conflict among members when it came to setting goals for the antlerless deer harvest each fall. Heated debate surrounded the application for DMAP tags and the number of antlerless deer to be taken in the 2025-ha QDM area. Even though taking does was restricted to the QDM area, and there were almost no antlerless deer harvested on 10,930 ha (84%) of the ALC property, this issue was very contentious each fall.

Rather than using science to justify the need for antlerless deer harvest, possibly it would have been more effective to use a messenger with greater credibility and a different case study from another hunting club. For example, a lead member from a hunt club in Pennsylvania where QDM was practiced at the Kinzua Deer Cooperative may have been more trusted than a scientist or agency deer manager. If a peer hunter reinforced the benefits of fewer deer (i.e., larger bucks, healthier deer, more wildflowers and biodiversity), possibly the message about increased doe harvest would have been better received at ALC. It is impossible to separate deer hunting culture and values from deer management, and science alone may not always "sell" a deer management program.

EDITOR COMMENTS

The wildlife and forestry consultants providing professional advice on managing deer density and impacts made exhaustive and comprehensive evaluation of the nutritional and other ecological aspects of deer management related to forage condition and status of tree regeneration. They were able to provide a scientific explanation for the lack of quality or high density of deer within the Quality Deer Management Area: there simply was an insufficient forage base to produce quality deer or sustain an abundant herd. However, the culture and history of club members, who remembered the days of high deer density following an enormous production of deer forage following a massive blowdown of overstory trees, prevailed over scientific fact and reason as they kept calling for more forest harvest to produce additional forage and increase deer density. The argument that managing for deer density at Maximum Sustained Yield (MSY), quality deer, and successful regeneration and recruitment of tree species was not achievable was ignored, as was the recommendation to harvest more antlerless deer to reduce pressure on understory vegetation and provide better distribution of forage to the deer herd. The inevitable result followed: further depletion and degradation of understory vegetation, continued poor quality and low abundance of deer, and continuing scarcity of understory herbaceous vegetation.

David S. deCalesta, editor

REFERENCES

Alerich, C. L. and D. A. Drake. 1995. *Forest Statistics for New York 1980 and 1993. Resource Bulletin NE-132.* Radnor PA. United States Department of Agriculture Forest Service Northeastern Experiment Station.

Brothers, A. and M. E. Ray, Jr. 1975. *Producing quality whitetails.* Laredo TX. Fiesta Publishing Company.

Brown, E. 1985. *The Forest Preserve of New York State.* Glens Falls NY. Adirondack Mountain.

Comstock, E., Jr. and M. C. Webster. 1990. *The Adirondack League Club, 1890–1990.* Old Forge NY. The Adirondack League Club.

Crimmins, S. M., J. W. Edwards, W. M. Ford et al. 2010. Browsing patterns of white-tailed deer following increased timber harvest and a decline in population density. *International Journal of Forestry Research* 2010:1–7.

Debbie, J. G. 1961. An evaluation of the white-tailed deer herd on the Adirondack League Club with respect to browse damage, hunter kill, and winter management. *M.S. Thesis*, Department of Conservation, Cornell University, Ithaca NY.

Hewitt, D. G. 2011. *Biology and management of white-tailed deer.* Boca Raton FL. CRC Press.

Horsley, S. B., S. L. Stout, and D. S. deCalesta. 2003. White-tailed deer impact on the vegetation dynamics of a northern hardwood forest. *Ecological Applications* 13:98–118.

Hurst, J. E. and W. F. Porter. 2008. Evaluation in shifts in white-tailed deer winter yards in the Adirondack region of New York. *Journal of Wildlife Management* 72:367–375.

Kobe, R.K., G. E. Likens, and C. Eagar. 2002. Tree seedling growth and mortality responses to manipulations of calcium and aluminum in a northern hardwood forest. *Canadian Journal of Forest Research* 32:954–966.

Mautz, W. W. 1978. Sledding on a bushy hillside: The fat cycle in deer. *Wildlife Society Bulletin* 6:88–90.

McMartin, B. 1994. *The great forest of the Adirondacks*. Utica NY. North Country Books.

Miller, B. F., T. A. Campbell, B. R. Laseter et al. 2009. White-tailed deer herbivory and timber harvesting rates: implications for regeneration success. *Forest Ecology and Management* 258:1067–1072.

Neth, P. C. 1956. The Adirondack League Club Deer Herd: An Evaluation of Winter Mortality, Hunting Kill, and Browse Survey Data. *Ph.D. Dissertation*, Department of Conservation, Cornell University, Ithaca NY.

New York State Energy and Research Development Authority. 2013. *Effects of acidic deposition and soil acidification on sugar maple trees in the western Adirondack Mountains.* Summary Report No. 13-04b. Albany NY.

Park, B. B. and R. D. Yanai. 2009. Nutrient concentrations in roots, leaves and wood of seedling and mature sugar maple and American beech at two contrasting sites. *Forest Ecology and Management* 258:1153–1160.

Parker, K. L., C. T. Robbins, and T. A. Hanley. 1984. Energy expenditures for locomotion by mule deer and elk. *Journal of Wildlife Management* 48:474–488.

Rawinski, T. J. 2014. White-tailed deer in northeastern forests: understanding and assessing impacts. *NA-IN-02-14.* Newtown Square PA. Northeastern Area State and Private Forestry, United States Department of Agriculture Forest Service Northeastern Experiment Station.

Rongstad, O. J. and J. R. Tester. 1969. Movements and habitat use of white-tailed deer in Minnesota. *Journal of Wildlife Management* 33:366–379.

Sage, Jr., R. W. 1987. Unwanted vegetation and its effects upon regeneration success. In: *Symposium of northern hardwood silviculture*, ed. R. Nyland, Radnor PA. USDA Forest Service.

Sage, R. W., W. F. Porter, and H. B. Underwood. 2003. Windows of opportunity: White-tailed deer and the dynamics of northern hardwood forests. *Journal of Nature Conservation* 10:1–8.

Saunders, D. A. 1988. *Adirondack mammals.* Syracuse NY. State University of New York, College of Environmental Science and Forestry.

Smith, M. 2005. *Cattaraugus #13 tract tour.* Falconer NY. Forecon, Inc.

Thompson, C. B., J. B. Holter, H. H. Hayes et al. 1973. Nutrition of white-tailed deer. I. Energy requirement of fawns. *Journal of Wildlife Management* 37:301–311.

Tierson, W. C. 1967. Influence of logging, beech control, and partial deer control on northern hardwood reproduction. *M. F. Thesis*, State University of New York, College of Environmental Science and Forestry, Syracuse NY.

Tierson, W. C., G. F. Mattfeld, R. W. Sage et al. 1985. Seasonal movements and home ranges of white-tailed deer in the Adirondacks. *Journal of Wildlife Management* 49:760–769.

Tilghman, N. G. 1989. Impacts of white-tailed deer on forest regeneration in northwestern Pennsylvania. *Journal of Wildlife Management* 53:524–532.

Tyndall, J. 1879. *Fragments of science, Vol. II.* London. Longmans, Green, and Co.

Waller, D. M. and W. S. Alverson. 1997. The white-tailed deer: A keystone herbivore. *Wildlife Society Bulletin* 25:217–226.

Webster, D. A. 1965. *The Adirondack League Club 75th anniversary.* Old Forge NY. The Adirondack League Club.

Appendix 1: Protocol for Estimating Deer Density by Pellet Group Counts, Deer Impact, and Deciduous and Coniferous Canopy Closure (for Deer Forage and Hiding Cover)

CONTENTS

A1.1 INTRODUCTION

Data for estimating deer density, deer impact (including proxies for multiple forest resources), and overstory canopy closure (as proxies for deer forage and hiding cover) may be collected simultaneously from a single system of plots arrayed along transect lines arranged in grids, resulting in considerable savings of financial and human resources. Because of the random nature by which transect lines and grids can be located, data derived from them may be analyzed for determination of significance of differences for individual parameters (e.g., deer density) identified among sites and/or among years. Data should be collected in spring, after snow melt and while wildflowers are in flower, but before green-up (proliferation of ferns after green-up will obscure pellet groups and wildflowers).

A1.2 REQUIREMENTS

Required financial, equipment, and human resources are minimal: crews of two observers can collect all data on two typical transect grids per 8-hour day, excluding time to travel to and from grids. Equipment includes data sheets, topographic maps of grids, a 1.2-m stick (mark distance on a walking stick used for maneuvering in the woods), clipboard, and compass. Because data are often collected during inclement weather, waterproof data sheets (available from forestry supply companies) should be used.

A1.3   ARRANGEMENT AND LAYOUT OF PLOTS, TRANSECT LINES, AND GRIDS

The system is that described by deCalesta (2013) and Pierson and deCalesta (2015) for estimating deer density and deer impact on forest resources. Plots are circular, with a radius of 1.2 m (4 feet) and area of 4.5 m² (50 feet²) and generally placed ∼30 m (100 feet) apart along transect lines. Transect lines are 1600 m (1 mile) long and spaced 300 m (1000 feet) apart, usually arranged in grids of five parallel transect lines (Figure A1.1).

The area enclosed by such grids is approximately 190 ha (∼470 acres or 0.75 miles²), the low end of average deer home range.

Forest properties rarely are perfect squares or rectangles. Rather, they are somewhat irregularly shaped, may be long and thin, and may consist of separate but nearby properties, especially when <500 ha in total. Accordingly, on these properties, a center transect line should pass through the middle of the property, aligned with the longer dimension. If additional transect lines can be fit within property boundaries, they should be spaced 300 m apart and parallel to the center transect line. On larger properties, multiple grids may be located representatively across the landscape to include different habitats/successional stages. There is no criterion for how many grids to place within larger landscapes in the hundreds of hectares. In the KQDC case history (Chapter 34), 26 grids were located within a 30,000-ha (113-mile²) area for a concentration of one 190-ha grid/1150 ha representing a sampling of ∼16% of the landscape. On larger landscapes, a network of lines 300 m apart placed perpendicularly to each other over maps of properties forms a sampling network. A random sample of the network intersections may be selected based on the number of grids desired. Grids should be centered on selected intersections (center of middle line in each grid forms center of grid) and identified by a number/letter placed on a prominent, permanent marker (large tree, large stake).

On disparate but nearby properties owned/managed by a single entity, permission may be sought to include and census non-owned properties within a contiguous landscape, and grids may be placed within the boundaries of such aggregated properties.

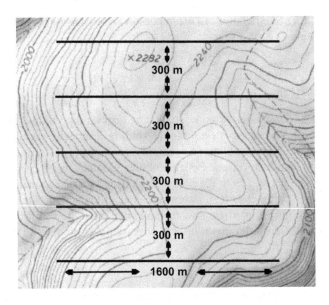

FIGURE A1.1   Layout of transect lines within a grid.

A1.4 COLLECTION OF DATA

A1.4.1 DEER DENSITY

Groups of deer pellets are counted if at least 50% of pellets fall within plots. Pellet groups must contain at least 10 pellets before being counted as a single group. Some plots may contain more than one pellet group, but differences in color and shape of individual pellets within different groups makes identification of separate pellet groups relatively easy. The number of plots assessed per transect line is recorded by dot tally (Figure A1.1). The number of deer pellets observed is recorded for each individual transect line, also by dot tally. For example, if only one pellet group is recorded for an entire transect line (which often occurs, as do zero counts) a single dot is recorded. Additional pellet groups are indicated by additional dots: four pellet groups are denoted by four dots arranged as the corners of a square. Additional pellet groups are denoted by drawing a line between neighboring/kitty-corner dots (Figure A1.2).

A1.4.2 DEER IMPACT ON SEEDLINGS

A small number of indicator seedlings should be selected: one or two highly preferred as deer forage, one or two of moderate preference, and one or two nonpreferred that seem mostly to be used as emergency forage during times of winter food scarcity. Twigs browsed by rabbits and hares may be distinguished from those browsed by deer by appearance. Rabbits and hares use their upper and lower incisors to create a sharply defined 45° clipped appearance; deer tear off terminal portions of twigs between their lower incisors and upper palate pad, producing a shredded effect (Figure A1.3).

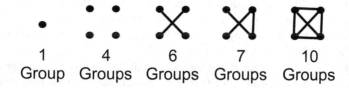

| 1 | 4 | 6 | 7 | 10 |
| Group | Groups | Groups | Groups | Groups |

FIGURE A1.2 Depiction of dot tally arrangement of dots and lines. After four pellet groups are recorded, additional groups are recorded by connecting dots. After the 10-dot figure, additional groups are recorded by starting over with a single and then additional dots and lines.

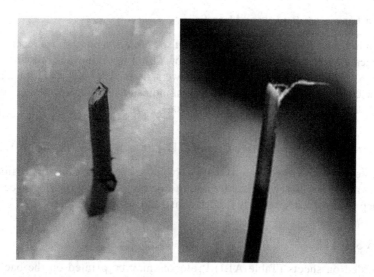

FIGURE A1.3 Appearance of twigs browsed by rabbits and hares (left panel) and deer (right panel). (Photos by David Jackson and David S. deCalesta.)

FIGURE A1.4   Browsing impact levels on seedlings, left to right: zero impact, light impact, moderate impact, heavy impact, severe impact.

For each indicator species, browse status per plot is recorded as: no impact (no twigs browsed), light impact ($> 0\%$ to $< 50\%$ of twigs browsed), moderate impact ($>50\%$ twigs browsed but not hedged), and heavy-severe impact ($>50\%$ twigs browsed and hedged) (Figure A1.4). Browse status is recorded for each indicator species by denoting level of deer impact. If more than one seedling of an indicator species is present within a plot, the level of deer impact exhibited by a majority of the seedlings is recorded (e.g., if one seedling has moderate impact and two have heavy-severe impact, the heavy-severe level is recorded for that plot—in the case of ties, the higher impact level is recorded). Thus, plots do not record number of seedlings browsed but rather level of impact representative of the majority of seedlings of that species.

If woody vegetation of any species is present within a plot but there is no deer browsing, an entry is made for *no impact* on regeneration for that plot. This accounts for *no browsing impact* that is indicated by no impact on any woody species.

The number of plots with observed levels of impact on indicator seedlings is recorded in the same manner as pellet groups—by dot tally.

A1.4.3   DEER IMPACT ON INDICATOR HERB SPECIES

Impact level on indicator herb species (e.g., *Trillium* sp., *Maianthemum* sp.) is recorded by dot tally and characterized as: 1 = wildflowers in flower, grow tall; 2 = wildflowers present but don't flower or grow tall; 3 = wildflowers absent; 4 = no wildflowers, invasive ferns and exotics dominate ground cover. As is the case with recording impact level on indicator seedling species, the dominant level of impact is recorded for each indicator herb per plot.

A1.4.4   PERCENT OVERSTORY CANOPY CLOSURE, DECIDUOUS AND CONIFEROUS SPECIES

Overstory/midstory canopy closure is recorded by dot tally as open (no overstory or midstory canopy closure directly above the plot) or closed by noting whether overstory/midstory trees/branches are directly over a plot, recorded separately for deciduous and coniferous trees.

A1.5   DATA SHEETS

Data are recorded on sheets (Table A1.1) Protocols may be printed on the back side of data sheets.

TABLE A1.1

Example Data Sheet for Recording Deer Density and Impact

Site:	Grid Number:	Date:	Weather Conditions:		Observer:	
Transect		Transect #1	Transect #2	Transect #3	Transect #4	Transect #5
Number plots						
# Pellet groups						
Plots no impact						
Plots no regen.						
Overstory Cover: Conifer						
Overstory Cover: Decid.						
Midstory Cover: Conifer						
Midstory Cover: Decid.						

Beech impact

0	L
M	H-S

Black cherry

0	L
M	H-S

Red Maple

0	L
M	H-S

Trillium

1	2
3	4

Maianthemum

1	2
3	4

Seedling impact: 0 = no browsing; L = <50% twigs browsed; M = >50% twigs browsed, no hedging; H-S = >50% twigs browsed, seedling hedged.

Herb impact for *Trillium* and *Maianthnemum:*   1 = wildflowers in flower, grow tall; 2 = wildflowers present but don't flower or grow tall; 3 = wildflowers absent; 4 = no wildflowers; invasive ferns and exotics dominate ground cover.

A1.6   PROTOCOLS FOR COLLECTING DEER DENSITY AND IMPACT (BACK SIDE OF DATA SHEET)

A1.6.1   PELLET GROUPS

1. Pellet groups are counted within 1.2-m-radius plots located at 30-m intervals along transect lines 1600 m long. The first plot taken is 30 m from the starting point from the beginning of each transect line, and the last plot is ~30 m from the end of the line. At the end of each transect, take a 90° bearing and travel 300 m to the starting point of the next transect. There

should be ∼52 plots per line (exceptions: plots inside fenced enclosures or that fall in bodies of water aren't counted). Distances may be measured by pacing or hip chain. Keep track of plots for each transect line by tallying each plot by "dot tally" (see instructions below).

2. There must be at least 10 pellets in a group before it is counted, at least half of the pellets must be within the 4-foot-radius plot, and pellets must be on top of leaves. Record the number of pellet groups by dot tally (see instructions for dot tallies below).

3. If fenced enclosures or ponds/lakes are encountered along the transect line, either climb over the fence/wade through the water and continue along the transect line, or take a sighting on the other side of the fence/water, walk around, and resume the line. Do not count pellet groups inside the fence/water. Deduct the number of plots that would have fallen inside the fence/water from the total of plots possible (52 for pellet group counts, 26 for impact counts) taken for each line. Try to keep the transect line as close to 1600 m as possible. Observers using hand-held GPS units can plot waypoints from one side of the fence to the other and walk around rather than climb over the fence—just be sure to note the length of transect line not walked inside the fence to be sure you only travel along 1600 m of transect line (including length inside fence).

4. Write the total number of plots and pellet groups for each transect line (top of form, other side of page).

5. Record numbers of dead deer observed along transect line.

A1.6.2 Deer Impact

(1) Record deer damage to seedlings within the 1.2-m-radius plot by each of five species (striped maple, beech, red maple, black cherry, hemlock). Record data for seedlings ≥5 cm tall. (2) Impact will be recorded at every other pellet group plot. If no regeneration exists on plot, tally in "plots without regeneration" box. If regeneration exists, but is not browsed, tally in "plots with regen, no browsing" box. If seedlings are browsed, record impact for each of five indicator species (beech, striped maple, black cherry, red maple, hemlock). Use the Impact Diagram (Figure A1.4) to characterize impact. Record impact in one of 5 categories: **0**—no impact; **light (L)**—0% to 50% of seedling stems are browsed; **moderate (M)**—more than 50% of stems are browsed but plant is not hedged; **heavy (H)**—more than 50% of stems are browsed and the plant is hedged (plant is browsed to a small ball of twigs); **severe (S)**—more than 50% of stems are browsed, the plant is hedged, and is less than 6 inches tall. There may be more than one seedling (including groups of seedlings as stump sprouts) for any species within a plot. Use your best judgment to characterize browsing across seedlings for each species within each plot; that is, if most are heavily browsed, record damage as heavy for the species for that plot. Record data with dot tally described above.

A1.7 ANALYSIS

A1.7.1 Deer Density: Small Properties

Because deer density data collected from properties smaller than 200 ha does not sample areas of typical deer home range size or larger, estimates of deer/km² cannot be calculated without assuming the result is representative of deer density that will impact the property. Rather, pellet group data from smaller properties should be used to estimate the number of deer using the property fall–spring (leaf-fall to green-up). Estimates of number of deer causing impact on smaller properties are useful as they can be used to determine how many deer should be removed by hunting/culling. Deer impact data collected from smaller properties do reflect deer impact regardless of property size and may be used as a surrogate for estimating deer/km².

Deer density/abundance estimates derived from pellet group counts require these assumptions/ constants:

1. The number of pellet groups deposited per day is the same for all sex and age classes of deer, and a reasonable approximation is 15 groups/deer/day (from Sawyer et al. 1990).
2. Pellet group counts are only for groups deposited from the preceding fall after leaf-off date (date when trees have shed their leaves the previous fall, covering previous years' pellet group but leaving current year's deposition uncovered). Counting groups after leaf-off date ensures that only pellet groups deposited by deer from fall–spring are counted. Number of days since leaf-off is the number of days between date of leaf-off and collection of pellet group data. On properties where there is no overstory cover of deciduous trees to deposit leaves over pellet groups from previous years (e.g., under coniferous cover, in fields and meadows without overstory deciduous trees), pellet groups from previous deposition periods are clearly more deteriorated and of paler color than pellet groups from the current year, and the assumption is that pellet group differences allow observers to differentiate and count only current period groups. Determining the number of days since leaf-off requires the assumption that appearance of pellet groups deposited prior to the typical fall leaf-off date (~early November) from previous years would be markedly different from that of the current year's groups due to heightened deterioration occurring over spring, summer, and fall of previous years when considerably warmer weather, including warm spring, summer, and fall rains, would result in much greater deterioration.

The calculation is straightforward : number of deer using the property

$$= \frac{\text{Number of pellet groups counted from leaf-off to day of count}}{\text{Deposit rate}\,(15\,\text{pellet groups/day})\times(\text{days since leaf-off})\times(\sum\text{plot area/property area})}$$

Example

Number pellet groups counted $= 22$
Days since leaf-off (November 15 to April 25) $= 161$
Plots $= 34$
Plot area $= \sum$ plots (34) $\times$ area/plot (1.2 m^2 $\times$ 3.14) $= 153$ m^2
Area of property $= 0.45$ km^2
Number of deer $= 16$ pellet groups/(15 groups/day $\times$ 161 days $\times$ 153 m^2/450,000 m^2)
   $= 19.5$ deer, round to nearest whole number $= 20$

Assuming the abundance estimate is representative of deer density impacting the property, one could estimate deer/km^2 by multiplying number of deer (20) by area of property/area of 0.45 km^2 (0.45 km^2 $\times$ 20) $= 12$ deer/km^2. A deer density this high would be associated with failure of seedling regeneration of preferred species, loss of herb and shrub understory, understory dominated by ferns or exotic herbs/shrubs, and deer of poor quality.

A1.7.2 Deer Density: Large Properties

Assumptions regarding pellet group deposition rate and pellet deposition period are the same as those for small properties. Calculating deer density rather than deer abundance produces values that can be compared with varying levels of carrying capacity and related to forest resources impacted. As for smaller properties, deer density estimates for large properties may also be expressed as total deer using property.

Calculation for deer/km^2

$$= \frac{\text{Number of pellet groups counted from leaf-off to day of count}}{\text{Deposit rate}\,(15\,\text{pellet groups/day})\times(\text{days since leaf-off})\times(\sum\text{plot area m}^2/\text{m}^2\text{ in 1 km}^2)}$$

Example

Number pellet groups counted = 18
Days since leaf-off (November 15 to April 25) = 161
$\sum$ plots = 250
Plot area = $\sum$ plots (250) $\times$ area/plot (4.5 m^2) = 1125 m^2
Area of property = 315 ha (3,150,000 m^2)
Number of deer = 18 pellet groups/(15 groups/day $\times$ 161 days $\times$ 1125 m^2/3,150,000 m^2)
= 20.9 deer/km^2, round up to nearest whole number = 21 deer/km^2

Deer density for an entire property is calculated by using plot totals summed across grids. However, if areas within properties have suspected differences in forage availability, deer harvest rates, or other factors suggesting deer density and impact should be calculated separately for each area, that may be accomplished by estimating deer density and impact (and percent canopy closure) by groups of grids that fall within the different areas.

A1.7.3   DEER IMPACT AND CANOPY CLOSURE

All the deer impact parameters—impact on seedlings, herbaceous plants, proxies—and canopy closure are calculated on a per plot basis, rendering the calculation easy. For example, if 150 plots were taken within a property, and 50 exhibited no impact on regeneration, plots with no impact would be calculated at 33.3%. Ditto plots exhibiting specific deer impact levels; for example, if 15 of 150 plots exhibited heavy-severe impact on red maple seedlings, the reported impact would be 10% of plots with heavy-severe impact on red maple seedlings.

A1.7.4   CALCULATING SIGNIFICANCE OF DIFFERENCES IN DEER
DENSITY/IMPACT/COVER AMONG AREAS/YEARS

Most private managers, particularly small woodlot owners, will not have the need to determine whether differences in estimates of deer density/impact/canopy/closure are significantly different through application of statistical analysis. However, managers of large properties, such as state and national forests or commercial timberlands, may wish to determine whether differences of values of interest (deer density/impact/canopy closure) among areas/years are of statistical significance. Generally, these managers have on staff persons who can provide the desired analysis. This can be accomplished by stratifying data by transect line within grids respectively for individual areas and/or years. Parameter values among transect lines within individual grids produce independent estimates of those values (deCalesta 2013). Transect lines within individual grids (of five transect lines) may be assigned a number 1–5 randomly for each grid. Calculating the values for density/impact/canopy closure separately by transect line among grids produces five replicate values for the parameters, which may then be subjected to analysis of variance by professional data analysts to determine the significance level of any differences.

REFERENCES

deCalesta, D. S. 2013. Reliability and precision of pellet-group counts for estimating landscape-level deer density. *Human-Wildlife Interactions* 7:60–68.

Pierson, T. G. and D. S. deCalesta. 2015. Methodology for estimating deer impact on forest resources. *Human-Wildlife Interactions* 9:67–77.

Sawyer, T. G., R. L. Marchinton, and W. M. Lentz. 1990. Defecation rates of female white-tailed deer in Georgia. *Wildlife Society Bulletin* 18:16–18.

Appendix 2: Protocol for Aging Deer by Tooth Wear and Eruption

Hunters are interested in the age structure of the herd they are hunting. More importantly, the age structure of the herd can be approximated by deer brought to checking stations (assuming that hunters bring harvested deer to checking stations every year in representative proportions by fawns, yearlings, young, and mature deer). Changes in the age structure over time reflect whether managers are affecting the deer herd through harvest in ways that contribute to achieving goals related to deer density and impact. Change in age structure of the herd from top-heavy older deer, especially does, to mostly younger deer (fawns, yearlings, and young adults) is a reliable indicator that herd density is being reduced by hunting.

Fawns are easily identified by their small size and presence of milk (temporary) teeth. Yearlings are larger than fawns and usually smaller than adults (2½ years and older), and their snouts are generally shorter relative to total head length. Antler points are not diagnostic of age in adult bucks: there is no reliable relationship between number of antler points and deer age in years.

Age of individual deer may be obtained accurately by counting growth rings (similar to the idea of counting tree rings to age a tree) from teeth (generally incisors) that have been extracted, sectioned, and examined under a microscope. Sectioning and counting teeth requires special equipment: if deer age is to be obtained in that manner, teeth are sent to specialists. Depending on number of teeth aged, costs may range from ~\$3 to ~\$75 per tooth; plus, there is generally a delay of several weeks or more before age estimates are provided.

However, placing deer into one of four age groups (fawn, yearling, young adult [2½ to 3½], and mature adult [4½ +]) provides managers with sufficient information to satisfy deer hunter requests for age structure of the herd. Generally, few deer survive beyond 5 years, especially antlered deer. For example, of 700 bucks harvested 2001–2012 and brought to checking stations in the Kinzua Quality Deer Cooperative (KQDC) case history (Chapter 34), 94.3% were aged (by wear and eruption technique) as younger than 4 years; of 438 does brought to the same checking station over the same time frame, 87.9% were younger than 4 years of age (deCalesta , personal communication). Characterizing deer health attributes (weight and antler measurements) by these four age classes (fawns, yearlings, young adults, and mature adults) over time allows managers to determine whether their deer management program affects deer quality.

Deer age can be placed reliably into four categories by distinctive wear and eruption patterns (Figure A2.1). Fawn deer have much smaller jaws and much smaller incisors, and the third premolar (third tooth beyond the gap after the incisors) has three distinctive segments, and only one molar has erupted through the gum line.

For yearlings, all three molars have erupted and the third premolar has three cusps (it's a milk tooth to be replaced by a permanent two-cusped tooth by the second year of life).

For deer over 1½ years of age, the three-cusped third premolar has been replaced by a permanent, two-cusped tooth. For 2½- and 3½-year-old deer (young adult deer), the last cusp on the last molar is still sharp, not yet having been ground down.

For deer 4½ years and older (mature adult deer), the last cusp on the third molar is no longer sharp, having been ground down. All teeth of mature adult deer display much more wear on all surfaces than those of 2½- and 3½-year-old young adult deer.

Outdoor supply companies sell nylon deer jawbone sets for aging deer: the sets depict wear appearance of deer teeth by year, from fawns to 8½-year-old deer.

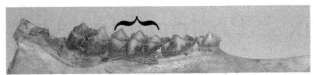

Fawn: 3rd pre-molar 3 sharp lingual* crests, 1 molar erupted

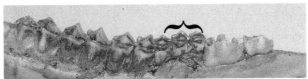

Yearling: 3rd pre-molar 3 dull lingual crests, 3 molars erupted

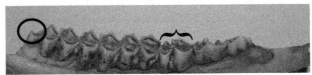

Young adult (2½ years old): last labial** crest last molar sharp; 3rd pre-molar 2 sharp lingual crests

Young adult (3½ years old): last labial crest last molar sharp; 3rd pre-molar 2 dull lingual crests

Mature adult (4½ years+): last labial crest 3rd molar flattened; all teeth heavily worn

FIGURE A2.1 Comparisons of wear and eruption patterns, fawn–4½⁺ year-old deer. *Lingual crest = tooth ridges on inside (lingual side) of teeth; **Labial crest = tooth ridges on outside (labial side) of teeth.

Examination of the jaws of deer brought to checking stations may be done by forcing the lower jaw open a sufficient distance to allow inspection of premolar and molar teeth in the lower jaw, if the deer has been recently harvested (within a few hours). For other deer, and especially larger deer, the cheek skin along the jaw must be slit to facilitate opening the jaws wide enough for a good look at the teeth. IMPORTANT! Before slitting the cheek skin, checking station operators must ask the hunter if it is all right to slit the skin. Hunters wishing to have their deer heads mounted will object to this action, as it will deface the head for mounting. In these cases, a jaw separator (Figure A2.2—available through forestry and wildlife catalogues) will have to be inserted (rounded end) between the upper and lower jaw just behind the incisor teeth and just before the premolar teeth and twisted 90° to force the upper and lower jaws sufficiently far apart to view molar teeth. Checking station operators may need to use a flashlight to see the molars, especially the third molar, for age estimation. Some technicians use tree loppers to cut the back of the jawbone where it curves up to the skull after the upper and lower jaws have been separated by an extractor, and then slip the loop of the jawbone extractor around and under the cut end of the jawbone and pull the loop toward the

FIGURE A2.2 Jawbone extractor.

front of the mouth. In most cases, this pulls the cut jawbone out of the mouth for easy examination. However, even skilled technicians have difficulty in extracting jawbones and often damage the cusps of molar teeth when cutting through the end of the jawbone, making age estimation chancy. It is far safer to use the extractor to open closed jaws as far as possible to examine the jawbone intact inside the mouth.

Index

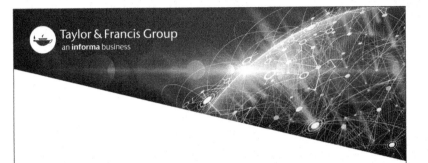

Printed in the United States
by Baker & Taylor Publisher Services